(continued on last page of book)

FUNCTIONS AND CHANGE

A Modeling Alternative
to College Algebra

Preliminary Edition

Bruce Crauder / *Benny Evans* / *Alan Noell*

Oklahoma State University

This project was supported, in part,
by the
National Science Foundation
Opinions expressed are those of the authors
and not necessarily those of the Foundation

HOUGHTON MIFFLIN COMPANY   BOSTON   NEW YORK

Gratitude and Thanks

We are deeply grateful for our families
who supported us and believed in this project.

For Douglas, Robbie, and, most of all, Anne—BCC
For Carla, Daniel, and Anna—BDE
For Evelyn, Philip, Laura, Stephen, and especially Liz—AVN

Editor-in-Chief: Charles Hartford
Senior Associate Editor: Maureen Ross
Editorial Assistant: Kathy Yoon
Senior Project Editor: Chere Bemelmans
Senior Cover Design Coordinator: Deborah Azerrad Savona
Senior Production/Design Coordinator: Carol Merrigan
Senior Manufacturing Coordinator: Marie Barnes
Marketing Manager: Michael Busnach

Cover Designer: Deborah Azerrad Savona
Cover Photographer: © Harvey Payne

Printed in the U.S.A.

Library of Congress Catalog Card No.: 98-72023

ISBN: 0-395-91158-3

123456789-CS-02 01 00 99 98

If you are interested in providing feedback on the *Preliminary Edition*,
contact your local Houghton Mifflin sales representative or
Michael Busnach at **michael_busnach@hmco.com**.

CONTENTS

4   EXPONENTIAL FUNCTIONS   209

PREFACE

To Instructors

Mathematics is among the purest and most powerful of sciences, an art form of unsurpassed beauty, and a descriptive language that codifies ideas from many areas, including business, engineering, and the natural, physical, and social sciences, alway showing that major concepts drawn from many different sources are in their essence the same. Mathematics pervades modern society and, to the knowledgeable eye, is everywhere evident in nature.

Practicing engineers, mathematicians, and scientists require a deep understanding of mathematics as well as a level of exactitude and facility with symbol manipulation that is sometimes difficult for entering students to master. For many, frustration with these aspects of mathematics obscures its power, beauty, and utility. But modern technology in the form of graphing calculators or computers can supplant much of the drudgery of mathematics, move the focus toward important concepts and ideas, and make mathematics more accessible.

> The goal of this text is to use technology and informal descriptions to empower entering students with mathematics as a descriptive problem-solving tool, and to reveal mathematics as an integral part of nature, science, and society.

Mathematics in context: style, pedagogy, and topics

This text differs from traditional textbooks in many ways. A quick glance through the book shows that there are more words and fewer formulas than usual. Also evident are the extensive examples and problems. The choices of examples and exercises are part of an important theme in the text: mathematics is easily learned from carefully chosen, realistic examples. In general, mathematical principles are developed through examples. Only after the examples give good intuition and understanding are more general and abstract notions made explicit. In practice, this style has worked very well at Oklahoma State University, as well as at the other schools class testing earlier drafts. Students are able to read the text and understand the examples. Students have opinions and bring their own independent understanding of the topics in the examples and exercises, so classroom discussions are more lively.

This text easily accommodates pedagogies other than traditional lecture. Because so much of the text focuses on examples, spending class time in discussions or working in groups is easy. The examples are realistic, which promotes bringing outside materials into the classroom and having students find examples from their other classes.

This course arose through an effort to provide students with the mathematical tools they will need in courses that traditionally require college algebra. The topics were selected after lengthy consultations with our colleagues from departments across the campus. The skills our colleagues wanted students to learn from an entry-level mathematics course included facility with graphs, tables of values, linear algebraic manipulations, and, most importantly, some level of confidence in relating sentences to formulas, tables, and graphs. Perhaps surprisingly, our colleagues also wanted a qualitative understanding of rates of change, leading us to make this one of the most inportant themes in the text. Rates of change is a unifying concept, following from the observation that knowledge of the initial value and how a function changes are sufficient to understand the function. This concept is fully developed and exploited for linear and exponential functions in Chapters 3 and 4 and is carried through the rest of the text.

Our consultations also indicated that these goals are met only marginally, if at all, by a traditional college algebra course. The most common perception we got from our colleagues in other areas is that even when entering students have some facility with elementary mathematics, they may be afraid to apply it and indeed may see no relation at all between mathematics that they have been exposed to and applications of mathematics in other areas. As a result, compared to traditional college algebra texts, our treatment of topics is more geared toward applications in other disciplines. For example, we often use data tables to display functions, and students become proficient in recovering the formula for the function from data.

One of the most important goals of this text is to provide students with the opportunity to succeed at sophisticated mathematics. Our experience with the material in its fourth year of use is that the appropriately used calculator gives students the power needed to perform significant mathematical analyses. Many of our students report that success in mathematics is a new and finally pleasing experience. We also wanted students to see mathematics as part of their life experience, and we have anecdotal stories to indicate some success in that area as well.

This text was not designed as a prerequisite for calculus, but, as it develops, the text provides an excellent preparation for some of the newer reformed applied calculus texts. We are using it as such at Oklahoma State University and are pleased with the results.

Graphing calculator use and reference

This text is designed to be used with a graphing calculator, and the calculator is an essential part of the presentation as well as the exercises. Throughout the text we employ TI-83 screens, but many graphing calculators and even some computer software can serve the purpose. Because the major graphing calculator producers have made their products remarkably powerful and easy to use, students should have no difficulty becoming proficient with the basic calculator operation. To ensure this, an accompanying *Keystroke Guide* provides TI-82 and

TI-83 keystrokes for creating tables and graphs, entering expressions, etc., in the Quick Reference section of the *Guide*. Keystrokes are cross-referenced using footnote boxes in the main text. For example, ▢3.4 on page 175 of the text indicates that the keystrokes needed to to create this graph are found in item number **3.4** of the Quick Reference pages of the *Guide*. We recommend removing the appropriate Quick Reference sheets from the *Keystroke Guide* for consultation while reading the text. The *Keystroke Guide* also includes generic instructions to help students become familiar with graphing calculators in general, along with related exercises.

Suggested paths through this text

The seven chapters in this text provide more than enough material for a one-semester course with three credit hours. Chapters 1 though 4 form the core material. In the context of rates of change, we examine functions from several points of view and study in some detail the important examples of linear and exponential functions. After covering the core material, instructors have some options, as shown below.

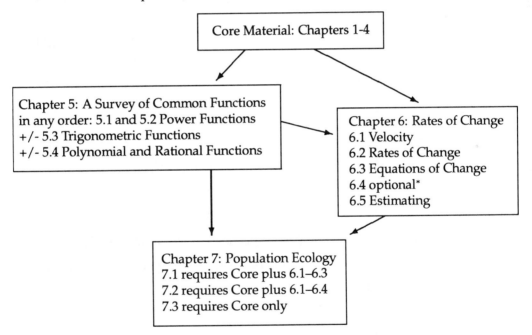

*Note: The discussion of graphical solutions for equations of change (i.e., differential equations) in Section 6.4 is the most challenging material in this text. (The authors believe it may also be the most rewarding.) It is necessary for parts of the presentation in Section 7.2 but otherwise may be omitted at the instructor's discretion. In particular, Section 6.5 on calculating rates of change does not depend on this discussion of equations of change. At Oklahoma State University we have found that students respond well to Chapter 6, and examining the texts used in courses from ecology to economics shows how useful the development of rates of change can be to students from a diverse collection of backgrounds.

Supplementary materials

- *Graphing Calculator Keystroke Guide and Drill Exercise Supplement* includes:

  - generic instructions to help students become familiar with their graphing calculator, along with related exercises.

  - TI-82 and TI-83 keystrokes for creating tables and graphs, entering expressions, etc., in the Quick Reference section of the *Guide*. Keystrokes are cross-referenced using footnote boxes in the main text. For example, $\boxed{3.4}$ on page 175 of the text indicates that the keystrokes needed to create this graph are found in item number **3.4** of the Quick Reference pages of the *Guide*.

  - supplementary drill exercises for each chapter for students needing extra practice.

- *Student Solutions Guide* includes complete solutions to all odd-numbered exercises.

- *Complete Solutions Guide* includes complete solutions to all exercises for instructors.

- *Instructor's Resource Guide with Tests* includes general teaching tips for adapting to this book's approach, specific section-by-section teaching tips, sample tests and quizzes, and transparency masters.

To Students

Every effort has been made to show mathematics as you are likely to encounter it in other courses as well as in daily life. Learning mathematics requires effort. But learning is also fun, and success in mathematics can be rewarding in terms of personal accomplishment. It can also facilitate understanding in other courses as well as in everyday experience, and it may be a key to attaining your career goals. We intend that you reap these and other benefits from your experience with this text.

How to learn with this text

Effective use of this text requires that you actively participate in the presentation. You should read with your graphing calculator turned on and with the Quick Reference pages from the *Keystroke Guide* handy. (Please see the description of this supplement above to see what it teaches and how it works.) It is not sufficient simply to read the examples in the text. Rather, you should work through each example yourself as it is presented, and when a calculator screen is shown, you should reproduce it on your own calculator.

As you begin, you will note that rarely is the final answer for an example presented simply as a number or a graph. Rather the answer is accompanied by sentences explaining how the answer was obtained and with appropriate conclusions. You should follow this pattern in solving the exercises. Your solution should include whatever calculations, graphs, or tables (copied from your calculator) you use as well as a clear statement of your conclusion accompanied by an explanation of your methods. A simple test of the clarity of your explanation is whether your peers can understand the solution by reading your work.

Our answers to the odd-numbered exercises are provided at the back of the text. They are of necessity brief and in general not acceptable as complete solutions. You should also be aware that for many of the exercises, there is no simple *right answer*. Instead, there is room for a number of conclusions, and any of them may be acceptable provided they are accompanied by a convincing argument. Sometimes you may have an answer that is different from that of one of your peers, and neither of your answers matches the one given in the back of the book, yet both of you may have correct solutions.

Mathematics is a tool that enhances your reasoning ability. It does not supplant that ability, and it is not a device that gives magical, unassailable answers. Whenever you are led to a conclusion that flies in the face of common sense, you should question the validity of your work and check carefully for a mistake.

A solicitation

This text is designed to be read by students, and while we are very much interested in input from instructors, the evolution of the text into its final form will be heavily influenced by what you have to say. We earnestly solicit any and all comments about the presentation. We would like your reactions to topics included or omitted as well as your estimation of the effectiveness of the presentation. We appreciate hearing about any errors, omissions, or inaccuracies in this preliminary edition. The best way to get information to us is through e-mail to `crauder`, `bevans`, or `noell`, each at `@math.okstate.edu`.

Thanks

Class Testers In addition to being used at our school, earlier drafts of *Functions and Change* were class tested at the following schools. We would like to thank the following instructors, some of whom have been using various drafts for several years, for being willing to try something different and for providing feedback based on their experiences. Thanks also to the students at these schools for their participation.

Bill Coberly	University of Tulsa
Doug Colbert	University of Nevada at Reno
Joel Haack	University of Northern Iowa
Joe Harkin	SUNY Brockport
John Lomax	Northeastern Oklahoma A & M

Reviewers The following reviewers also deserve our thanks for their encouragement, criticisms, and suggestions.

Judith Ahrens	Pellissippi State Technical Community College
Daniel Alexander	Drake University
Doug Colbert	University of Nevado at Reno
Dewey Furness	Ricks College
Joel Haack	University of Northern Iowa
John Haverhals	Bradley University
Judith Hector	Walters State Community College
Miles Hubbard	St. Cloud State University
Gina Kietzmann	Elmhurst College
Jerry Kissick	Portland Community College
John Lomax	Northeastern Oklahoma A & M
Patty Monroe	Greenville Technical College
Bernd Rossa	Xavier University
Mary Jane Sterling	Bradley University
Jeremy Underwood	Clayton College and State University

In writing this book, we have also relied on the help of other mathematicians as well as specialists from agriculture, biology, business, chemistry, ecology, economics, engineering, physics, political science, and zoology. We offer our thanks to Bruce Ackerson, Brian Adam, Robert Darcy, Joel Haack, Stanley Fox, Adrienne Hyle, Smith Holt, Jerry Johnson, Lionel Raff, Scott Turner, and Gary Young. Errors and inaccuracies in applications are due to the authors' misrepresentation of correct information provided by our able consultants. We are grateful to the National Science Foundation for its foresight and support of initial development and to Oklahoma State University for its support. We very much appreciate Charles Hartford's continuing patience and good humor through some trying times.

The most important participants in the development of this work are the students at Oklahoma State University, particularly those in the fall of 1995 and spring of 1996, who suffered through very early versions of this text, and whose input has shaped the current version. This book is written for entering mathematics students, and further student reaction will direct the evolution of this preliminary edition into a better product. Students and teachers at Oklahoma State University have had fun and learned with this material. We hope the same happens for others.

BRUCE CRAUDER
BENNY EVANS
ALAN NOELL

A Modeling Alternative
to College Algebra

Preliminary Edition

PROLOGUE *Calculator Arithmetic*

Graphing calculators provide powerful tools for mathematical analysis, and this power has profound effects on how modern mathematics and its applications are done. Many mathematical applications which traditionally required sophisticated mathematical development can now be successfully analyzed at an elementary level. Indeed, modern calculating power places entering students in a position to attack problems that in the past would have been considered too complicated. The first step is to become proficient with arithmetic on the calculator. In this chapter we discuss key mathematical ideas associated with calculator arithmetic. Chapter 1 of the *Keystroke Guide* is intended to provide additional help for those who may be new to the operation of the calculator, or who need a brief refresher on arithmetic operations.

Typing mathematical expressions

When we write expressions such as $\frac{71}{7} + 3^2 \times 5$ using pen and paper, the paper serves as a two-dimensional display, and we can express fractions by putting one number on top of another and exponents by using a superscript. When such expressions are entered on a computer, calculator, or typewriter, they must be written on a single line using special symbols and often additional parentheses. The *caret* symbol $\wedge$ is commonly used to denote an exponent, so in *typewriter notation* $\frac{71}{7} + 3^2 \times 5$ comes out as

$$71 \div 7 + 3 \wedge 2 \times 5 \, .$$

In Figure 0.1, we have entered $\boxed{0.1}$ this expression, and the resulting answer 55.14285714 is shown in Figure 0.2. You should use your calculator to verify that we did it correctly. (Since this is its first occurrence, we will point out that the footnote symbol in a box $\boxed{0.1}$ indicates that the exact keystrokes for doing this are shown on the Quick Reference pages of the *Keystroke Guide*.)

Figure 0.1: *Entering* $\frac{71}{7} + 3^2 \times 5$

Figure 0.2: *The value of* $\frac{71}{7} + 3^2 \times 5$

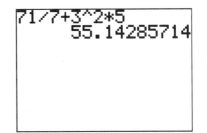

Rounding

When a calculation yields a long answer such as 55.14285714 in Figure 0.2, we will commonly shorten it to a more manageable size by *rounding*. Rounding means that we keep a few of the digits after the decimal point, possibly changing the last one, and discard the rest. There is no set rule for how many digits after the decimal point you should keep; in practice it depends on how much accuracy you need in your answer as well as the accuracy of input data. As a general rule in this text we will round to two decimal places. Thus for

$$\frac{71}{7} + 3^2 \times 5 = 55.14285714$$

we would report the answer as 55.14.

In order to make the abbreviated answer more accurate, it is standard practice to increase the last decimal entry by one if the following entry is 5 or greater. Verify with your calculator that

$$\frac{58.7}{6.3} = 9.317460317 \, .$$

In this answer the next digit after 1 is 7, which indicates that we should round up, and so we would report the answer rounded to two decimal places as 9.32. Note that in reporting 55.14 as the rounded answer above, we followed this same rule. The next digit after 4 in 55.14285714 is 2, which does not indicate that we should round up.

To provide additional emphasis for this idea, we have shown in Figure 0.3 a calculation where rounding does not change the last reported digit, and in Figure 0.4 a calculation where rounding requires that the last reported digit be changed.

Figure 0.3: An answer that will be reported as 1.71

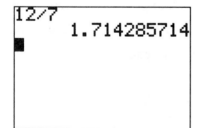

Figure 0.4: An answer that will be reported as 1.89

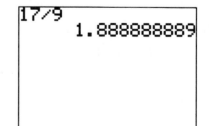

KEY IDEA 0.1: ROUNDING

When reporting complicated answers, we will adopt the convention of keeping two places beyond the decimal point. The last digit is increased by one if the next following number is 5 or greater.

Scientific notation

It is cumbersome to write down all the digits of some very large or very small numbers. A prime example of such a large number is *Avogadro's number*, which is the number of atoms in 12 grams of carbon 12. Its value is about

$$602,000,000,000,000,000,000,000.$$

An example of a small number which is awkward to write is the mass in grams of an electron:

$$0.000\,000\,000\,000\,000\,000\,000\,000\,000\,000\,911 \text{ grams.}$$

Scientists and mathematicians usually express such numbers in a more compact form using *scientific notation*. In this notation, numbers are written in a form with one nonzero digit to the left of the decimal point times a power of 10. Examples of numbers written in scientific notation are 2.7×10^4 and 2.7×10^{-4}. The power of 10 tells how the decimal point should be moved in order to write out the number longhand. The 4 in 2.7×10^4 means that we should move the decimal point 4 places to the right. Thus

$$2.7 \times 10^4 = 27,000$$

since we move the decimal point four places to the right. When the exponent on 10 is negative, the decimal point should be moved to the left. Thus

$$2.7 \times 10^{-4} = 0.00027$$

since we move the decimal point four places to the left. With this notation, Avogadro's number comes out as 6.02×10^{23}, and the mass of an electron as 9.11×10^{-31} grams.

Many times calculators display numbers like this but use a different notation for the power of 10. For example, Avogadro's number 6.02×10^{23} is displayed as 6.02E23, and the mass in grams of an electron 9.11×10^{-31} is shown as 9.11E-31. In Figure 0.5 we have calculated 2^{50}. The answer reported by the calculator written in longhand is 1,125,899,907,000,000. In presenting the answer in scientific notation, it would in many settings be appropriate to round to two decimal places as 1.13×10^{15}. In Figure 0.6 we have calculated $\frac{7}{3^{20}}$. The answer reported there equals 0.000 000 002 007 580 394. If we write it in scientific notation and round to two decimal places, we get 2.01×10^{-9}.

Figure 0.5: *Scientific notation for a large number*

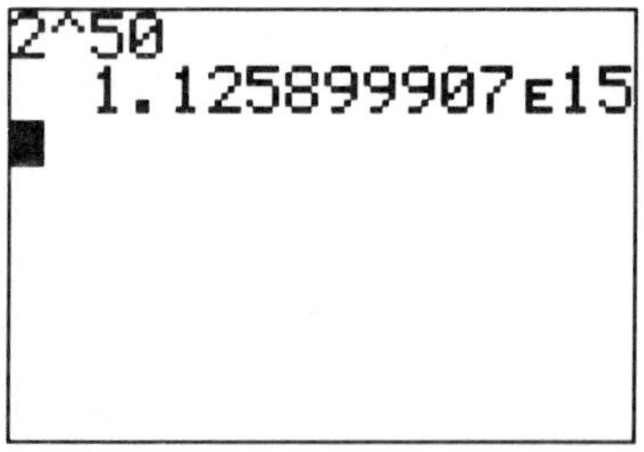

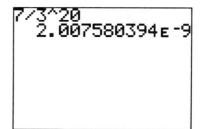

Figure 0.6: *Scientific notation for a small number*

```
7/3^20
    2.007580394E-9
```

Parentheses and grouping

When parentheses appear in a calculation, then the operations inside are to be done first. Thus $4(2 + 1)$ means that we should first add up 2+1 and then multiply the result by 4, getting an answer of 12. This is correctly entered ⌐0.2⌐ and calculated in Figure 0.7. Where parentheses appear, their use is essential. If we had entered the expression as $4 \times 2 + 1$, thus leaving out the parentheses, the calculator would think we meant first to multiply 4 times 2 and then to add 1 to the result, giving an incorrect answer of 9. This incorrect entry is shown in Figure 0.8.

Figure 0.7: *A correct calculation of* $4(2 + 1)$ *when parentheses are properly entered*

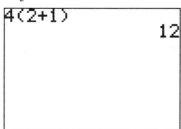

Figure 0.8: *An incorrect calculation of* $4(2 + 1)$ *caused by omitting parentheses*

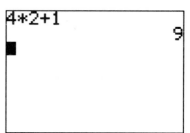

Sometimes parentheses do not appear, but we must supply them. For example, $\dfrac{17}{5 + 3}$ means $17 \div (5 + 3)$. The parentheses are there to show that the whole expression $5 + 3$ goes in the denominator. To do this on the calculator, we must supply ⌐0.3⌐ these parentheses. Figure 0.9 shows the result. If the parentheses are not used, and $\dfrac{17}{5 + 3}$ is entered as $17 \div 5 + 3$, the calculator will think that only the 5 goes in the denominator of the fraction. This error is shown in Figure 0.10. Similarly $\dfrac{8 + 9}{7 + 2}$ means $(8 + 9) \div (7 + 2)$; the parentheses around $8 + 9$ indicate that the entire expression goes in the numerator, and the parentheses around $7 + 2$ indicate that the entire expression goes in the denominator. Enter ⌐0.4⌐ the expression on your calculator and check that the answer rounded to two places is 1.89. The same problem

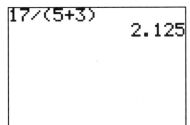

Figure 0.9: Proper use of parentheses in the calculation of $\dfrac{17}{5+3}$

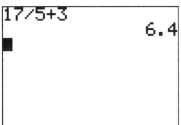

Figure 0.10: An incorrect calculation of $\dfrac{17}{5+3}$ caused by omitting parentheses

can occur with exponents. For example $3^{2.7 \times 1.8}$ in typewriter notation is $3 \wedge (2.7 \times 1.8)$. Check $\boxed{0.5}$ to see that the answer rounded to two places is 208.36.

In general, we advise that if you have trouble entering an expression into your calculator, or if you get an answer that you know is incorrect, go back and re-enter the expression after first writing it out in typewriter notation, and be careful to supply all needed parentheses.

Minus signs

The minus sign used in arithmetic calculations actually has two different meanings. If you have \$9 in your wallet and spend \$3, then you will have $9 - 3 = 6$ dollars left. Here the minus sign means that we are to perform the operation of subtracting 3 from 9. Suppose in another setting that you receive news from the bank that your checking account is overdrawn by \$30, and so your balance is -30 dollars. Here the minus sign is used to indicate that the number we are dealing with is negative; it does not signify an operation between two numbers. In everyday usage, the distinction is rarely emphasized and may go unnoticed. But most calculators actually have different keys $\boxed{0.6}$ for the two operations, and they cannot be used interchangeably. Thus differentiating between the two operations becomes a crucial factor in correct calculator operation.

Once the problem is recognized, it is usually easy to spot when the minus sign denotes subtraction (when two numbers are involved) and when it indicates a change in sign (when only one number is involved). The following examples should help clarify the situation:

$$-8 - 4 \quad \text{means} \quad \textit{negative 8 subtract 4} \; \boxed{0.7}$$

$$\frac{3 - 7}{-2 \times 3} \quad \text{means} \quad \frac{3 \; \textit{subtract} \; 7}{\textit{negative} \; 2 \times 3} \; \boxed{0.8}$$

$$2^{-3} \quad \text{means} \quad 2^{\textit{negative} \; 3}.$$

The calculation [0.9] of 2^{-3} is shown in Figure 0.11. If we try to use the calculator's subtraction key [0.10], the calculator will not understand the input and will produce an error message such as the one in Figure 0.12.

Figure 0.11: Calculation of 2^{-3} *using the negative key*

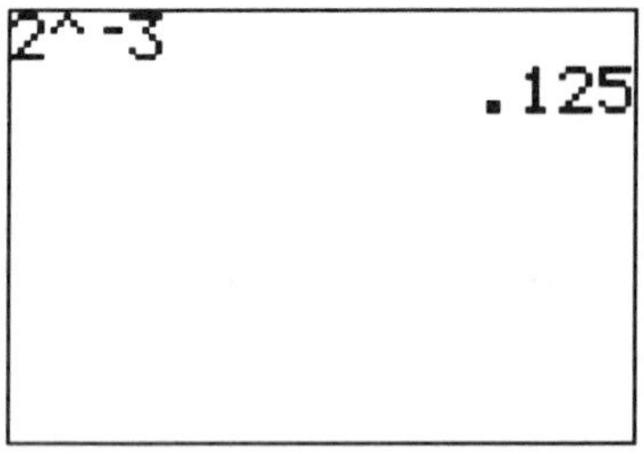

Figure 0.12: Syntax error when subtraction operation is used in 2^{-3}

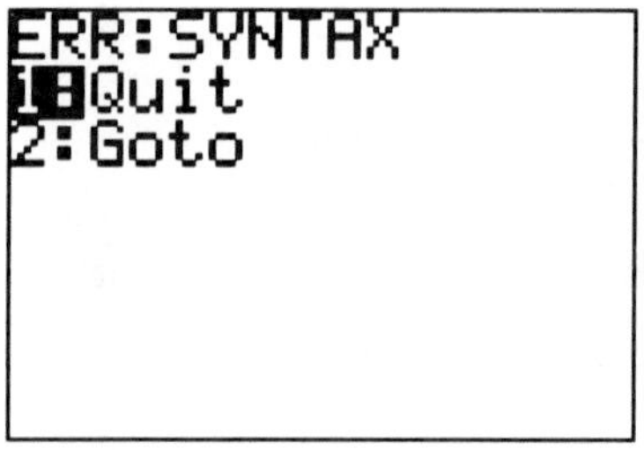

EXAMPLE 0.1 *Some Simple Calculations*

Make the following calculations and report the answer rounded to two digits beyond the decimal point.

1. $\dfrac{\sqrt{11.4 - 3.5}}{26.5}$
2. $\dfrac{7 \times 3^{-2} + 1}{3 - 2^{-3}}$

Solution to Part 1: To make sure everything we want is included under the square root symbol, we need to use parentheses. In typewriter notation, this looks like

$$\sqrt{\ }(11.4 - 3.5) \div 26.5 \; .$$

We have calculated [0.11] this in Figure 0.13. Since the third digit beyond the decimal point, 6, is five or larger, we report the answer as 0.11.

Solution to Part 2: We need to take care to use parentheses to insure that the numerator and denominator are right, and we must use the correct keys for negative signs and subtraction. In expanded typewriter notation,

$$\frac{7 \times 3^{-2} + 1}{3 - 2^{-3}} = (7 \times 3 \wedge \; \textit{negative}\; 2 + 1) \div (3 \; \textit{subtract} \; 2 \wedge \; \textit{negative}\; 3) \; .$$

The result [0.12] 0.6183574879 is shown in Figure 0.14. We round this to 0.62.

Figure 0.13: *Solution to Part 1*

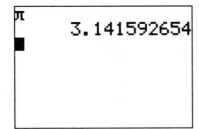

Figure 0.14: *Solution to Part 2*

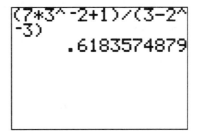

Special numbers π and e

Two numbers, π and e, occur so often in mathematics and its applications that they deserve special mention. The number π is familiar from the formulas for the circumference and area of a circle:

$$\text{Area of a circle of radius } r \;=\; \pi r^2$$

$$\text{Circumference of a circle of radius } r \;=\; 2\pi r \,.$$

The approximate value of π is 3.14159, but its exact value cannot be expressed by a simple decimal, and that is why it is normally written using a special symbol. Most calculators allow you to enter ⟨0.13⟩ the symbol π directly as shown in Figure 0.15. When we ask the calculator for a numerical answer ⟨0.14⟩ we get the decimal approximation of π shown in Figure 0.15.

The number e may not be as familiar as π, but it is just as important. Like π, it cannot be expressed exactly as a decimal, but its approximate value is 2.71828. In Figure 0.16 we have entered ⟨0.15⟩ e, and the calculator responded with the decimal approximation shown.

Figure 0.15: *A decimal approximation of π*

Figure 0.16: *A decimal approximation of e*

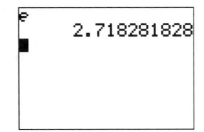

Chain calculations

Some calculations are most naturally done in stages. Many calculators have a special key $\boxed{0.16}$ that accesses the result of the last calculation, allowing you to enter your work in pieces. To show how this works, let's look at

$$\left(\sqrt{13}-\sqrt{2}\right)^3 + \frac{17}{2+\pi}\;.$$

We will make the calculation in pieces. First we calculate $\left(\sqrt{13}-\sqrt{2}\right)^3$. Enter $\boxed{0.17}$ this to get the answer in Figure 0.17. To finish the calculation we need to add this answer to $\dfrac{17}{2+\pi}$:

$$\left(\sqrt{13}-\sqrt{2}\right)^3 + \frac{17}{2+\pi} = \text{First answer} + \frac{17}{2+\pi}\;.$$

In Figure 0.18 we have used the answer $\boxed{0.18}$ from Figure 0.17 to complete the calculation.

Figure 0.17: The first step in a chain calculation

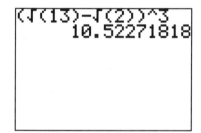

Figure 0.18: Completing a chain calculation

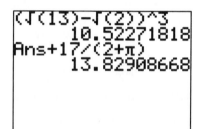

Accessing the results of one calculation for use in another can be particularly helpful if the same thing appears several times in an expression. For example, let's calculate

$$\frac{3^{\frac{7}{9}} + 2^{\frac{7}{9}}}{5^{\frac{7}{9}} - 4^{\frac{7}{9}}}\;.$$

Since $\dfrac{7}{9}$ occurs several times, we have calculated it first in Figure 0.19. Then we have used the results to complete $\boxed{0.19}$ the calculation. We would report the final answer rounded to two decimal places as 7.30.

There is an additional advantage to accessing directly answers of previous calculations. It might seem reasonable to calculate $\dfrac{7}{9}$ first, then round it to two decimal places, and then use that to complete the calculation. Thus we would be calculating

$$\frac{3^{0.78} + 2^{0.78}}{5^{0.78} - 4^{0.78}}\;.$$

This is done in Figure 0.20, which shows the danger in this practice. We got an answer rounded to two decimal places of 7.27, somewhat different from the more accurate answer 7.30 which we got earlier. In many cases errors caused by early rounding can be much more severe than is shown by this example. In general, if you are making a calculation in several steps, you should not round until you get the final answer. An important exception to this general rule occurs in applications where the result of an intermediate step must be rounded because of the context. For example, in a financial computation dollar amounts would be rounded to two decimal places.

Figure 0.19: *Accessing previous results to get an accurate answer*

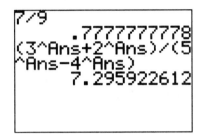

Figure 0.20: *Inaccurate answer caused by early rounding*

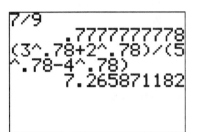

EXAMPLE 0.2   *Compound Interest and APR*

There are a number of ways in which lending institutions report and charge interest.

1. Paying *simple interest* on a loan means that you wait until the end of the loan before calculating or paying any interest. If you borrow $5000 from a bank that charges 7% simple interest, then after t years you will owe

$$5000 \times (1 + 0.07t) \text{ dollars.}$$

Under these conditions, how much money will you owe after 10 years?

2. Banks more commonly *compound the interest*. That is, at certain time periods the interest you have incurred is calculated and added to your debt. From that time on you incur interest not only on your principal (the original debt) but on the added interest as well. Suppose the interest is compounded yearly, but you make no payments and there are no finance charges. Then, again with a principal of $5000 and 7% interest, after t years you will owe

$$5000 \times 1.07^t \text{ dollars.}$$

Under these conditions, how much will you owe after 10 years?

3. For many transactions such as automobile loans or home mortgages, interest is compounded monthly rather than yearly. In this case, the amount owed is calculated each month using the *monthly interest rate*. If r (as a decimal) is the monthly interest rate, then after m months, the amount owed is

$$5000 \times (1 + r)^m \text{ dollars,}$$

assuming the principal is $5000.

The value of r is usually not apparent from the loan agreement. But lending institutions are required by the *Truth in Lending Act* to report the *annual percentage rate* or APR in a prominent place on all loan agreements. The same statute requires that the value of r be calculated using the formula

$$r = \frac{\text{APR}}{12} .$$

If the annual percentage rate is 7%, what is the amount owed after 10 years?[1]

Solution to Part 1: To get the amount owed after 10 years, we use $t = 10$ to get

$$5000 \times (1 + 0.07 \times 10) .$$

Enter $\boxed{0.20}$ this on the calculator as we have done in Figure 0.21. We find that the amount owed in 10 years will be $8500.

Solution to Part 2: This time we use $\boxed{0.21}$

$$5000 \times 1.07^{10} .$$

From Figure 0.22 we see that, rounded to the nearest cent, the amount owed will be $9835.76. Comparison with Part 1 shows the effect of compounding interest. We should note that at higher interest rates the effect is more dramatic.

---

[1] Many consider the relationship between the monthly interest rate and the APR mandated by the Truth in Lending Act to be misleading. If for example you borrow $100 at an APR of 10%, then if no payments are made you may expect to owe $110 at the end of one year. If interest is compounded monthly, then you will in fact owe somewhat more. For more information see the discussion in Section 5.6 of *Fundamentals of Corporate Finance* by S. Ross, R. Westerfield, and B. Jordan, published by Richard D. Irwin Inc., 1995. See also Exercise 9.

Figure 0.21: *Balance after 10 years using simple interest*

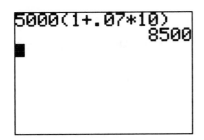

Figure 0.22: *Balance after 10 years using yearly compounding*

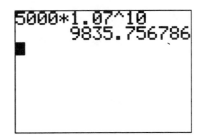

Solution to Part 3: The first step is to use the formula

$$r = \frac{\text{APR}}{12} = \frac{0.07}{12}$$

to get the value of r as we have done in Figure 0.23. Ten years is 120 months, and this is the value we use for m. Using this value for m and incorporating the value of r that we just calculated, we entered $\boxed{0.22}$ $5000 \times (1 + r)^{120}$ in Figure 0.24, and we conclude that the amount owed will be \$10,048.31. Comparing this with the answer from Part 2, we see that the difference between yearly and monthly compounding is significant. It is important that you know how interest on your loan is calculated, and this may not be easy to find out from the paperwork you get from a lending institution. The APR will be reported, but the compounding periods may not be shown at all.

Figure 0.23: *Getting the monthly interest rate from the APR*

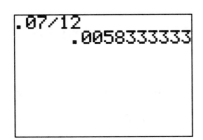

Figure 0.24: *Balance after 10 years using monthly compounding*

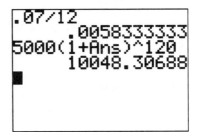

Prologue Exercise Set

1. **Arithmetic:** Calculate the following and report the answer rounded to two decimal places. For some of the calculations you may wish to use the chain calculation facility of your calculator to help avoid errors.

   (a) $(4.3 + 8.6)(8.4 - 3.5)$

   (b) $\dfrac{2^{3.2} - 1}{\sqrt{3} + 4}$

   (c) $\sqrt{2^{-3} + e}$

   (d) $(2^{-3} + \sqrt{7} + \pi)\left(e^2 + \dfrac{7.6}{6.7}\right)$

   (e) $\dfrac{17 \times 3.6}{13 + \frac{12}{3.2}}$

2. **A good investment:** You have just received word that your original investment of $850 has increased in value by 13%. What is the value of your investment today?

3. **A bad investment:** You have just received word that your original investment of $720 has decreased in value by 7%. What is the value of your investment today?

4. **An uncertain investment:** Suppose you invested $1300 in the stock market two years ago. During the first year, the value of the stock increased by 12%. During the second year, the value of the stock decreased by 12%. How much money is your investment worth at the end of the two-year period? Did you earn money or lose money? (<u>Note</u>: The answer to the first question is *not* $1300.)

5. **Pay raise:** You receive a raise in your hourly pay from $7.25 per hour to $7.50 per hour. What percent pay raise does this represent?

6. **Heart disease:** The number of deaths in a certain county due to heart disease decreased from 235 in one year to 221 in the next year. What percent decrease in deaths due to heart disease does this represent?

7. **Future value:** Business and finance texts refer to the value of an investment at a future time as its *future value*. If an investment of P dollars is compounded yearly at an interest rate of r as a decimal, then the value of your investment after t years is given by

$$\text{Future value} = P \times (1 + r)^t .$$

In this formula, $(1 + r)^t$ is known as the *future value interest factor,* so the formula above can also be written

$$\text{Future value} = P \times \text{Future value interest factor} .$$

Financial officers normally calculate this (or look it up in a table) first.

(a) What future value interest factor will make an investment double? Triple?

(b) If you have an investment which is compounded yearly at a rate of 9%, find the future value interest factor for a 7 year investment.

(c) Use the results from Part (b) to calculate the 7 year future value if your initial investment is $5000.

8. **The Rule of 72:** *This is a continuation of Exercise 7.* Financial advisors sometimes use a rule of thumb known as the *Rule of 72* to get a rough estimate of the time it takes for an investment to double in value. For an investment which is compounded yearly at an interest rate of $r\%$, this rule says it will take about $\dfrac{72}{r}$ years for the investment to double. In this calculation r is the integer interest rate rather than a decimal. Thus if the interest rate is 8% we would use $\dfrac{72}{8}$ rather than $\dfrac{72}{0.08}$.

For the remainder of this exercise we will consider an investment which is compounded yearly at an interest rate of 13%.

(a) According to the Rule of 72 how long will it take the investment to double in value?

Parts (b) and (c) of this exercise will check to see how accurate this estimate is for this particular case.

(b) Using the answer you got from Part (a) of this exercise, calculate the future value interest factor (as defined in Exercise 7). Is it exactly the same as your answer to the first question in Part (a) of Exercise 7?

(c) If your initial investment was $5000, use your answer from Part (b) to calculate the future value. Did your investment exactly double?

9. **The Truth in Lending Act:** Many lending agencies compound interest more often than yearly, and, as was noted in Example 0.2, they are required to report in a prominent place on the loan agreement the annual percentage rate or APR. Furthermore, they are required to calculate the APR in a specific way. If r is the monthly interest rate then the APR is calculated using

$$\text{APR} = 12 \times r .$$

   (a) Suppose a credit card company charges a monthly interest rate of 1.9%. What APR are they required to report?

   (b) If you borrow $6000 from a credit card company which quotes an APR of 22.8%, and if no payments are made, the phrase *Annual Percentage Rate* leads some to believe that at the end of one year interest would be calculated as 22.8% simple interest on $6000. How much would you owe at the end of a year if interest is calculated in this way?

   (c) If interest is compounded monthly (which is common), the actual amount you would owe is given by

$$6000 \times 1.019^{12} .$$

   What is the actual amount you would owe at the end of a year?

10. **The size of the Earth:** The radius of the Earth is approximately 4000 miles.

   (a) How far is it around the equator? (<u>Hint</u>: You are looking for the circumference of a circle.)

   (b) What is the volume of the Earth? (<u>Note</u>: The volume of a sphere of radius r is given by $\frac{4}{3}\pi r^3$.)

   (c) What is the surface area of the Earth? (<u>Note</u>: The surface area of a sphere of radius r is given by $4\pi r^2$.)

11. **When the radius increases:**

   (a) A rope is wrapped tightly around a wheel with a radius of 2 feet. If the radius of the wheel is increased by 1 foot to a radius of 3 feet, by how much must the rope be lengthened to fit around the wheel?

   (b) Consider a rope wrapped around the equator of the Earth. We noted in Exercise 10 that the radius of the Earth is about 4000 miles. That is 21,120,000 feet. Suppose now that the rope is to be suspended exactly one foot above the equator. By how much must the rope be lengthened to accomplish this?

12. **The length of Earth's orbit:** The Earth is approximately 93 million miles from the sun. For this exercise we will assume that the Earth's orbit is a circle.[2]

    (a) How far does the Earth travel in a year?

    (b) What is the velocity in miles per year of the Earth in its orbit? (<u>Hint</u>: Recall that $\text{Velocity} = \dfrac{\text{Distance}}{\text{Time}}$.)

    (c) How many hours are there in a year? (<u>Note</u>: For this problem ignore leap years.)

    (d) What is the velocity in miles per hour of the Earth in its orbit?

13. **A population of bacteria:** Some populations, such as bacteria, can be expected under the right conditions to show *exponential growth*. If 2000 bacteria of a certain type are incubated under ideal conditions, then after t hours we expect to find 2000×1.07^t bacteria present. How many bacteria would we expect to find after 8 hours? How many after two days?

PROLOGUE SUMMARY

Modern graphing calculators are well designed for ease of use, but care must be taken when entering expressions. The most common errors occur when parentheses are omitted or misused. Also, when you use a calculator, rounding and scientific notation are significant concepts. The special numbers e and π are important.

Typing Expressions and Parentheses

When entering an expression in the calculator, you must enter it not as one would write it on paper, but rather in *typewriter notation*. If you have trouble getting an expression into the calculator properly, first write it out on paper in typewriter notation and then enter it into the calculator. Parentheses are essential when you need to tell the calculator that a certain operation is to be applied to a group of numbers.

Rounding

In order to do accurate calculations, the calculator uses decimals with many digits. Often only a few digits after the decimal point are needed for the final answer. This is accomplished by *rounding*. There is no set number of digits used in rounding; that depends on the accuracy of data entered and on the accuracy needed for the answer. In general, however, answers are reported in this text rounded to two decimal places.

---

[2]The orbit of the Earth is in fact an ellipse, but for many practical applications the assumption that it is a circle yields reasonably accurate results.

Rounding Convention for This Text: Unless otherwise specified, answers should be rounded to two decimal places. If the third digit beyond the decimal point is less than 5, all digits beyond the second are discarded. If the third digit is 5 or larger, we increase the second digit by one before discarding additional digits.

Scientific Notation

Some numbers use so many digits that it is more convenient to express them in *scientific notation*. This simply means to write the number using only a few digits and multiply by a power of 10 which tells how the decimal point should be adjusted. The adjustment required depends on the sign on the power of 10. Scientific notation can be entered in the calculator using 10 to a power, but when the calculator reports an answer in scientific notation, a special notation is common.

Entry	Calculator display	Meaning
$number \times 10^{+k}$	*number* **E+k**	Move decimal point k places right
$number \times 10^{-k}$	*number* **E-k**	Move decimal point k places left

Special numbers

There are two special numbers, e and π, which occur so often in mathematics and its applications that their use cannot be avoided. Modern calculators allow for their direct entry. The number π is the familiar ratio of the circumference of a circle to its diameter. The number e is perhaps less familiar but is just as important, and it often arises in certain exponential contexts. Neither of these numbers can be expressed exactly as a finite decimal, but their approximate values are given below.

$$\pi \approx 3.14159$$
$$e \approx 2.71828$$

CHAPTER 1 *Functions*

A fundamental idea in mathematics and its applications is that of a *function*, which tells how one thing depends on others. This idea is so basic that it is impossible to say where and when it originated, and it was almost certainly conceived independently by any number of people at different times and places. The idea remains today as the cornerstone to understanding and using mathematics.

In applications of mathematics, functions are often representations of real phenomena or events. Thus we say that they are *models*. Obtaining a function or functions to act as a model is commonly the key to understanding physical, natural, and social science phenomena. This applies to business and many other areas as well.

1.1 FUNCTIONS GIVEN BY FORMULAS

There are a number of common ways in which functions are presented and used. We look first at functions given by formulas since this provides a natural context for explaining how a function works.

Functions of one variable

If your job pays $7.00 per hour, then the money M (in dollars) that you make depends on the number of hours h that you work, and the relationship is given by a simple formula:

$$\text{Money} = 7 \times \text{Hours worked}$$
$$M = 7h \text{ dollars.}$$

The formula $M = 7h$ shows how the money M that you earn depends on the number of hours h that you work, and we say that M *is a function of* h. In this context we are thinking of h as a *variable* whose value we may not know until the end of the week. Once the value of h is known, the formula $M = 7h$ can be used to calculate the value of M. To emphasize that M is a function of h it is common to write $M = M(h)$ and to write the formula as $M(h) = 7h$.

Functions given in this way are very easy to use. For example, if you work 30 hours, then in *functional notation*, $M(30)$ is the money you would earn. To calculate that, you need only replace h in the formula by 30:

$$M = M(30) = 7 \times 30 = 210 \text{ dollars.}$$

It is important to remember that h is measured in hours and M is measured in dollars. You will not be very happy if your boss makes a mistake and pays you $7 \times 10 = 70$ cents for 10 hours worked. You may be happier if she pays you $7 \times 30 = 210$ dollars for 30 minutes of work, but both calculations are of course incorrect. The formula is not useful unless you state in words the units you are using. A proper presentation of the formula for this function would be $M = 7h$, where h is measured in hours and M is measured in dollars. The words giving the units are as important as the formula.

We should also note that you can use different letters for variables if you want. Whatever letters you use, it is critical that you explain in words what the letters mean. We could for example use the letter t instead of h to represent the number of hours worked. If we do that, we would emphasize the functional relationship with $M = M(t)$ and present the formula as $M = 7t$, where t is the number of hours worked, and M is the money earned in dollars.

Functions of several variables

Sometimes functions depend on more than one variable. Your grocery bill G may depend on the number a of apples you buy, the number s of sodas you buy, and the number p of frozen pizzas you put in your basket. If apples cost 60 cents each, sodas cost 50 cents each, and pizzas cost \$3.25 each, then we can express $G = G(a, s, p)$ as

$$\text{Grocery bill} \quad = \quad \text{Total cost of apples} + \text{Total cost of sodas} + \text{Total cost of pizzas}$$
$$G \quad = \quad 0.6a + 0.5s + 3.25p \,,$$

where G is measured in dollars. The notation $G = G(a, s, p)$ is simply a way of emphasizing that G is a function of the variables a, s, and p, or that the value of G depends on a, s, and p. We could also give a correct formula for the function as $G = 60a + 50s + 325p$, where G is measured this time in cents. Either expression is correct as long as we explicitly say what units we are using.

EXAMPLE 1.1   *A Grocery Bill*

Suppose your grocery bill is given by the function $G = G(a, s, p)$ above (with G measured in dollars).

1. Use functional notation to show the cost of buying 4 apples, 2 sodas, and 3 pizzas, and then calculate it.

2. Explain the meaning of $G(2, 6, 1)$.

3. Calculate the value of $G(2, 6, 1)$.

Solution to Part 1: Since we are buying 4 apples, we use $a = 4$. Similarly, we are buying 2 sodas and 3 pizzas, so that $s = 2$ and $p = 3$. Thus, in functional notation our grocery bill is $G(4, 2, 3)$. To calculate this we use the formula $G = 0.6a + 0.5s + 3.25p$, replacing a by 4, s by 2, and p by 3:

$$
\begin{aligned}
G(4, 2, 3) &= 0.6 \times 4 + 0.5 \times 2 + 3.25 \times 3 \\
&= 13.15 \text{ dollars.}
\end{aligned}
$$

Thus the cost is $13.15.

Solution to Part 2: The expression $G(2, 6, 1)$ is the value of G when $a = 2$, $s = 6$ and $p = 1$. It is your grocery bill when you buy 2 apples, 6 sodas, and 1 frozen pizza.

Solution to Part 3: We calculate $G(2, 6, 1)$ just as we did in Part 1, but this time we use $a = 2$, $s = 6$, and $p = 1$:

$$
\begin{aligned}
G(2, 6, 1) &= 0.6 \times 2 + 0.5 \times 6 + 3.25 \times 1 \\
&= 7.45 \text{ dollars.}
\end{aligned}
$$

Thus the cost is $7.45.

Even when the formula for a function is complicated, the idea of how you use it remains the same. Let's look for example at $f = f(x)$ where f is determined as a function of x by the formula

$$
f = \frac{x^2 + 1}{\sqrt{x}} \ .
$$

The value of f when x is 3 is expressed in functional notation as $f(3)$. To calculate $f(3)$, we simply replace x in the formula by 3:

$$
f(3) = \frac{3^2 + 1}{\sqrt{3}} \ .
$$

You should check $\boxed{1.1}$ to see that the calculator gives an answer of 5.773502692, which we round to 5.77. Do not allow formulas such as this one to intimidate you. With the aid of the calculator it is easy to deal with them.

EXAMPLE 1.2   *Borrowing Money*

When you borrow money to buy a home or a car, you pay off the loan in monthly payments, but interest is always accruing on the outstanding balance. This makes the determination of your monthly payment on a loan more complicated than you may expect. If you borrow P dollars at a monthly interest rate[1] of r (as a decimal) and wish to pay off the note in t months, then your monthly payment $M = M(P, r, t)$ in dollars can be calculated using

$$M = \frac{Pr(1 + r)^t}{(1 + r)^t - 1} .$$

1. Explain the meaning of $M(7800, 0.0067, 48)$ and calculate its value.

2. Suppose you borrow $5000 to buy a car and wish to pay it off over 3 years. Take the prevailing monthly interest rate to be 0.58%. (That is an APR of $12 \times 0.58 = 6.96\%$.) Use functional notation to show your monthly payment and then calculate its value.

Solution to Part 1: The expression $M(7800, 0.0067, 48)$ gives your monthly payment on a $7800 loan that you pay off in 48 months (4 years) at a monthly interest rate of 0.67%. (That is an APR of $12 \times 0.67 = 8.04\%$.) To get its value, we use the formula above, putting 7800 in place of P, 0.0067 in place of r, and 48 in place of t:

$$M(7800, 0.0067, 48) = \frac{7800 \times 0.0067 \times 1.0067^{48}}{1.0067^{48} - 1} .$$

This can be entered all at once on the calculator, but to avoid typing errors we show the calculation in pieces. The calculation $\boxed{1.2}$ of the denominator $1.0067^{48} - 1$ is shown in Figure 1.1. To complete the calculation $\boxed{1.3}$ we need to get

$$\frac{7800 \times 0.0067 \times 1.0067^{48}}{\text{Answer from first calculation}} .$$

We round the answer shown in Figure 1.2 to get the monthly payment of $190.57.

---

[1] Here we are assuming monthly payment and interest compounding. If you use the annual percentage rate or APR reported on your loan agreement, then you have $r = \dfrac{\text{APR}}{12}$. See also Exercise 9 at the end of the Prologue.

Figure 1.1: *The first step in calculating a loan payment*

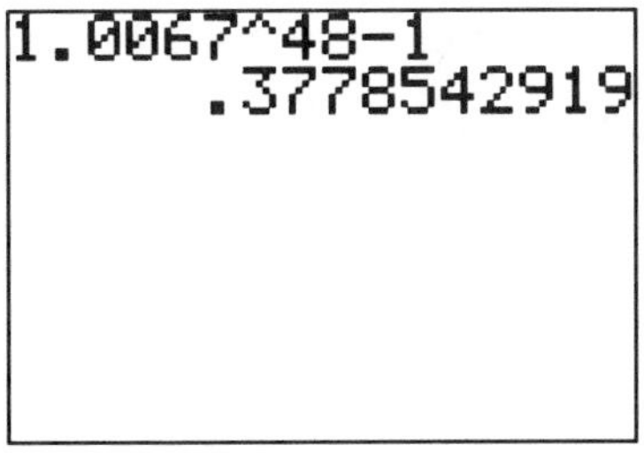

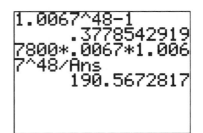

Figure 1.2: *Completing the calculation*

Solution to Part 2: We borrow $5000, so we use $P = 5000$. The monthly interest rate is 0.58%, so we use $r = 0.0058$, and we pay the loan off in 3 years or 36 months, making $t = 36$. In functional notation, the monthly payment is $M(5000, 0.0058, 36)$. To calculate it we use

$$M(5000, 0.0058, 36) = \frac{5000 \times 0.0058 \times 1.0058^{36}}{1.0058^{36} - 1}.$$

Once again we make the calculation in two stages. First $\boxed{1.4}$ we get $1.0058^{36} - 1$ as shown in Figure 1.3. As before we use this answer to complete the calculation $\boxed{1.5}$:

$$\frac{5000 \times 0.0058 \times 1.0058^{36}}{\text{Answer from the first calculation}}.$$

The result in Figure 1.4 shows that we will have to make a monthly payment of $154.29.

Figure 1.3: *The first step in calculating the payment on a $5000 loan*

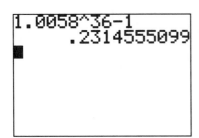

Figure 1.4: *Completing the calculation*

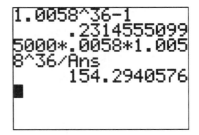

Exercise Set 1.1

1. **Practicing calculations with formulas:** For each of the following functions $f = f(x)$ find the value of $f(3)$, reporting the answer rounded to two decimal places. Use your calculator where it is appropriate.

    (a) $f = 3x + \dfrac{1}{x}$

    (b) $f = 3^{-x} - \dfrac{x^2}{x+1}$

    (c) $f = \sqrt{2x + 5}$

2. **Flushing chlorine:** City water, which is slightly chlorinated, is being used to flush a tank of heavily chlorinated water. The concentration $C = C(t)$ of chlorine in the tank t hours after flushing begins is given by

$$C = 0.1 + 2.78e^{-0.37t} \text{ milligrams per gallon.}$$

    (a) What is the initial concentration of chlorine in the tank?

    (b) Express the concentration of chlorine in the tank after 3 hours using functional notation, and then calculate its value.

3. **A population of deer:** When a breeding group of animals is introduced into a restricted area such as a wildlife reserve, the population can be expected to grow rapidly at first but level out when the population grows to near the maximum that the environment can support. Such growth is known as *logistic population growth*, and ecologists sometimes use a formula to describe it. The number N of deer present at time t (measured in years since the herd was introduced) on a certain wildlife reserve has been determined by ecologists to be given by the function

$$N = \frac{12.36}{0.03 + 0.55^t} \ .$$

    (a) How many deer were initially on the reserve?

    (b) Calculate $N(10)$ and explain the meaning of the number you have calculated.

    (c) Express the number of deer present after 15 years using functional notation and then calculate it.

    (d) How much increase in the deer population do you expect from the tenth to the fifteenth year?

4. **A car that gets 32 miles per gallon:** The cost C of operating a certain car which gets 32 miles per gallon is a function of the price g in dollars per gallon of gasoline and the distance d in miles that you drive. The formula for $C = C(g, d)$ is $C = \dfrac{gd}{32}$ dollars.

   (a) Use functional notation to express the cost of operation if gasoline costs 98 cents per gallon and you drive 230 miles. Calculate the cost.

   (b) Calculate $C(1.03, 172)$ and explain the meaning of the number you have calculated.

5. **Radioactive substances** change form over time. For example, carbon 14, which is important for radiocarbon dating, changes through radiation into nitrogen. If we start with 5 grams of carbon 14, then the amount $C = C(t)$ of carbon 14 remaining after t years is given by

$$C = 5 \times 0.5^{\frac{t}{5730}}.$$

   (a) Express the amount of carbon 14 left after 800 years in functional notation, and then calculate its value.

   (b) How long will it take before half of the carbon 14 is gone? Explain how you got your answer. (<u>Hint</u>: You might use trial and error to solve this, or you might solve it by looking carefully at the exponent.)

6. **A roast** is taken from the refrigerator (where it had been for several days) and placed immediately in a preheated oven to cook. The temperature $R = R(t)$ of the roast t minutes after being placed in the oven is given by

$$R = 325 - 280e^{-0.005t} \text{ degrees Fahrenheit.}$$

   (a) What is the temperature of the refrigerator?

   (b) Express the temperature of the roast 30 minutes after being put in the oven in functional notation, and then calculate its value.

   (c) By how much did the temperature of the roast increase during the first 10 minutes of cooking?

   (d) By how much did the temperature of the roast increase from the first hour to 10 minutes after the first hour of cooking?

7. **What if interest is compounded more often than monthly?** Some lending institutions compound interest daily or even *continuously*. (The term continuous compounding means that interest is being added at each instant in time.) The point of this exercise is to show that, for most consumer loans, the answer you get with monthly compounding is very close to the right answer even if the lending institution compounds more often. In Part 1

of Example 1.2 we showed that if you borrow $7800 from an institution which compounds monthly using an APR of 8.04%, then in order to pay off the note in 4 years, you have to make a payment of $190.57.

(a) Would you expect your monthly payment to be higher or lower if interest is compounded daily rather than monthly? Explain why.

(b) Which would you expect to result in a larger monthly payment, daily compounding or continuous compounding? Explain your reasoning.

(c) If interest is compounded continuously using an APR of a (as a decimal), then you can get the *effective annual rate* (or *EAR*) r as a decimal using $r = e^a - 1$. Calculate the EAR in this case if the APR is 0.0804. (Round your answer here to four places beyond the decimal point.)

(d) When interest is compounded continuously using an EAR of r, you can calculate your monthly payment $M = M(P, r, Y)$ for a loan of P dollars to be paid off in Y years using

$$M = \frac{Pr}{12(1 - e^{-rY})} \text{ dollars} .$$

Calculate the monthly payment on a loan of $7800 to be paid off over 4 years using this formula. You should use the value of r you got in Part (c). How does this answer compare with the result in Example 1.2?

8. **Present value:** The amount of money originally put into an investment is known as the *present value P* of the investment. For example if you buy a $50 U.S. Savings Bond which matures in 10 years, the present value of the investment is the amount of money you have to pay for the bond today. The value of the investment at some future time is known as the *future value F*. Thus if you buy the savings bond mentioned above, its future value is $50.

If the investment pays an interest rate of r (as a decimal) compounded yearly, and if we know the future value F for t years in the future, then the present value $P = P(F, r, t)$, the amount we have to pay today, can be calculated using

$$P = F \times \frac{1}{(1 + r)^t}$$

if we measure F and P in dollars. The term $\frac{1}{(1 + r)^t}$ is known as the *present value factor* or the *discount rate*, so the formula above can also be written as

$$P = F \times \text{discount rate} .$$

(a) Explain in your own words what information the function $P(F, r, t)$ gives you.

For the remainder of this problem, we will deal with an interest rate of 9% compounded yearly and a time t of eighteen years in the future.

(b) Calculate the discount rate.

(c) Suppose you wish to put money into an account which will provide $40,000 to help your child attend college eighteen years from now. How much money would you have to put into savings today in order to attain that goal?

9. **How much can I borrow?** The function in Example 1.2 can be rearranged to show the amount of money $P = P(M, r, t)$ in dollars you can afford to borrow at a monthly interest rate of r (as a decimal) if you are able to make t monthly payments of M dollars:

$$ P = M \times \frac{1}{r} \times \left(1 - \frac{1}{(1+r)^t} \right). $$

Suppose you can afford to pay $350 per month for four years.

(a) How much money can you afford to borrow for the purchase of a car if the prevailing monthly interest rate is 0.75%? (That is 9% APR.) Express the answer in functional notation, and then calculate it.

(b) Suppose your car dealer can arrange a special monthly interest rate of 0.25% (or 3% APR). How much can you afford to borrow now?

(c) Even at 3% APR you find yourself looking at a car you can't afford, and you consider extending the period you are willing to make payments to five years. How much can you afford to borrow under these conditions?

10. **Financing a new car:** You are buying a new car and you plan to finance your purchase with a loan you will repay over 48 months. The car dealer offers two options: either dealer financing with a low APR, or a $2000 rebate on the purchase price. If you use dealer financing, you will borrow $14,000 at an APR of 3.9%. If you take the rebate, you will reduce the amount you borrow to $12,000 but you will have to go to the local bank for a loan at an APR of 8.85%. Would you take the dealer financing or the rebate? How much would you save over the life of the loan by taking the option you chose?

To answer the first question you may need the formula

$$ M = \frac{Pr(1+r)^{48}}{(1+r)^{48} - 1}. $$

Here M is your monthly payment (in dollars) if you borrow P dollars with a term of 48 months at a monthly interest rate of r (as a decimal), and $r = \dfrac{\text{APR}}{12}$.

1.2 FUNCTIONS GIVEN BY TABLES

Long before the idea of a function was formalized, it was used in the form of tables of values. Some of the earliest surviving samples of mathematics are from Babylon and Egypt and date from 2000 to 1000 B.C. They contain a variety of tabulated functions such as tables of squares of numbers. Functions given in this way almost always leave gaps and are incomplete. In this respect they may appear less useful than functions given by formulas. On the other hand, tables often clearly show trends that are not easily discerned from formulas, and in many cases tables of values are much easier to obtain than is a formula.

Reading tables of values

The population N of the United States depends on the date d. That is, $N = N(d)$ is a function of d. Table 1.1 was taken from the 1995 edition of the *Statistical Abstract of the United States*. It shows the population in millions each decade from 1950 through 1990. This is a common way to express functions when data is gathered by *sampling*, in this case by census takers.

Table 1.1: Population of the United States

$d = $ Year	1950	1960	1970	1980	1990
$N = $ Population in millions	151.87	179.98	203.98	227.23	249.40

In order to get the population in 1980, we look at the column corresponding to $d = 1980$ and read $N = 227.23$ million people. In functional notation this is $N(1980) = 227.23$ million people. Similarly, we read from the table that $N(1970) = 203.98$, indicating that the U.S. population in 1970 was 203.98 million people.

Filling gaps by averaging

Functions given by tables of values have their limitations in that they almost always leave gaps. Sometimes it is appropriate to fill these gaps by *averaging*. For example, Table 1.1 does not give the population in 1975. In the absence of further information, a reasonable guess for the value of $N(1975)$ would be the average of the populations in 1970 and 1980:

$$\frac{N(1970) + N(1980)}{2} = \frac{203.98 + 227.23}{2} = 215.61 \text{ million people.}$$

Thus the population in 1975 was approximately 215.61 million people. U.S. census data records the actual population of the United States in 1975 as 215.47 million people, and so it seems that the idea of averaging to estimate the value of $N(1975)$ worked pretty well in this case.

We would emphasize here that in the absence of further data we have no way of determining the exact value of $N(1975)$, and so you should not be misled to believe that the answer we gave is the only acceptable one. If, for example, you had reason to believe that the population of the United States grew faster in the earlier part of the decade of the 70's than it did in the later part, then it would be reasonable for you to give a larger value for $N(1975)$. Such an answer supported by an appropriate argument has as much validity as the one given here.

Average rates of change

A key tool in the analysis of functions is the idea of an *average rate of change*. To illustrate the idea, let's get the best estimate we can for $N(1972)$, the population of the U.S. in 1972. Since 1972 is not halfway between 1970 and 1980, it does not make sense here to average the two as we did above to estimate the population in 1975, but a simple extension of this idea will help. From 1970 to 1980 the population increased from 203.98 million to 227.23 million people. That is an increase of $227.23 - 203.98 = 23.25$ million people in ten years. Thus, on average, during the decade of the 70's the population was increasing by $\frac{23.25}{10} = 2.325$ million people per year. This is the *average yearly rate of change in N* during the 70's, and there is a natural way to use it to estimate the population in 1972. In 1970 the population was 203.98 million people, and during this period the population was growing by about 2.325 million people per year. Thus in 1972 the population $N(1972)$ was approximately

$$
\begin{aligned}
\text{Population in 1972} \ &= \ \text{Population in 1970 } + \text{Two years of growth} \\
&= \ 203.98 + 2 \times 2.325 = 208.63 \text{ million.}
\end{aligned}
$$

KEY IDEA 1.1: EVALUATING FUNCTIONS GIVEN BY TABLES

You can evaluate functions given by tables by locating the appropriate entry in the table. It is sometimes appropriate to fill in gaps in the table using averages or average rates of change.

EXAMPLE 1.3 *Women Employed Outside the Home*

Table 1.2 shows the number W of women in the United States employed outside the home as a function of the date d. It was taken from the *1995 Statistical Abstract of the United States.*

Table 1.2: Number of women employed outside the home in the U.S.

d = Year	1960	1970	1980	1990
W = Number in millions	21.3	29.6	41.9	53.5

1. Explain the meaning of $W(1970)$ and give its value.

2. Explain the meaning of $W(1975)$ and estimate its value.

3. Express the number of women employed outside the home in 1972 in functional notation and use the average yearly rate of change from 1970 to 1980 to estimate its value.

Solution to Part 1: The expression $W(1970)$ represents the number of women in the United States employed outside the home in 1970. Consulting the second column of the table, we find that $W(1970) = 29.6$ million women.

Solution to Part 2: The expression $W(1975)$ represents the number of women in the United States employed outside the home in 1975. This value is not given in the table, so we must estimate it. Since 1975 is halfway between 1970 and 1980, it is reasonable to estimate $W(1975)$ as the average of $W(1970)$ and $W(1980)$:

$$\frac{W(1970) + W(1980)}{2} = \frac{29.6 + 41.9}{2} = 35.75 \text{ million.}$$

Solution to Part 3: In functional notation the number of women employed outside the home in 1972 is $W(1972)$. To estimate this function value, we first calculate the average rate of change during the decade of the 70's. From 1970 to 1980 the number of women employed outside the home increased from 29.6 million to 41.9 million, an increase of 12.3 million over the 10 year period. Thus in the decade of the 70's the number increased on average by $\frac{12.3}{10} = 1.23$ million per year. In the two year period from 1970 to 1972 the increase was about $2 \times 1.23 = 2.46$ million. We estimate that

$$
\begin{aligned}
W(1972) &= W(1970) + \text{ two years' growth} \\
&= W(1970) + 2.46 \\
&= 29.6 + 2.46 \\
&= 32.06 \text{ million.}
\end{aligned}
$$

There are features of this problem that are worth emphasis. The things we do with mathematics are neither, as some may have been led to believe, magic nor beyond the understanding of the average citizen. They are common sense ideas which are used on a regular basis. Secondly, many times applications of mathematics do not involve a set way to solve the problem, and often there is not a simple "right answer." What is important is clear thinking leading to helpful information.

Spotting trends

In some situations, tables of values show clear trends or *limiting values* for functions. A good example of this is provided by the famous[2] yeast experiment of Tor Carlson in the early 1900's. In this experiment, a population of yeast cells growing in a confined area was carefully monitored as it grew. Each day, the amount of yeast[3] was found and recorded. We let t be the number of hours since the experiment began and $N = N(t)$ the amount of yeast present. Carlson only presented data through 18 hours of growth. In Table 1.3 we have partially presented Carlson's data and added data which modern studies indicate he might have recorded if the experiment were continued. Thus, for example, $N(5) = 119$ is the amount of yeast present 5 hours after the experiment began.

Table 1.3: Amount of yeast

Time t	0	5	10	15	20	25	30
Amount of yeast N	10	119	513	651	662	664	665

We want to estimate the value of $N(35)$, that is, the amount of yeast present after 35 hours. Before we look at the data, let's think how our everyday experience tells us that a similar, but more familiar, situation would progress. The population of mold on a slice of bread[4] will show behavior much like that of yeast. On bread, the mold spreads rapidly at first, but

---

[2]This experiment is described by R. Pearl in "The Growth of Populations," *Quart. Rev. Biol* **2**(1927), 532-548. It served to illustrate the now widely used *logistic population growth model* which we will examine in more detail in subsequent chapters.

[3]The actual unit of measure in unclear from the original reference. It is simply reported as *amount of yeast*.

[4]See also Exercise 11.

eventually the bread is covered and no room is left for further population growth. We would expect the amount of mold to stabilize at a *limiting value*. The same sort of growth seems likely for yeast growing in a confined area. In terms of the function N, that means we expect to see its values increase at first but eventually stabilize, and this is borne out by the data in Table 1.3, where evidently the yeast population is leveling out at about 665. Thus, it is reasonable to expect that $N(35)$ is about 665.

We note here that the significance of 665 is not just that it is the last entry in the table. Rather, this entry is indicative of the trend shown by the last few entries, 662, 664, and 665, of the table. We also emphasize that we did not propose a value for $N(35)$ by looking at the data divorced from its meaning. Before we looked at the data, the physical situation led us to believe that N would level out at *some value*, and our everyday experience with moldy bread tells us that once mold begins to grow, this level will be reached in a relatively short time. The data simply provided us with the ability to make a good guess at the limiting value which we expected to see.

KEY IDEA 1.2: LIMITING VALUES

Information about physical situations can sometimes show that limiting values are to be expected for functions which describe them. Under such conditions, the limiting value may be estimated from a trend established by the data.

The next example shows how using rates of change can confirm the expectation that there is a limiting value.

EXAMPLE 1.4   *A Skydiver*

During the period of a skydiver's free fall, he is pulled downward by gravity but his velocity is retarded by air resistance, which physicists believe increases as velocity does. Table 1.4 shows the downward velocity $v = v(t)$ in feet per second of an average size man t seconds after jumping from an airplane.

Table 1.4: Velocity in feet per second t seconds into free fall

Time t	0	10	20	30	40	50	60
Velocity v	0	147	171	175	175.8	176	176

1. From the table, describe in words how the velocity of the skydiver changes with time.

2. Use functional notation to give the velocity of the skydiver 15 seconds into the fall. Estimate its value.

3. Explain how the physical situation leads you to believe that the function v will approach a limiting value.

4. Make a table showing the average rate of change per second in velocity v over each of the ten-second intervals in Table 1.4. Explain how your table confirms the conclusion of Part 3.

5. Estimate the *terminal velocity* of the skydiver, that is, the greatest velocity that the skydiver can attain.

Solution to Part 1: During the first part of the fall the velocity increases rapidly, but the rate of increase slows as the fall progresses.

Solution to Part 2: In functional notation, the velocity 15 seconds into the fall is $v(15)$. It is not given in Table 1.4, so we must estimate its value. One reasonable estimate is the average of the velocities at 10 and 20 seconds into the fall:

$$\frac{v(10) + v(20)}{2} = \frac{147 + 171}{2} = 159 \text{ feet per second.}$$

With the information we have, this is a reasonable estimate for $v(15)$, but a little closer look at the data might lead us to adjust this. Notice that velocity increases very rapidly (by $147 - 0 = 147$ feet per second) during the first ten seconds of the fall but only by $171 - 147 = 24$ feet per second during the next ten seconds. It appears that the rate of increase in velocity is slowing dramatically. Thus it is reasonable to expect that the velocity increased more from 10 to 15 seconds than it did from 15 to 20 seconds. We might be led to make an upward revision in our estimate of $v(15)$. It turns out that the actual velocity at $t = 15$ is 164.44 feet per second, but, with the information we have, we have no way of finding that out. What is important here is to obtain an estimate that is reasonable and is supported by an appropriate argument.

Solution to Part 3: The force of gravity causes the skydiver's velocity to increase, but this is retarded by air resistance, which increases along with velocity. When downward velocity reaches a certain level, we would expect the force of retardation to match the downward pull of gravity. From that point on, velocity will not change.

Solution to Part 4: Over the interval from $t = 0$ to $t = 10$ the change in v is $147 - 0 = 147$ feet per second, so the average rate of change per second in v over this ten-second interval is $\frac{147}{10} = 14.7$ feet per second per second. From $t = 10$ to $t = 20$ the change in v is

$171 - 147 = 24$ feet per second, so the average rate of change per second in v over this interval is $\dfrac{24}{10} = 2.4$ feet per second per second. If we repeat this computation for each of the intervals, we get the values shown in Table 1.5. Here the average rate of change is measured in feet per second per second.

Table 1.5: *Average rate of change in velocity v over each ten-second interval*

Time interval	0 to 10	10 to 20	20 to 30	30 to 40	40 to 50	50 to 60
Average rate of change in v	14.7	2.4	0.4	0.08	0.02	0

Note that Table 1.5 is consistent with our observations in Part 1 and Part 2. The table shows that the average rate of change in velocity decreases to zero as time goes on. This confirms our conclusion in Part 3 that velocity will not change after a time.

Solution to Part 5: From Part 3 and Part 4, we have good reason to believe that the data should show a limiting value for velocity. From $t = 30$ to $t = 60$ the velocity seems to be inching up toward 176 feet per second, and it appears to level out there. This is confirmed by the fact that the average rate of change in velocity decreases to zero. The limiting value of 176 for v is the terminal velocity of the skydiver. You may be interested to know that this is about 120 miles per hour.

Exercise Set 1.2

1. **The American food dollar:** The following table shows the percentage $P = P(d)$ of the American food dollar that was spent on eating away from home (at restaurants for example) as a function of the date d.

d= year	1970	1980	1988
P=percent spent away from home	24%	29%	33%

   (a) Find $P(1980)$ and explain what it means.

   (b) What does $P(1975)$ mean? Estimate its value.

   (c) What is the average increase per year in percentage of the food dollar spent away from home for the period from 1980 to 1988?

   (d) What does $P(1982)$ mean? Estimate its value. (Hint: Your calculation in Part (c) should be useful.)

   (e) Estimate the value of $P(1990)$ and explain how you made your estimate.

2. **Cable TV:** The following table gives the total dollars $C = C(t)$ in billions spent by Americans on cable TV in year t.

t = year	1984	1989	1994
C = billions of dollars	6.98	11.71	16.55

   (a) Find $C(1989)$ and explain what it means.

   (b) Find the average rate of increase per year during the period from 1984 to 1989.

   (c) Estimate the value of $C(1986)$. Explain how you got your answer.

3. **A cold front:** At 4 p.m. on a winter day, an arctic air mass moved from Kansas into Oklahoma, causing temperatures to plummet. The temperature $T = T(h)$ in degrees Fahrenheit h hours after 4 p.m. in Stillwater, Oklahoma, on that day is recorded in the following table.

h = hours since 4 p.m.	0	1	2	3	4
T = temperature	62	59	38	26	22

   (a) Use functional notation to express the temperature in Stillwater at 5:30 p.m., and then estimate its value.

   (b) What was the average decrease per minute in temperature between 5 p.m. and 6 p.m.?

   (c) Estimate the temperature at 5:12 p.m.

(d) At about what time did the temperature reach the freezing point? Explain your reasoning.

4. **A troublesome snowball:** One winter afternoon a child brings, unbeknownst to his mom, a snowball into the house, lays it on the floor, and then goes to watch TV. Let $W = W(t)$ be the volume of dirty water that has soaked into the carpet t minutes after the snowball was deposited on the floor. Explain in practical terms what the limiting value of W represents, and tell what physically has happened when this limiting value is reached.

5. **Falling with a parachute:** If an average size man jumps from an airplane with a properly opening parachute, his downward velocity $v = v(t)$ in feet per second t seconds into the fall is given by the following table.

t = seconds into the fall	0	1	2	3	4
v = velocity	0	16	19.2	19.84	19.97

(a) Explain why you expect v to have a limiting value and what physically this limiting value represents.

(b) Estimate the terminal velocity of the parachutist.

6. **Carbon 14:** Carbon 14 is a radioactive substance which decays over time. One of its important uses is in dating relatively recent archaeological events. In the following table, time t is measured in thousands of years, and $C = C(t)$ is the amount in grams of carbon 14 remaining.

t = thousands of years	0	5	10	15	20
C = grams remaining	5	2.73	1.49	0.81	0.44

(a) What is the average yearly rate of decay of carbon 14 during the first 5000 years?

(b) How many grams of carbon 14 would you expect to find remaining after 1236 years?

(c) What would you expect to be the limiting value of C?

7. **Newton's law of cooling** says that a hot object cools rapidly when the difference between its temperature and that of the surrounding air is large, but it cools more slowly when the object nears room temperature. Suppose a piece of aluminum is removed from an oven and left to cool. The following table gives the temperature $A = A(t)$ in degrees Fahrenheit of the aluminum t minutes after it is removed from the oven.

t = minutes	0	30	60	90	120	150	180	210
A = temperature	302	152	100	81	75	73	72	72

(a) Explain the meaning of $A(75)$ and estimate its value.

(b) Find the average decrease per minute of temperature during the first half-hour of cooling.

(c) Find the average decrease per minute of temperature during the first half of the second hour of cooling.

(d) Explain how Parts (b) and (c) support Newton's law of cooling.

(e) Use functional notation to express the temperature of the aluminum after one hour and 13 minutes. Estimate the temperature at that time. (Note: Your work in Part (c) should be helpful.)

(f) What is the temperature of the oven? Express your answer using functional notation and give its value.

(g) Explain why you would expect the function A to have a limiting value.

(h) What is room temperature? Explain your reasoning.

8. **Effective percentage rate for various compounding periods:** We have seen that the annual percentage rate APR does not, in spite of its name, generally tell directly how much interest accrues on an investment in a year. That value, known as the *effective annual rate* or *EAR*, depends on how often the interest is compounded. Consider an investment with an annual percentage rate of 12%. The following table gives the EAR, $E = E(n)$, if interest is compounded n times each year. For example, there are 8760 hours in a year, so that column corresponds to compounding each hour.

$n =$ compounding periods	1	2	12	365	8760	525,600
$E = $ EAR	12%	12.36%	12.683%	12.747%	12.749%	12.749%

(a) State in everyday language the type of compounding that each column represents.

(b) Explain in practical terms what $E(12)$ means and give its value.

(c) Use the table to calculate the interest accrued in one year on an $8000 loan if the APR is 12% and interest is compounded daily.

(d) Estimate the EAR if compounding is done continuously, that is, if interest is added at each moment in time. Explain your reasoning.

9. **Growth in height:** The following table gives for a certain man his height $H = H(t)$ in inches at age t in years.

t = age (years)	0	5	10	15	20	25
H = height (inches)	21.5	42.5	55.0	67.0	73.5	74.0

(a) Use functional notation to express the height of the man at age 13, and then estimate its value.

(b) Now we study the man's growth rate.

    i. Make a table showing, for each of the five-year periods, the average yearly growth rate, that is, the average yearly rate of change in H.

    ii. During which five-year period did the man grow the most in height?

    iii. Describe the general trend in the man's growth rate.

(c) What limiting value would you estimate for the height of this man? Explain your reasoning in physical terms.

10. **Growth in weight:** The following table gives for a certain man his weight $W = W(t)$ in pounds at age t in years.

t = age (years)	4	8	12	16	20	24
W = weight (pounds)	36	54	81	128	156	163

(a) Make a table showing, for each of the four-year periods, the average yearly rate of change in W.

(b) Describe in general terms how the man's gain in weight varied over time. During which four-year period did the man gain the most in weight?

(c) Estimate how much the man weighed at age 30.

(d) Estimate how much he weighed at birth.

11. **A home experiment:** In our discussion of Carlson's experiment with yeast, we indicated that there should be similarities between the growth of yeast and the growth of mold on a slice of bread. In this exercise, you will verify that. Begin with a slice of bread which has a few moldy spots on it. Put it in a plastic bag and leave it on a warm place such as your kitchen counter. Estimate the percentage of the bread surface that is covered by mold two or three times each day until the bread is covered with mold. (This may take several days to a week.) Record your data and provide a written report describing the growth of the mold.

1.3 FUNCTIONS GIVEN BY GRAPHS

The idea of picturing a function as a graph is usually credited to the 17th century mathematicians Fermat and Descartes. But some give credit to Nicole Oresme, who preceded Fermat and Descartes by 300 years. In his study of velocity he hit upon the idea of representing the velocity of an object using two dimensions. He drew a *base line* to represent time and then added *perpendiculars* from this line whose length represented the velocity at each instant. The velocity, he said, "cannot be known any better, more clearly, or more easily than by such mental images and relations to figures."[5] Oresme was expressing his belief, shared today by modern mathematicians, that the graph provides deeper insight into a function and makes mathematics easier. It is one more illustration of the old cliché that "A picture is worth a thousand words." One of the best features of a graph is that it provides an overall view of a function which makes it easy to deduce important properties.

Reading graphs

Figure 1.5 shows the percentage of unemployment $U = U(d)$ in Canada as a function of the date d. It is common to describe this as a graph of U *versus* d or as a graph of U *against* d. Either of these indicates that the horizontal axis corresponds to d and the vertical axis corresponds to U. To find the unemployment in 1970, we locate 1970 on the horizontal axis, move along the vertical line shown in Figure 1.6 up to the graph, and then move horizontally to locate the corresponding percentage on the vertical axis. In this case we see that it is about 5%, so in functional notation $U(1970) = 5\%$. As we see in this example, graphical presentations

Figure 1.5: Unemployment in Canada

Figure 1.6: Getting U(1970) from the graph

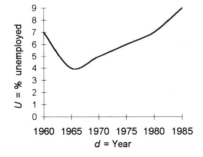

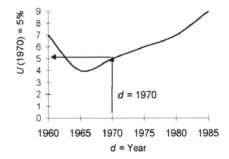

sometimes allow only rough approximation of function values, but interesting features of a function are often easy to spot when it is presented graphically. For example, in Figure 1.7 we

---

[5]From Marshall Clagett, transl. and ed., *Nicole Oresme and the Medieval Geometry of Qualities and Motions*, University of Wisconsin Press, Madison, 1968.

have marked the *minimum*, or lowest point on the graph, as well as regions where the graph is *decreasing* and where it is *increasing*. This shows clearly that the unemployment rate declined from 1960 until it reached a minimum of about 4% in 1965 and that it generally increased during the ensuing 20 years. This is typical of the shape of a graph near a minimum value; the graph decreases down to the minimum value and increases after that. Similarly, the highest point, or *maximum*, typically occurs where the graph changes from increasing to decreasing. Sometimes, however, maximum or minimum values may occur at the ends of a graph. For example, we see in Figure 1.5 that, during the period from 1960 to 1985, unemployment in Canada reached a maximum of about 9% in 1985, at the right-hand end of the graph.

A great deal more information is available from the graph. Suppose for example we want to know when unemployment reached 8%. In functional notation, we want to solve for d the equation $U(d) = 8$ percent. To do so, we locate 8 on the vertical axis and move horizontally until we get to the graph. From that point we drop down to the horizontal axis as illustrated in Figure 1.8, and we see that unemployment reached 8% around 1982. In functional notation this is $U(1982) = 8\%$. We might also ask for the average yearly increase during the decade of the 70's. From the graph we read $U(1970) = 5\%$ and $U(1980) = 7\%$. That is an increase of 2 percentage points over the ten year period, so the average yearly increase is 0.2 percentage points per year.

Figure 1.7: *Regions of decrease, increase, and the minimum value*

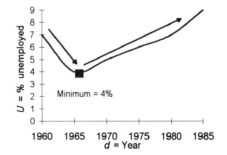

Figure 1.8: *Finding when unemployment reached 8%*

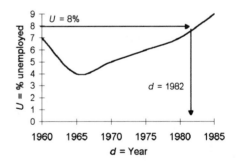

KEY IDEA 1.3: EVALUATING FUNCTIONS GIVEN BY GRAPHS

To evaluate a function given by a graph, locate the point of interest on the horizontal axis, move vertically to the graph, and then move horizontally to the vertical axis. The function value is the location on the vertical axis.

EXAMPLE 1.5 *Currency Exchange*

Figure 1.9 shows the value $G = G(d)$ in American cents of the German mark as a function of the date d.

Figure 1.9: *The value of the German mark in American cents*

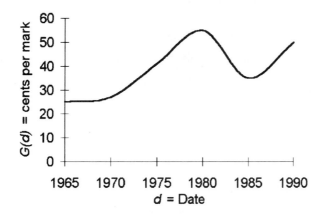

1. Explain the meaning of $G(1970)$ and estimate its value.

2. From 1965 through 1990 what was the largest value the German mark attained? When did that happen?

3. What was the average yearly increase in the value of the mark from 1975 to 1980?

4. During which year was the mark decreasing most rapidly in value?

5. During the 1965 to 1990 period, what would have been the ideal buy and sell dates for an American investing in German marks? Explain your reasoning.

Solution to Part 1: The expression $G(1970)$ is the value in cents of the German mark in 1970. From the graph we read that $G(1970)$ is about 26 cents.

Solution to Part 2: The mark reaches its highest value where the graph reaches its highest point. That appears to be about 55 cents in 1980.

Solution to Part 3: From 1975 to 1980 the mark increased in value from $G(1975) = 40$ cents to $G(1980) = 55$ cents. That is an increase of 15 cents over the five year period. Thus the value of the mark increased by about $\dfrac{15}{5} = 3$ cents per year.

Solution to Part 4: The value of the mark is most rapidly decreasing where the graph is going downward the most steeply. That is difficult to discern exactly from the graph, but it surely occurred sometime between 1980 and 1985. The year 1982 appears to be fairly close.

Solution to Part 5: There is more than one reasonable answer for this part. The ideal buy and sell dates for an American trading marks depend on the goals of the investment. If the goal was simply to get the maximum cumulative return on investment, the American should have bought the mark at its cheapest, about 25 cents in 1965, and sold it when it achieved its maximum value of 55 cents in 1980.

Many times investors are concerned not only with cumulative profit but also with how long it takes the profit to accrue. With this in mind one might prefer to have bought in 1970 for about 26 cents and sold in 1980 for 55 cents. There are a number of other answers that are acceptable if properly explained.

Concavity and rates of change

Important features of a graph include places where it is increasing or decreasing, maxima and minima, and places where the graph may cross the horizontal axis. But a feature called *concavity* is often as important as any of these. A graph may be *concave up*, meaning that it has the shape of a wire whose ends are bent upward, or it may be *concave down*, meaning that it has the shape of a wire whose ends are bent downward. Figures 1.10 and 1.11 show graphs that are concave up, and Figures 1.12 and 1.13 show graphs that are concave down.

Figure 1.10: *A graph that is concave up and increasing*	***Figure 1.11:*** *A graph that is concave up and decreasing*

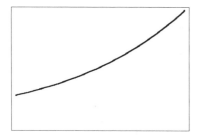

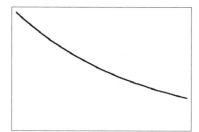

Note that the concavity of a graph does not determine whether it is increasing or decreasing, but it does give additional information about the rate of increase or decrease. The graphs in Figures 1.10 and 1.12 are both increasing, but they increase in different ways. As we move to the right in Figure 1.10 the graph gets steeper. That is, the graph is increasing at a faster rate. When we move to the right in Figure 1.12 the graph becomes less steep, showing that the rate of increase is slowing. This is typical. Increasing graphs which are concave up increase at an increasing rate, while increasing graphs which are concave down increase at

Figure 1.12: A graph that is concave down and increasing

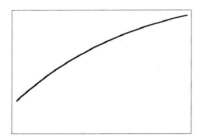

Figure 1.13: A graph that is concave down and decreasing

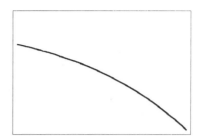

a decreasing rate. A similar analysis for Figures 1.11 and 1.13 shows that decreasing graphs which are concave up decrease at a decreasing rate, while decreasing graphs which are concave down decrease at an increasing rate. Concavity, if it persists, can have important long-term effects. For example, if prices were increasing, it would be a good deal more troubling in the long term if the graph of prices against time were concave up than if it were concave down.

EXAMPLE 1.6   *Concavity and Currency Exchange*

Consider once more the value of the German mark as given by the graph in Figure 1.9.

1. During what years is the graph concave up? Explain in practical terms what this means about the value of the mark during these periods.

2. During what years is the graph concave down? Explain in practical terms what this means about the value of the mark during this period.

Solution to Part 1: The graph is clearly concave up from 1965 to about 1975. Since the graph is increasing and concave up, the value of the mark is increasing at an increasing rate during this period.

The graph is also concave up from about 1982 until 1990. We divide our analysis here into the places where the graph is decreasing and increasing. From 1982 to 1985 the graph is decreasing and concave up. The fact that it is concave up means that as we near 1985, the graph decreases at a slower rate. From 1985 to 1990 the graph is increasing and concave up. Thus as we approach 1990, the rate of increase is getting larger.

Solution to Part 2: The graph is concave down only from about 1975 to 1982. As in Part 1, we divide our analysis into the places where the graph is increasing and where it is decreasing. From 1975 to 1980, the graph is increasing and concave down. That means the value of the mark was increasing, but at a decreasing rate. From 1980 to 1982, the graph is decreasing and concave down. Thus during this period, the value of the mark decreases more rapidly as we approach 1982.

Inflection points

We refer to points where concavity changes, up to down or down to up, as *inflection points*. In Figure 1.14 we have shown a graph where concavity changes. These changes in concavity determine the two inflection points marked in Figure 1.15. The first inflection point in

---

Figure 1.14: *A graph with changing concavity*

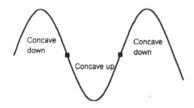

Figure 1.15: *Inflection points*

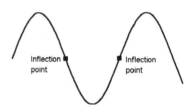

Figure 1.15 is where the graph changes from concave down to concave up, and the second is where the graph changes from concave up to concave down. Notice that the first inflection point in Figure 1.15 is also the point where the graph is decreasing most rapidly. Similarly, the second inflection point is where the graph is increasing most rapidly. Points of maximum increase or decrease are often found at inflection points, and in many applications this is where we should look. For example, in Part 4 of Example 1.5 we found that the value of the mark was most rapidly decreasing in 1982. Also in Example 1.6 we identified this same point as the place where concavity changes from down to up, that is, as an inflection point. We can also observe that the graph in Figure 1.9 is increasing most rapidly at about 1975, and this corresponds to an inflection point where concavity changes from up to down.

EXAMPLE 1.7   *Driving Down a Straight Road*

You leave home at noon and drive your car down a straight road for a visit to a friend's house. Your distance D (in miles) from home as a function the time t (in hours) since noon is shown in the graph in Figure 1.16.

1. During what time period are you at your friend's house?

2. Over what time period is the graph decreasing? Explain in practical terms what this portion of the graph represents.

Figure 1.16: *Distance from home versus time*

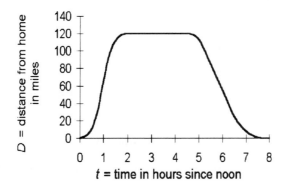

3. At what value of *t* is the graph rising most steeply? Explain in terms of your speedometer what your answer means.

4. At what time do you get back home?

Solution to Part 1: During the time we are at the friend's house our distance from home is unchanging, or constant, so the graph of D versus t is level over that period. Since the graph is level between about $t = 2$ and $t = 4\frac{1}{2}$ hours after noon, we are at the friend's house between about 2:00 p.m. and 4:30 p.m. Note that the constant value of D over this period is about 120, so the friend lives about 120 miles from our home.

Solution to Part 2: The graph is decreasing from around $t = 4\frac{1}{2}$ to around $t = 7\frac{3}{4}$, or roughly from 4:30 p.m. to 7:45 p.m. Since D is our distance from home, this portion of the graph represents our trip back home.

Solution to Part 3: The steepest rise in the graph occurs at the first inflection point, where the graph changes from concave up to concave down. Examination of the graph shows that this point occurs at about $t = 1$. Before $t = 1$, the graph is concave up, so the distance D is increasing at an *increasing* rate; for a time after $t = 1$, the graph is concave down and the distance is increasing at a *decreasing* rate. This means that the greatest rate of increase in distance occurs at $t = 1$. Now on the way to our friend's house the speedometer reading tells us how fast we are going away from home; in other words, it tells us the rate of increase in distance. So we interpret the time when the graph is rising most steeply as the time on our way to the friend's house when the speedometer reading is the greatest. In short, on the first part of our trip we reach our maximum speed at about 1:00 p.m.

Solution to Part 4: When we get back home the distance D reaches 0, so the graph touches the horizontal axis. This occurs at around $t = 7\frac{3}{4}$ hours after noon, or 7:45 p.m.

Drawing a graph

A graph is a pictorial representation of a function, and it is generally not difficult to make such pictures. Here is an example. According to the *1996 Information Please Almanac*, in 1900 about 9.3 Americans per thousand were married each year. This number reached an all-time high of about 12.2 per thousand in 1945 (in the aftermath of World War II). This number then decreased and reached a low of 8.5 per thousand in 1960. This information describes a function $M = M(d)$ which gives the number of marriages per thousand Americans in year d, and we want to draw its graph.

To make a graph of the function M means that we graph M against d. That is, the horizontal axis will correspond to the date and the vertical axis will correspond to the number of marriages per thousand. These axes are so labeled in Figure 1.17. We note that the verbal description of the function gave us three data points, which we express in functional notation:

$$M(1900) = 9.3$$
$$M(1945) = 12.2$$
$$M(1960) = 8.5 .$$

The next step is to mark these points on the graph as shown in Figure 1.17. To complete the graph, we join these points with a curve incorporating the additional information we have. In particular, the verbal description of M tells us that the graph should decrease after 1945. Our completed graph is in Figure 1.18.

Figure 1.17: Locating given points for M versus d

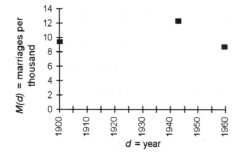

Figure 1.18: Marriages per thousand Americans

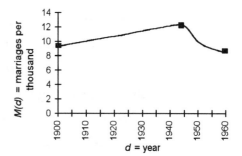

It is important to note that there was a certain amount of guesswork done to produce the graph in Figure 1.18. The points we located in Figure 1.17 are not guesses, but the way we chose to join these points to complete the graph in Figure 1.18 does involve a good deal of guessing. For example, we were told that the marriage rate per thousand decreased from 1945 to 1960, but we were not told how it decreased. The true graph might for example be concave down in this region, or it might be a jagged line rather than the smooth curve we have drawn. When graphs are used to represent functions given by sampling data it is inevitable that some guessing is involved in connecting known data points. You should be aware that artistic flair can dramatically affect the appearance of such graphs. For example, if we wanted to emphasize the high marriage rate at the end of World War II, we might have drawn the graph as in Figure 1.19. If we wanted to de-emphasize it, we might have made the graph as in Figure 1.20. Note that both of these graphs agree with the verbal description we were given, but they convey quite different information to the eye. Further emphasis or de-emphasis of particular features of a function can be attained by adjusting the scale on the horizontal or vertical axis.

Figure 1.19: *A representation of the marriage rate which emphasizes the peak in 1945*	*Figure 1.20:* *A representation of the marriage rate which de-emphasizes the peak in 1945*

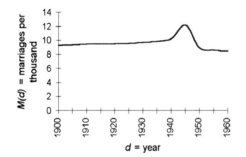

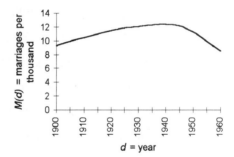

How do we know which of the pictures, if any, among Figures 1.18, 1.19, and 1.20 provide a true representation of the function M? The answer is that graphs show a more accurate representation of a function when more data points are used, and if your interest is in seeing the true nature of the American marriage rate in the first half of the twentieth century, you should consult the almanac referred to above, where a good deal more information is available. See also Exercise 12 at the end of this section. As a citizen, you should be especially leery of free-hand graphs which do not provide references for data sources or for which the artist may be an advocate of a particular point of view.

EXAMPLE 1.8 Education in the United States

According to United States census figures, in 1940 about 62% of Americans in the 25 to 29 year age group had completed less than 12 years of school. This figure decreased to about 25% in 1970. From that point on, the decrease was much less pronounced, dropping to about 12% in 1990. Let $E = E(d)$ be the percentage of Americans in the 25 to 29 year age group who have completed less than 12 years of school in year d.

1. Three exact data points were given. Express them in functional notation.

2. Draw a graph of E against d. Be sure to label your axes and make the graph as accurate as the data allows.

Solution to Part 1: We are told that 62% of the target population had less than 12 years of school in 1940. Thus $E(1940) = 62\%$. Similarly $E(1970) = 25\%$, and $E(1990) = 12\%$.

Solution to Part 2: To graph E against d means that the date d goes on the horizontal axis and the percentage E goes on the vertical axis. In Figure 1.21 we have labeled both years and percentage points in 10 year increments and noted what the numbers on the axes represent. Next we located the three given function values in Figure 1.21. To complete the graph as shown in Figure 1.22, we follow the verbal description and draw a decreasing curve through these points. Note that the fact that the decrease was less pronounced after 1970 is reflected by the fact that the graph does not go down so steeply from that point on. Remember though that there are a number of related graphs all of which could be correct.

Figure 1.21: *Locating three data points*

Figure 1.22: *Percentage of 25 to 29 year age group with less than 12 years of school*

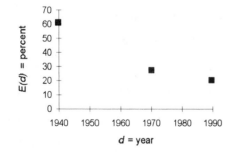

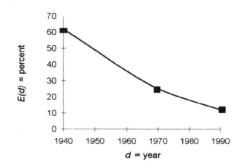

Exercise Set 1.3

1. **Sketching a graph with given concavity:**

   (a) Sketch a graph that is always decreasing but starts out concave down and then changes to concave up. There should be a point of inflection in your picture. Mark and label it.

   (b) Sketch a graph that is always decreasing but starts out concave up and then changes to concave down. There should be a point of inflection in your picture. Mark and label it.

2. **An investment with continuous compounding:** In 1995 an investor put money into a fund that pays 8% (EAR) yearly interest compounded continuously. The graph in Figure 1.23 shows the value $v = v(d)$ of the investment (in dollars) as a function of the date d.

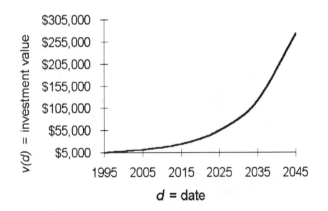

Figure 1.23: *An investment*

   (a) Express the original investment using functional notation and give its value.

   (b) Is the graph concave up or concave down? Explain how this affects the growth in value of the account.

   (c) When will the value of the investment reach $55,000?

   (d) What is the average yearly increase from 2035 to 2045?

   (e) Which is larger, the average yearly increase from 2035 to 2045 or the average yearly increase from 1995 to 2005? Explain your reasoning.

3. **A stock market investment:** A stock market investment of $10,000 was made in 1960. During the decade of the 60's, the stock lost half its value. Beginning in 1970, the value increased until it reached $35,000 in 1980. After that its value has remained stable. Let $v = v(d)$ denote the value of the stock in dollars as a function of the date d.

   (a) What are the values of $v(1960)$, $v(1970)$, $v(1980)$, and $v(1990)$?

   (b) Make a graph of v against d. Label the axes appropriately.

   (c) Estimate the time when your graph indicates that the value of the stock was most rapidly increasing.

4. **The value of the Canadian dollar:** The value $C = C(d)$ in American dollars of the Canadian dollar is given by the graph in Figure 1.24.

Figure 1.24: *The value in American dollars of the Canadian dollar*

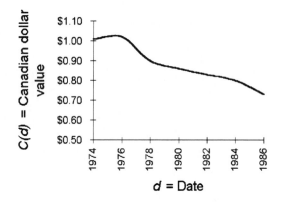

   (a) Describe how the value of the Canadian dollar fluctuated from 1974 to 1986. Give specific function values in your description where they are appropriate.

   (b) When was the Canadian dollar worth 90 American cents?

   (c) What was the average yearly decrease in the value of the Canadian dollar from 1976 to 1978?

   (d) For an American investor trading in Canadian dollars, what would have been ideal buy and sell dates? Explain your answer.

5. **Logistic population growth:** The graph in Figure 1.25 shows the population $N = N(d)$ in thousands of animals of a particular species on a protected reserve as a function of the date d. Ecologists refer to growth of the type shown here as *logistic population growth*.

Figure 1.25: *A population (in thousands) showing logistic population growth*

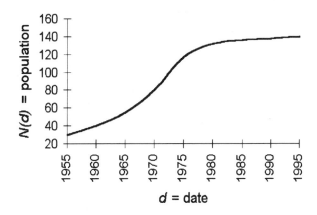

(a) Explain in general terms how the population N changes with time.

(b) When did the population reach 80 thousand?

(c) Your answer in Part (b) is the solution of an equation involving $N(d)$. Which equation?

(d) During which period is the graph concave up? Explain what this means about population growth during this period.

(e) During which period is the graph concave down? Explain what this means about population growth during this period.

(f) When does a point of inflection occur on the graph? Explain how this point may be interpreted in terms of the growth rate.

(g) What is the *environmental carrying capacity* of this reserve for this species of animal? That is, what is the maximum number of individuals that the environment can support?

(h) Make a new graph of $N(d)$ under the following new scenario. The population grew as in Figure 1.25 until 1980, when a natural disaster caused half of the population to die. After that the population resumed logistic growth.

6. **Cutting trees:** In forestry management it is important to know the *net stumpage value* of a stand (i.e., a group) of trees. This is the commercial value of the trees minus the costs of felling, hauling, etc. The graph in Figure 1.26 shows the net stumpage value V (in dollars per acre) of a Douglas-fir stand in the Pacific Northwest as a function of the age t (in years) of the stand.[6]

Figure 1.26: *Net stumpage value of a Douglas-fir stand*

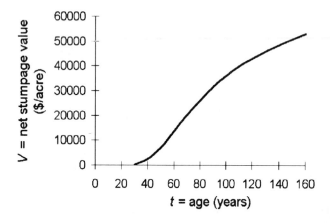

(a) Estimate the net stumpage value of a Douglas-fir stand which is 60 years old.

(b) Estimate the age of a Douglas-fir stand whose net stumpage value is $40,000 per acre.

(c) At what age does the commercial value of the stand equal the costs of felling, hauling, etc.?

(d) At what age is the net stumpage value increasing the fastest?

(e) This graph shows V only up to age $t = 160$ years, but the Douglas-fir lives for hundreds of years. Draw a graph to represent what you expect for V over the life span of the tree. Explain your reasoning.

7. **Tornados in Oklahoma:** The graph in Figure 1.27 shows the number $T = T(d)$ of tornados reported by the Oklahoma Climatological Survey.

(a) When were the most tornados reported? How many were reported in that year?

(b) When were the fewest tornados reported? How many were reported in that year?

(c) What was the average yearly rate of decrease in tornadic activity from 1986 to 1987?

---

[6]The normal-yield table used here (with site index 140) is from a study by R. E. McArdle et al., as presented by Thomas E. Avery and Harold E. Burkhart, *Forest Measurements*, 4th edition, 1994, McGraw-Hill, New York.

Figure 1.27: *Tornados in Oklahoma*

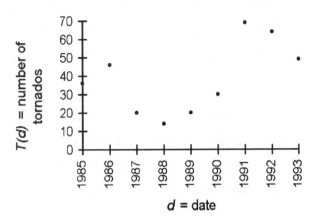

(d) What was the average yearly rate of increase in tornadic activity from 1990 to 1991?

(e) What was the average yearly *rate of change* (that is, the average yearly rate of total increase or total decrease) in tornadic activity from 1987 to 1989?

8. **Wind chill:** The graph in Figure 1.28 shows the temperature $T = T(v)$ adjusted for wind chill as a function of the velocity v of the wind when the thermometer reads 30 degrees Fahrenheit. The adjusted temperature T shows the temperature with an equivalent cooling power if there were no wind.

Figure 1.28: *Temperature adjusted for wind chill when the thermometer reads 30 degrees*

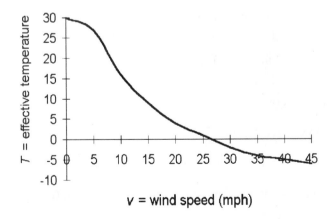

(a) At what wind speed is the temperature adjusted for wind chill equal to 0?

(b) Your answer in Part (a) is the solution of an equation involving $T(v)$. Which equation?

(c) At what value of v would a small increase in v have the greatest effect on $T(v)$? In other words, at what wind speed could you expect a small increase in wind speed to cause the greatest change in wind chill? Explain your reasoning.

(d) Suppose the wind speed is 45 miles per hour. Judging from the shape of the graph, how significant would you expect the effect on $T(v)$ to be if the wind speed increases?

9. **Inflation:** During a period of high inflation, a political leader was up for re-election. Inflation had been increasing during his administration, but he announced that the *rate of increase* of inflation was decreasing. Draw a graph of inflation versus time which illustrates this situation. Would this announcement convince you that economic conditions were improving?

10. **Walking to school:** You walk from home due east to school to get a book and then you walk back west to visit a friend. The graph in Figure 1.29 shows your distance D (in yards) east of home as a function of the time t (in minutes) since you left home.

Figure 1.29: Distance east of home versus time

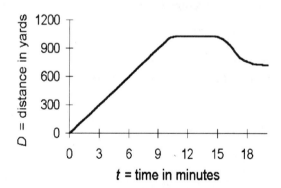

(a) How far away is school?

(b) At what time do you reach school?

(c) At what time(s) are you a distance of 900 yards from home?

(d) Compute the average rate of change in D over the interval from $t = 0$ to $t = 3$, from $t = 3$ to $t = 6$, and from $t = 6$ to $t = 9$.

(e) What does your answer to Part (d) tell you about how fast you are walking to school? How is this related to the shape of the graph over the interval from $t = 0$ to $t = 9$?

(f) At what time are you walking back west the fastest?

11. **Driving a car:** You are driving a car. The graph in Figure 1.30 shows your distance D (in miles) from home as a function of the time t (in minutes) since 1:00 p.m. Make up a driving story that matches the graph. Be sure to explain how your story incorporates the times when the graph is increasing, decreasing, or constant.

Figure 1.30: Distance from home versus time

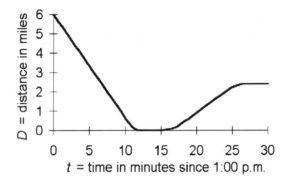

12. **A research project:** In the discussion following Figures 1.19 and 1.20 we pointed out that the right way to get an accurate graph of marriage rates was to include more data points. On page 835 of the *1996 Information Please Almanac*, the marriage rate per thousand Americans is given in five year periods. This information will also be available in other reference sources in your library. Pursue an appropriate reference and settle the question of which (if any) of the graphs we presented is an accurate representation of American marriage rates by producing your own graph.

1.4 FUNCTIONS GIVEN BY WORDS

Many times mathematical functions are described verbally. In order to work effectively with such functions, it is essential that the verbal description be clearly understood. Sometimes this is enough, but more often it is necessary to supplement the verbal description with a formula, graph, or table of values.

Comparing formulas and words

For example, there are initially 2000 bacteria in a petri dish. The bacteria reproduce by cell division, and each hour the number of bacteria doubles. This is a verbal description of a function $N = N(t)$, where N is the number of bacteria present at time t. It is common in situations like this to begin at time $t = 0$. Thus $N(0)$ is the number of bacteria we started with, 2000. One hour later the population doubles, and so $N(1) = 4000$. In one more hour the population doubles again, giving $N(2) = 8000$. Continuing, we see that $N(3) = 16,000$.

Notice that, while we can calculate $N(4)$ or $N(5)$, it is not so easy to figure out $N(4.5)$, the number of bacteria we have after four hours and 30 minutes. To make such a calculation, we need another description of the function, a formula. This situation is an example of *exponential growth*, which we will study in more detail in Chapter 4; there we will learn how to find formulas to match verbal descriptions of this kind. For now we only want to make a comparison to indicate that the verbal description matches the formula $N = 2000 \times 2^t$ without showing where the formula came from. That is, we want to show that the formula gives the same answers as the verbal description. You can verify the following calculations by hand or using your calculator:

$$N(0) = 2000 \times 2^0 = 2000 \times 1 = 2000$$
$$N(1) = 2000 \times 2^1 = 2000 \times 2 = 4000$$
$$N(2) = 2000 \times 2^2 = 2000 \times 4 = 8000$$
$$N(3) = 2000 \times 2^3 = 2000 \times 8 = 16,000 .$$

Notice that these are the same values we got using the function's verbal description. You may wish to do further calculations to satisfy yourself that the formula agrees with the verbal description. This formula allows us to calculate the number of bacteria present after four hours thirty minutes. Since that is 4.5 hours after the experiment began, we use $t = 4.5$:

$$N(4.5) = 2000 \times 2^{4.5} = 45,255 .$$

We have rounded here to the nearest whole number since we don't expect to see fractional parts of bacteria.

We should point out that comparisons via the few calculations we made cannot establish for a certainty that the formula and verbal descriptions match. In fact, such a certain conclusion could not be drawn no matter how many individual calculations we made. But such calculations can establish patterns and provide partial evidence of agreement, and that is what we are after here.

EXAMPLE 1.9 *Purifying Water*

Water which is initially contaminated with a concentration of 9 milligrams of pollutant per liter of water is subjected to a cleaning process. The cleaning process is able to reduce the pollutant concentration by 25% each hour. Let $C = C(t)$ denote the concentration in milligrams per liter of pollutant in the water t hours after the purification process begins.

1. What is the concentration of pollutant in the water after 3 hours?

2. Compare the results predicted by the verbal description at the end of each of the first three hours with those given by the formula $C = 9 \times 0.75^t$.

3. The formula given in Part 2 is in fact the correct one. Use it to find the concentration of pollutant after four and one quarter hours of cleaning.

Solution to Part 1: There is initially 9 milligrams per liter of pollutant in the water. After one hour, this is reduced by 25%. That is a reduction of $0.25 \times 9 = 2.25$ milligrams per liter, leaving $9 - 2.25 = 6.75$ milligrams per liter in the water. During the second hour the level is reduced by 25% again, leaving $6.75 - 0.25 \times 6.75 = 5.0625$ milligrams per liter. After one more hour there will be $5.0625 - 0.25 \times 5.0625 = 3.80$ milligrams per liter (rounded to two decimal places).

Solution to Part 2: We have calculated the appropriate concentrations in Part 1. Use your calculator to verify the following and check Part 1 for agreement:

$$9 \times 0.75^1 = 6.75$$
$$9 \times 0.75^2 = 5.0625$$
$$9 \times 0.75^3 = 3.80 \text{ (rounded to two decimal places)}.$$

Solution to Part 3: To get the concentration after four and one quarter hours, we put 4.25 into the formula 9×0.75^t:

$$9 \times 0.75^{4.25} = 2.65 \text{ milligrams per liter.}$$

EXAMPLE 1.10 *A Risky Stock Market Investment*

An entrepreneur invested $2000 in a risky stock. Unfortunately, over the next year the value of the stock decreased by 7% each month. Let $V = V(t)$ denote the value in dollars of the investment t months after the stock purchase was made.

1. Make a table of values showing the initial value of the stock and its value at the end of each of the first five months.

2. Make a graph of investment value versus time.

3. Verify that the formula $V = 2000 \times 0.93^t$ gives the same values you found in your table above.

4. The formula given in Part 3 is in fact a valid description of the function V. What is the value of the investment after nine and a half months?

Solution to Part 1: We use the verbal description to find the needed function values. The original investment was 2000, and so $V(0) = 2000$. During the first month, the investment lost 7% of its value. That is, its value decreased by $0.07 \times 2000 = 140$ dollars. Thus $V(1) = 2000 - 140 = 1860$ dollars. During the second month, the value again decreased by 7%. That is $0.07 \times 1860 = 130.20$. So $V(2) = 1860 - 130.20 = 1729.80$ dollars.

Continuing the calculation in this manner, we get

$$
\begin{aligned}
V(3) &= \$1608.71 \\
V(4) &= \$1496.10 \\
V(5) &= \$1391.38 \,.
\end{aligned}
$$

Arranging these in a table of values, we get

t=Months	0	1	2	3	4	5
$V(t)$=Investment value	$2000	$1860	$1729.80	$1608.71	$1496.10	$1391.38

Solution to Part 2: Since we are graphing investment value versus time, we use time for the horizontal axis and investment value for the vertical axis. The first step is to locate the data points from the table as in Figure 1.31. Next we join up the dots as in Figure 1.32.

Figure 1.31: *Plotting points*

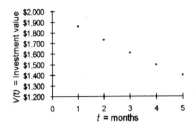

Figure 1.32: *Completing the graph*

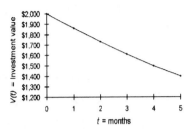

Solution to Part 3: We use $t = 0, 1, 2, 3, 4, 5$ to make the calculations:

$$V(0) = 2000 \times 0.93^0 \quad = \quad 2000 \times 1 = 2000$$

$$V(1) = 2000 \times 0.93^1 \quad = \quad 2000 \times 0.93 = 1860$$

$$V(2) = 2000 \times 0.93^2 \quad = \quad 1729.80$$

$$V(3) = 2000 \times 0.93^3 \quad = \quad 1608.71$$

$$V(4) = 2000 \times 0.93^4 \quad = \quad 1496.10$$

$$V(5) = 2000 \times 0.93^5 \quad = \quad 1391.38 \,.$$

We note that these calculations agree with the table of values we calculated in Part 1 using the verbal description of the function. This is evidence that $V = 2000 \times 0.93^t$ gives a correct formula for this function.

Solution to Part 4: We use the formula from Part 3 with $t = 9.5$:

$$V(9.5) = 2000 \times 0.93^{9.5} = 1003.73 \text{ dollars.}$$

Getting formulas from words

Many times verbal descriptions can be directly translated into formulas for functions. Suppose for example that an engineering firm invests $78,000 in the design and development of a more efficient computer hard disk drive. For each disk drive sold, the engineering firm makes a profit of $98. We want to get a formula that shows the company's net profit as a function of the number of disk drives sold, taking into account the initial investment. This is not a difficult problem, and you may find that you can do it without further help, but we want to show a general method that may be helpful for more complicated descriptions.

Step 1: Identify the function and the things on which it depends, and write the relationships you know in a formula using words. The function we are interested in is net profit. That is the profit on disk drives sold minus money spent on initial development. The profit on drives sold in turn depends on the number of sales:

$$\text{Net profit} \quad = \quad \text{Profit from sales} \; - \; \text{Initial investment} \tag{1.1}$$

$$= \quad \text{Profit per item} \; \times \; \text{Number sold} \; - \; \text{Initial investment.} \tag{1.2}$$

Step 2: Select and record letter names for the function and for each of the variables involved, and state their units. We can use any letters we want as long as we identify them and state clearly what they mean. To say that we want to express net profit as a function of the number of disk drives sold means that net profit is the function and the number of disk drives is the variable. We will use N to denote our function, the net profit in dollars, and d to represent the variable, the number of disk drives sold:

$$N \quad = \quad \text{Net profit in dollars}$$

$$d \quad = \quad \text{Number of drives sold.}$$

Step 3: Replace the words in Step 1 by the letters identified in Step 2 and appropriate information from the verbal description. Here we simply replace the words *Net profit* in Equation (1.2) by N, *Profit per item* by 98, *Number sold* by d, and *Initial investment* by 78,000:

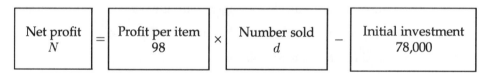

| Net profit N | = | Profit per item 98 | × | Number sold d | − | Initial investment 78,000 |

Thus we get the formula as $N = 98d - 78,000$, where d is the number of disk drives sold and N is the net profit in dollars. We can use this formula to provide information about the engineering firm's disk drive project. For example, in functional notation $N(1200)$ represents the net profit in dollars if 1200 disk drives are sold. We can calculate its value by replacing d by 1200: $N(1200) = 98 \times 1200 - 78,000 = 39,600$ dollars.

EXAMPLE 1.11 *Cutting a Diamond*

The total investment a jeweler has in a gem-quality diamond is the price paid for the rough stone plus the amount paid to work the stone. Suppose the gem cutter earns $40 per hour.

1. Choose variable and function names, and give a formula for a function that shows the total investment in terms of the cost of the stone and the number of hours required to work the stone.

2. Use functional notation to express the jeweler's investment in a stone that costs $320 and requires 5 hours and 15 minutes of labor by the gem cutter.

3. Use the formula you made in Part 1 to calculate the value from Part 2.

Solution to Part 1: To say that we want the total investment as a function of the cost of the stone and the hours of labor means that *total investment* is our function, and *cost of stone* and *hours of labor* are our variables. The total investment is the cost of the stone plus the cost of labor, which in turn depends on the number of hours needed to work the stone. We write out the relationship using a formula with words. We will accomplish this in two steps:

$$\text{Investment} \quad = \quad \text{Cost of stone } + \text{ Cost of labor} \tag{1.3}$$

$$= \quad \text{Cost of stone } + \text{ Hourly wage } \times \text{ Hours of labor.} \tag{1.4}$$

The next step is to choose letters to represent the function and the variables. We let c be the cost in dollars of the rough stone and h the number of hours of labor required. (It is fine to use any letters you like as long as you say in words what they represent.) We let $I = I(c, h)$ be the total investment in dollars in the diamond. To get the formula we want, we use Equation (1.4), replacing *Investment* by I, *Cost of stone* by c, *Hourly wage* by 40, and *Hours of labor* by h:

$$\boxed{\begin{array}{c}\text{Investment}\\ I\end{array}} = \boxed{\begin{array}{c}\text{Cost of stone}\\ c\end{array}} + \boxed{\begin{array}{c}\text{Hourly wage}\\ 40\end{array}} \times \boxed{\begin{array}{c}\text{Hours of labor}\\ h\end{array}}$$

The result is $I = c + 40h$ where I is the total investment in dollars, c is the cost in dollars of the rough stone, and h is the number of hours required to work the stone.

Note that in our final presentation of the formula we were careful to identify the meaning of each letter, noting the appropriate units. Such descriptions are as important as the formula. Also, in this case you may find it easy enough to bypass much of what we presented here and go directly to the final answer. You are very much encouraged, however, to put in the intermediate step using words just as it appears here. We will encounter many *word problems*, and this practice will make things much easier for you as the course progresses.

Solution to Part 2: The cost of the rough stone is $320, so we use $c = 320$. To get the value of h we need to remember that h is measured in hours. Five hours and fifteen minutes is five and a quarter hours, so $h = 5.25$ hours, and that is what we will put in for h. Thus, in functional notation the total investment is $I(320, 5.25)$.

Solution to Part 3: In the formula we put 320 in place of c and 5.25 in place of h:

$$I(320, 5.25) = 320 + 40 \times 5.25 = 530 \text{ dollars.}$$

oportion

It is common in many applications of mathematics to use the term *proportional* to indicate that one thing is a multiple of another. For example, the total salary of an hourly wage earner can be calculated using

$$\text{Salary} = \text{Hourly wage} \times \text{Hours worked}.$$

Another way of expressing this is to say that salary is *proportional to*[7] the number of hours worked. If we use S to denote the salary, w the hourly wage, and h the hours worked, then the formula is

$$S = wh\ .$$

Many texts use the special symbol $\propto$ to denote a proportionality relation. Thus $S \propto h$ is shorthand notation for the phrase "S is proportional to h." In this context, the hourly wage w would be termed the *constant of proportionality*. It is what we need to multiply by in order to change the proportionality relation into an equation. The statement that S is proportional to h usually carries with it the implicit understanding that h is the variable but that the constant of proportionality w does not change. Thus we would be discussing a worker with a fixed hourly wage but whose working hours may differ from pay period to pay period.

Since the salary S is a multiple of the hourly wage w, it is also correct to say that S is proportional to w, or in symbols that $S \propto w$. Looking at it this way, we are thinking of w as the variable and h (the number of hours worked) as the constant of proportionality. Implicit in this description is that h does not change, but w does. We might for example be discussing the long-term salary of someone who always works 40 hours each week but is anticipating raises.

When a function is given verbally, it can sometimes be a challenge to find a formula which represents it. But when such verbal descriptions are in terms of a proportion, it is always an easy task to get the formula. This can be done even without understanding the meaning of the letters involved. For example if you are told that one quantity denoted by the letter T is proportional to a second quantity denoted by z, then you know that $T \propto z$ and that there is a constant of proportionality k that makes the equation $T = kz$ true. Determining the value of k and making sense of the equation do of course depend on an understanding of what the letters mean.

---

[7]Sometimes the phrase *directly proportional to* is used.

EXAMPLE 1.12   *Density*

The total weight of a rock depends on its size and is proportional to the *density*. In this context, density is the weight per cubic inch. Let w denote the weight of the rock in pounds, s the size of the rock in cubic inches, and d the density of the rock in pounds per cubic inch.

1. What is the total weight of a 3 cubic inch rock which weighs 2 pounds per cubic inch?

2. Write an equation which shows the proportionality relation. What is the constant of proportionality?

3. Use the equation you got in Part 2 to find the total weight of a 14 cubic inch rock with density 0.3 pounds per cubic inch.

Solution to Part 1: A 3 cubic inch rock which weighs 2 pounds per cubic inch clearly weighs a total of 6 pounds. Notice that we got that by multiplying 3 times 2. This is the point of this part of the example: to illustrate that you get total weight by multiplying density times size.

Solution to Part 2: The weight w is proportional to the density d. That is, $w \propto d$. We noted in Part 1 that we get total weight by multiplying density times size. In letters this is $w = sd$. Thus we are thinking of s, the size of the rock, as the constant of proportionality.

Solution to Part 3: We use the formula we got in Part 2, putting in 14 for s and 0.3 for d:

$$w = sd = 14 \times 0.3 = 4.2 \text{ pounds.}$$

Exercise Set 1.4

1. **United States population growth:** In 1960 the population of the U.S. was about 180 million. Since that time the population has increased by approximately 1.2% each year. This is a verbal description of the function $N = N(t)$, where N is the population, in millions, and t is the number of years since 1960.

   (a) Express in functional notation the population of the U.S. in 1963. Calculate its value.

   (b) Use the verbal description of N to make a table of values that shows U.S. population in millions from 1960 through 1965.

   (c) Make a graph of U.S. population versus time. Be sure to label your graph appropriately.

   (d) Verify that the formula 180×1.012^t million people, where t is the number of years since 1960, gives the same values as those you found in the table in Part (b). (<u>Note</u>: Since t is the number of years since 1960, you would use $t = 2$ to get the population in 1962.)

   (e) Assuming the population has been growing at the same percentage rate since 1960, what value does the formula above give for the population in 1995? (<u>Note</u>: The actual population in 1995 was about 263 million.)

2. **Education and income:** According to the *1995 Digest of Education Statistics*, the median (middle of the range) annual income of a high school graduate in 1993 was about $27,000 per year. Assuming it takes 4 years to earn a bachelor's degree and two additional years to earn a master's degree, the median annual income increases by 12.1% each year an individual attends college. This describes a function $I = I(y)$, where I is the median income in 1993 for a person with y years of college education.

   (a) Express in functional notation the median income in 1993 of an individual with two years of college. Calculate the value.

   (b) Make a table of values that shows the median income in 1993 for individuals completing 0 through 4 years of college.

   (c) Make a graph of median income versus years of college completed. Be sure to provide appropriate labels.

   (d) Verify that the formula $I = 27 \times 1.121^y$ thousand dollars, where y is the number of years of college completed, gives the same values as those you found in the table you made.

(e) Using the formula given in Part (d), find the median income in 1993 of an individual holding a master's degree.

(f) Assuming it takes 3 years beyond the master's degree to complete a Ph.D., and assuming the formula in Part (d) applies, what was the median income of a Ph.D. in 1993?

(g) In fact the same rate of increase does not apply for the years spent on a Ph.D. The actual median income for a Ph.D. in 1993 was $63,149. Does that mean that the increase in median income for years spent earning a Ph.D. is higher or lower than 12.1% per year?

3. **A rental car:** A rental car agency charges $29.00 per day and 6 cents per mile.

   (a) Calculate the rental charge if you rent a car for 2 days and drive 100 miles.

   (b) Use a formula to express the cost of renting a car as a function of the number of days you keep it and the miles you drive. Identify the function and each variable you use, and state the units.

   (c) It is about 250 miles from Dallas to Austin. Use functional notation to express the cost to rent a car in Dallas, drive it to Austin, and return it in Dallas one week later. Use the formula from Part (b) to calculate the cost.

4. **Preparing a letter:** You pay your secretary $6.25 per hour. A stamped envelope costs 38 cents, and paper costs 3 cents per page.

   (a) How much does it cost to prepare and mail a 3 page letter if your secretary spends 2 hours on typing and corrections?

   (b) Use a formula to express the cost of preparing and mailing a letter as a function of the number of pages of the letter and the time it takes your secretary to type it. Identify the function and each of the variables you use, and state the units.

   (c) Use the function you made in Part (b) to find the cost of preparing and mailing a two-page letter that it takes your secretary 25 minutes to type. (Note: 25 minutes is $\frac{25}{60}$ hour.)

5. **A car that gets m miles per gallon:** The cost of operating a car depends on the gas mileage m that your car gets, the cost g per gallon of gasoline, and the distance d that you drive.

   (a) How much does it cost to drive 100 miles if your car gets 25 miles per gallon and gasoline costs 99 cents per gallon?

   (b) Find a formula that gives the cost C as a function of m, g, and d. Be sure to state the units of each variable.

   (c) Use functional notation to show the cost of driving a car that gets 28 miles per gallon a distance of 138 miles if gasoline costs $1.17 per gallon. Use the formula from Part (b) to calculate the cost.

6. **Renting motel rooms:** You own a motel with 30 rooms and have a pricing structure that encourages rentals of rooms in groups. One room rents for $85.00, two for $83.00 each, and in general the group rate per room is found by taking $2 off the base of $85 for each extra room rented.

   (a) How much money do you charge per room if a group rents 3 rooms? What is the total amount of money you take in?

   (b) Use a formula to give the rate you charge for each room if you rent n rooms to an organization.

   (c) Find a formula for a function $R = R(n)$ that gives the total revenue from renting n rooms to a convention host.

   (d) Use functional notation to show the total revenue from renting a block of 9 rooms to a group. Calculate the value.

7. **A cattle pen:** A rancher has 160 feet of fence, and he intends to use all of it as an enclosure for a rectangular cattle pen.

   (a) Suppose he decides to make one side of the pen 30 feet long. Draw a picture and label the length of each side of the pen. What is the length of each side? What is the area of this enclosure?

   (b) What would be the area of the pen if it were a square?

   (c) Can two rectangles with the same perimeter have different areas?

   (d) Find a formula for a function $A = A(l)$ which gives the area in square feet of the cattle pen in terms of the length l in feet of one of its sides. (<u>Hint</u>: First draw a picture of the pen and label one side l. Next figure out the lengths of the other sides in terms of l.)

8. **A car:** The distance d in miles that a car travels on a three hour trip is proportional to its speed s (which we assume remains the same throughout the trip) in miles per hour.

   (a) What is the constant of proportionality in this case?

   (b) Write a formula which expresses d as a function of s.

9. **Production rate:** The total number t of items that a manufacturing company can produce is directly proportional to the number n of employees.

   (a) Choose a letter to denote the constant of proportionality and write an equation that shows the proportionality relation.

   (b) What in practical terms does the constant of proportionality represent in this case?

10. **The $3x + 1$ problem:** Here is a mathematical function $f(n)$ that applies only to whole numbers n. If a number is even, divide it by 2. If it is odd, triple it and add 1. For example, 16 is even, so we divide by 2: $f(16) = \dfrac{16}{2} = 8$. On the other hand 15 is odd, so we triple it and add one: $f(15) = 3 \times 15 + 1 = 46$.

    (a) Apply the function f repeatedly beginning with $n = 1$. That is, calculate $f(1)$, f(the answer from the first part), f(the answer from the second part), and so on. What pattern do you see?

    (b) Apply the function f repeatedly beginning with $n = 5$. How many steps does it take to get to 1?

    (c) Apply the function f repeatedly beginning with $n = 7$. How many steps does it take to get to 1?

    (d) Try several other numbers of your own choosing. Does the process always take you back to 1? (<u>Note</u>: We can't be sure what your answer will be here. Every number that anyone has tried so far leads eventually back to 1, and it is conjectured that this happens no matter what number you start with. This is known to mathematicians as the $3x + 1$ *conjecture*, and it is, as of the writing of this book, an unsolved problem. If you can find a starting number that does not lead back to 1, or if you can somehow show that the path *always* leads back to 1, you will have solved a problem that has eluded mathematicians for a number of years. Good hunting!)

11. **Catering a dinner:** You are having a dinner catered. You pay a rental fee of $150 for the dining hall, plus you pay the caterer $10 for each person who attends the dinner.

    (a) Suppose you just want to break even.

         i. How much should you charge *per ticket* if you expect 50 people to attend?

        ii. Use a formula to express the amount you should charge per ticket as a function of the number of people attending. Be sure to explain the meaning of the letters you choose and the units.

       iii. You expect 65 people to attend the dinner. Use your answer to Part (ii) above to express in functional notation the amount you should charge per ticket, and then calculate that amount.

    (b) Suppose now that you want to make a profit of $100 from the dinner. Use a formula to express the amount you should charge per ticket as a function of the number of people attending. Again, be sure to explain the meaning of the letters you choose and the units.

1.5 CHAPTER SUMMARY

The idea of a *function* is as old as mathematics itself, and it is central to mathematics and applications. It is certainly a key topic in this text, where it occurs in one form or another throughout. A function is nothing more than a clear description of how one thing depends on another (or on several other things), and functions are presented in various ways. The most common ways of presenting functions are the section topics of this chapter.

Functions Given by Formulas

This is perhaps the way in which most people think of a function, and it is an important one. An example of a function given by a formula is

$$M = 7h,$$

where h is the number of hours an employee may work, and M is the money earned, in dollars. The formula simply says that one can calculate the money earned by multiplying the number of hours worked by 7. In other words, the formula describes the pay of an employee who earns $7.00 an hour. Sometimes formulas for functions are quite complicated. The calculator makes functions given by formulas easy to deal with in spite of their apparent complexity.

In applications of mathematics, functions are often representations of real phenomena or events. Thus we say that they are *models*. Obtaining a function or functions to act as a model is commonly the key to understanding physical, natural, and social science phenomena. This applies to business and many other areas as well.

Functions Given by Tables

One of the most common ways in which functions are encountered in daily life is in terms of *tables of values*. Such tables can be found everywhere: census data, payment schedules, college enrollment statistics, lists of species occupying a certain region, and a myriad of other familiar data tables. An example is the U.S. census data appearing in the following table.

d = Year	1950	1960	1970	1980	1990
N = Population in millions	151.87	179.98	203.98	227.23	249.40

This table gives U.S. population $N = N(d)$ as a function of the date. Tables are almost always incomplete; that is, some information is left out of the table. In this case, the population in 1954 is not reported. A common way of estimating function values not given in a table is by using the *average rate of change*. For example, from 1950 to 1960, the U.S. population grew from 151.87 million to 179.98 million. That is an increase of 28.11 million over a 10 year period. Thus from 1950 to 1960, the U.S. population grew at a rate of approximately

$$\frac{28.11}{10} = 2.811 \text{ million per year.}$$

This is the average yearly rate of change during the 1950's. It is reasonable to estimate the population in 1954 by

$$N(1950) = 151.87 + 4 \times 2.811 = 163.114 \text{ million.}$$

Sometimes, a function has a *limiting value* which may be estimated from the tabular form of a function. For example, the following table shows the amount of yeast present in an enclosed area t hours after observations began.

Time t	0	5	10	15	20	25	30
Amount of yeast N	10	119	513	651	662	664	665

Since an enclosed area is being observed, we expect that there is a limit to the amount of yeast which will ever be present. Looking at the table, it is reasonable to expect that this limiting value is about 665.

Functions Given by Graphs

Another way in which functions are commonly presented is with graphs. Generally, it is more difficult to get exact function values from a graph than from a formula or table, but a graph has the advantage of showing clearly certain overall features of a function. It clearly shows, for example, when a function is increasing or decreasing, when it reaches maxima or minima, the concavity of the graph, and often limiting values of the function. These function properties are displayed by the graph according to the following table.

Function property	Graph display
Increasing	Rising graph
Decreasing	Falling graph
Maximum	Graph reaches a peak
Minimum	Graph reaches a valley
Concave up	Graph is bent upward (holds water)
Concave down	Graph is bent downward (spills water)
Has a limiting value	Graph levels out on right side

The concavity of a graph gives important information about the rate of change of the function. Increasing graphs which are concave up represent functions which increase at an increasing rate, while increasing graphs which are concave down represent functions which increase at a decreasing rate. Similarly, decreasing graphs which are concave up represent functions which decrease at a decreasing rate, while decreasing graphs which are concave down represent functions which decrease at an increasing rate.

Functions Given by Words

Very often a function is presented with a verbal description, and the key to understanding it may well be to translate this verbal description into a formula, table, or graph. For example, suppose a company invests $78,000 in the design and development of a more efficient computer hard drive, and for each drive sold, the firm makes a profit of $98. This can be thought of as a verbal description of the net profit $N = N(d)$ as a function of the number d of drives sold. Furthermore, it is not difficult to translate this verbal description into a formula:

$$\text{Net profit} \quad = \quad \text{Profit from sales} \; - \; \text{Initial investment}$$
$$N \quad = \quad 98d - 78,000 \text{ dollars.}$$

A common way in which verbal descriptions of functions are given is in terms of *proportion*. This simply indicates that one thing is a multiple of another. For example, money earned by a wage employee is proportional to the number of hours worked. In terms of a formula, this proportionality statement means

$$\text{Money earned } = \text{Hourly wage } \times \text{ Hours worked.}$$

In this context, the hourly wage is known as the *proportionality constant*.

CHAPTER 2 *Graphical and Tabular Analysis*

Each type of function presentation that we studied in Chapter 1 has advantages and disadvantages, and one of the most useful analysis methods is to look at functions in more than one way. Tables and graphs often show information that is difficult to obtain directly from formulas. The graphing calculator makes it easy to go from a formula to a table or graph and hence becomes a tool for solving significant problems.

2.1 TABLES AND TRENDS

The advantage of functions given by formulas is that they allow for the calculation of any function value. Tables of values always leave gaps, but they may be more helpful than formulas for seeing trends, predicting future values, or discerning other interesting information about the function.

Getting tables from formulas

Let's look at an example to illustrate this. Proper management of wildlife depends on the ability of ecologists to monitor and predict population growth. In many situations it is reasonable to expect animal populations to exhibit *logistic growth*. A special formula that is studied extensively by ecologists describes this type of growth.

The circumstances surrounding the *George Deer Reserve* in Michigan have made it particularly easy for ecologists to monitor accurately the growth of the deer population on the reserve and to develop a logistic growth formula for the number $N = N(t)$ of deer expected to be present after t years:[1]

$$N = \frac{6.21}{0.035 + 0.45^t} \text{ deer.}$$

When a breeding group of animals is introduced into a limited area, one expects that it will over time grow to the largest size that the environment can support. Wildlife managers refer to this as the *environmental carrying capacity*. Let's find the carrying capacity for deer of the George Reserve. That is, we want to know the deer population after a long period of time.

---

[1]Dale R. McCullough, *The George Reserve Deer Herd*, The University of Michigan Press, Ann Arbor, 1979.

This is not an easy question to answer by looking at the formula, but if we use the formula to make a table of values, then the trend will become apparent. In the table below, we have calculated $N(0), N(5), N(10), \dots, N(30)$. (The initial population, the population after 5 years, the population after 10 years, etc.)

Year t	$N(t)$ Population in year t
0	6
5	116.18
10	175.72
15	177.4
20	177.43
25	177.43
30	177.43

Scanning down the right-hand column of the table, we see the growth of the deer population with time. From the first row of the table, we see that there were initially 6 deer on the reserve. During the first 10 years, the population increases rapidly, but the rate of increase slows down dramatically after that. It appears that after about 20 years, the population levels out at approximately 177 deer. (We have rounded to the nearest whole number since we don't expect to see parts of deer on the reserve.) This is the carrying capacity of the reserve.

The key idea here is that many times needed information is difficult to obtain directly from a formula. Supplementing the formula with a table of values can provide deeper insight into what is happening.

Tables of values can be made by calculating each wanted function value one at a time, but many calculators have a built-in feature which automatically generates such tables. Making tables of values using a calculator is a skill that will be needed often in what follows, and so you are strongly encouraged to consult the *Keystroke Guide* for instructions on how to do this. To become familiar with the procedure, you should work through the practice problems that are presented there.

When you make a table of values with a calculator, there are three key bits of information that you must input. You must tell your calculator which function you want to use, you must decide on a place to begin the table, the *table starting value*, and you must decide on the periods, the *table increment value*, when you want to see additional data. In making the table above for deer population, we used the function $\dfrac{6.21}{0.035 + 0.45^t}$ with a table starting value of

$t = 0$, and we viewed the data in five-year periods. That is, we used a table increment value of 5. In what follows, we will refer to these latter two items as the *table setup*. Thus, if we wanted to instruct you to make the table above in exactly the same way we did, we would say "Enter ⌐2.1¬ the function $\dfrac{6.21}{0.035 + 0.45^t}$, and for table setup ⌐2.2¬ use a starting value of $t = 0$ and a table increment value of 5." The function entry screen on a graphing calculator will typically appear as in Figure 2.1 and the completed function entry as in Figure 2.2, but displays will vary from calculator to calculator. The table setup will typically appear as in Figure 2.3, and

---

Figure 2.1: *A typical function entry screen*

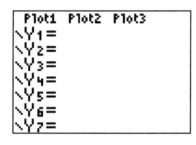

Figure 2.2: *The properly entered function*

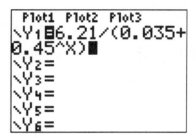

---

the completed table of values will typically appear as in Figure 2.4.

Figure 2.3: *A typical* TABLE SETUP *menu*

Figure 2.4: *A table of values for the deer population*

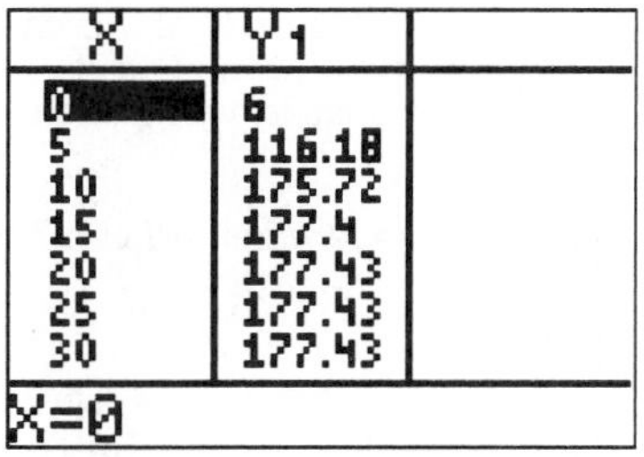

We are working with a function whose name is N and with a variable t, but as we see in Figures 2.2 and 2.4, the calculator has chosen its own name Y_1 for the function N and X for the variable t. Your calculator may use other letters to represent the function and the variable, but whatever letters your calculator uses, it is important to keep track of the proper associations. It is good practice to write down the correspondence, and we will always do that in the examples we present:

$$Y_1 \quad = \quad N, \text{ population} \tag{2.1}$$

$$X \quad = \quad t, \text{ time in years.} \tag{2.2}$$

Now if we want to use the table in Figure 2.4 to find the value of N when t is 20, we note from Equation (2.2) that we should look in the X column to find 20 and from Equation (2.1) that the corresponding Y_1 value 177.43 is the function value for N. That is, $N(20) = 177.43$.

EXAMPLE 2.1   *A Skydiver*

A falling object is pulled downward by gravity, but its fall is retarded by air resistance, which under appropriate conditions is directly proportional to velocity. When a skydiver jumps from an airplane her downward velocity $v = v(t)$ before she opens her parachute is given by

$$v = 176(1 - 0.834^t) \text{ feet per second,}$$

where t is the number of seconds that have elapsed since she jumped from the airplane.

1. Express the velocity of the skydiver 2 seconds into the fall using functional notation and calculate its value.

2. Describe how the velocity of the skydiver changes with time. Include in your description the average rate of increase in velocity during the first five seconds and the average rate of increase in velocity during the next five seconds.

3. What is the *terminal velocity*? That is, what is the maximum speed the skydiver can attain?

4. How long does it take the skydiver to reach 99% of terminal velocity?

Solution to Part 1: In functional notation, the downward velocity 2 seconds into the fall is $v(2)$. To make the calculation we put 2 in for t:

$$v(2) = 176(1 - 0.834^2) = 53.58 \text{ feet per second.}$$

Solution to Part 2: We want to see how the velocity increases with time. This is difficult to see from the formula, but a table of values showing velocity in five-second intervals will give us the information we need. Thus we want to enter $\boxed{2.3}$ the function and we want a table setup $\boxed{2.4}$ with a starting value of 0 and an increment of 5. The correctly entered function is in Figure 2.5, and the correctly configured table setup menu is in Figure 2.6.

Figure 2.5: *Entering the function for velocity*

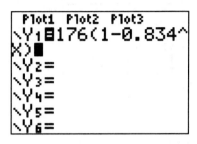

Figure 2.6: *Setting up the table*

Once again, the calculator is using its own choices of letters, and so we record the appropriate correspondences:

$$Y_1 = v, \text{ velocity}$$
$$X = t, \text{ time in seconds.}$$

When we view $\boxed{2.5}$ the table shown in Figure 2.7 we can read down the right-hand column to see that the velocity increases rapidly to begin with but seems to be leveling out near 30 seconds into the fall. It appears that the downward pull of gravity makes the skydiver accelerate rapidly at first, but air resistance seems to have a greater effect at high velocities. This is a consequence of the fact that air resistance is in this case directly proportional to velocity.

The table shows that the velocity increased from 0 to 104.99 feet per second during the first five seconds of the fall. Thus, during this period, velocity increased at an average rate of $\frac{104.99}{5} = 21$ feet per second per second. During the next five seconds of the fall, the velocity increased by $147.35 - 104.99 = 42.36$ feet per second. That gives an average increase in velocity of $\frac{42.36}{5} = 8.47$ feet per second per second. As we should have expected, the rate of increase in velocity is much less during the second five-second period than in the first five seconds. What would this say about the concavity of the graph of velocity against time?

Solution to Part 3: To get the terminal velocity, we want to know what the velocity of the skydiver would be if she continued free fall for a long time without opening her parachute.

That is, we want to look at the table for large values of t. Your graphing calculator may have a feature that lets you extend [2.6] the table without going back to the table setup menu. We have shown the table for $t = 35$ seconds to $t = 65$ seconds in Figure 2.8. This

Figure 2.7: *A table of values for velocity*

X	Y1	
0	0	
5	104.99	
10	147.35	
15	164.44	
20	171.34	
25	174.12	
30	175.24	

X=0

Figure 2.8: *Extending the table*

X	Y1	
35	175.69	
40	175.88	
45	175.95	
50	175.98	
55	175.99	
60	176	
65	176	

X=65

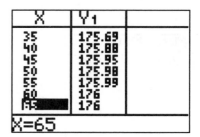

table shows clearly that velocity levels out at 176 feet per second. (Look further down the table for more evidence of this.) This is the terminal velocity, where the downward pull of gravity matches air resistance.

Solution to Part 4: Now 99% of terminal velocity is $0.99 \times 176 = 174.24$ feet per second. Consulting the table in Figure 2.7, we see that this velocity is reached about 25 seconds into the fall. You may wish to improve the accuracy of this answer by changing the table setup so that the increment is 0.5.

We should also note that questions about terminal velocity in particular and limiting values of functions in general can lead to unexpected difficulties which require more advanced mathematical analysis than is appropriate here. See Exercise 11 at the end of this section.

It is worth emphasizing the significance of what we have done here. The table of values allowed us to make an in-depth analysis of what happens in the free fall period of a skydiver's fall. None of the conclusions we drew in Parts 2 and 3 are apparent from the formula, but they are easily discernible from the table of values. The technology we are using gives us the power to attack and resolve real problems.

Optimizing with tables of values

We can also use tables of values to find maximum and minimum values for functions. To illustrate this, let's suppose we roll several dice hoping to get exactly 3 sixes, not more or less. The probability that this will occur depends on how many dice we roll. If we use only 4 dice, then it seems unlikely that we would get as many as 3 sixes. If we roll 100 dice then we would expect to get more than 3 sixes. Elementary probability theory can be used to show that if we roll N dice, then the probability $p = p(N)$ of getting exactly 3 sixes is given by the formula

$$p = \frac{N(N-1)(N-2)}{750} \times \left(\frac{5}{6}\right)^N .$$

Thus if we want to know the probability of getting exactly 3 sixes when we roll 10 dice, we put $N = 10$ into the formula for p:

$$p(10) = \frac{10 \times 9 \times 8}{750} \times \left(\frac{5}{6}\right)^{10} = 0.15505 .$$

If we round this to two decimal places, we get $p = 0.16$. This means that if you roll 10 dice you can expect to get exactly 3 sixes about 16 times in 100 rolls.

How many dice should we roll so that we have the best possible chance of getting exactly 3 sixes? To answer this, we want to make a table of values showing the probability of getting exactly 3 sixes for various values of N. First we enter $\boxed{2.7}$ the function as shown in Figure 2.9. We record the letter correspondences:

$$\mathsf{Y_1} \;\; = \;\; p, \text{ probability of exactly 3 sixes}$$
$$\mathsf{X} \;\; = \;\; N, \text{ number of dice.}$$

Next we set up the table using a starting value $\boxed{2.8}$ of 1 and an increment of 1. The correctly configured table setup menu is in Figure 2.10. When we view the table of values, we

---

Figure 2.9: *Entering a probability function*

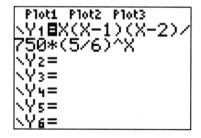

Figure 2.10: *Setting up the table*

---

get the display in Figure 2.11. We see that if we use 1 or 2 dice, the probability is 0, indicating as we expected that it is impossible to get 3 sixes by rolling fewer than 3 dice. If we roll 3 dice, we see that the probability of getting exactly 3 sixes is 0.00463, or 0.0046 rounded to two decimal places. This means that we would expect to get 3 sixes only 46 times out of 10,000 rolls. As the number of dice increases from 3 up through 7, the probability of getting exactly 3 sixes gets larger. In Figure 2.12 we have extended $\boxed{2.9}$ the table so that we can see what happens when we roll between 15 and 21 dice. This table shows us that the probability of 3 sixes increases as the number of dice increases up to 17 but decreases for more than 18 dice. The decrease occurs because if we roll too many dice, we would expect to get more than 3 sixes. Thus the probability is at its largest, 0.2452, if we use either 17 or 18 dice. You should look both forward and backward in the table to insure that there are not larger values outside the ranges shown in Figures 2.11 and 2.12.

Figure 2.11: Probability of 3 sixes when 1 through 7 dice are used

X	Y1	
1	0	
2	0	
3	.00463	
4	.01543	
5	.03215	
6	.05358	
7	.07814	

X=3

Figure 2.12: Probability of 3 sixes when 15 through 21 dice are used

X	Y1	
15	.23626	
16	.24231	
17	.2452	
18	.2452	
19	.24264	
20	.23789	
21	.23128	

X=21

EXAMPLE 2.2 Renting Canoes

A small business has 20 canoes which it rents for float trips down the Illinois River. The pricing structure offers a discount for group rentals. One canoe rents for \$35, two rent for \$34 each, and in general the group rate per canoe is found by taking \$1 off the base of \$35 for each extra canoe rented.

1. How much money is taken in if 3 canoes are rented to a group?
2. Write a formula that gives the price charged for each canoe if n canoes are rented to a group.
3. Find a formula $R = R(n)$ that shows how much money is taken in from renting n canoes to a group.
4. How large a rental to a single group will bring the most income?

Solution to Part 1: One canoe rents for \$35. Two rent for \$34 each, and three rent for \$33 each. Three are being rented, so $3 \times 33 = 99$ dollars is taken in.

Solution to Part 2: We want a formula expressing the amount charged per canoe in terms of the number of canoes rented to a group. If the business rents 1 canoe, the charge is $35 = 36 - 1$ dollars per canoe. If the group rental is 2 canoes, the charge is $34 = 36 - 2$ dollars per canoe; for 3, the charge is $33 = 36 - 3$ dollars per canoe. If n canoes are rented to a group, the charge is $36 - n$ dollars per canoe.

Solution to Part 3: We know that n canoes are rented at the rate we found in Part 2. Thus if the business rents n canoes, then it charges $36 - n$ dollars for each canoe. That means $n(36 - n)$ dollars are taken in:

$$R(n) = n(36 - n).$$

Solution to Part 4: We want a table that shows how much money is taken in from renting canoes in groups. Thus we need a table of values for R showing money taken in for $n = 1$ through $n = 20$. We enter $\boxed{2.10}$ the function and record the correspondence for function and variable names:

$$Y_1 \; = \; R, \text{ revenue}$$
$$X \; = \; n, \text{ canoes rented.}$$

Next we set up the table $\boxed{2.11}$ using a table starting value of 1 and an increment value of 1. The tables for 7 through 13 and 14 through 20 canoe rentals are in Figure 2.13 and Figure 2.14. Since the business only has 20 canoes, we need not look further down the table. It shows that the most money, \$324, is taken in if 18 canoes are rented to a group. You should view the entire table from $n = 1$ through $n = 20$ to make sure that nothing has been overlooked.

Figure 2.13: *Renting to groups that number 7 to 13*

X	Y1	
7	203	
8	224	
9	243	
10	260	
11	275	
12	288	
13	299	
X=7		

Figure 2.14: *Renting to groups that number 14 to 20*

X	Y1	
14	308	
15	315	
16	320	
17	323	
18	324	
19	323	
20	320	
X=20		

Exercise Set 2.1

1. **APR and EAR:** Recall that the APR (the *annual percentage rate*) is the percentage rate on a loan that the Truth in Lending Act requires lending institutions to report on loan agreements. It does not tell directly what the interest rate really is. If you borrow money for one year and make no payments, then in order to calculate how much you owe at the end of the year, you must use another interest rate, the EAR (the *effective annual rate*), which is not normally reported on loan agreements. The calculation is made by adding the interest indicated by the EAR to the amount borrowed.

   The relationship between the APR and the EAR depends on how often interest is compounded. If you borrow money at an annual percentage rate APR (as a decimal), and if interest is compounded n times per year, then the effective annual rate EAR (as a decimal) is given by

   $$\text{EAR} = \left(1 + \frac{\text{APR}}{n}\right)^n - 1.$$

   For the remainder of this problem, we will assume an APR of 10%. Thus in the formula above, we would use 0.1 in place of APR.

   (a) Would you expect a larger or smaller EAR if interest is compounded more often? Explain your reasoning.

   (b) Make a table that shows how the EAR depends on the number of compounding periods. Use your table to report the EAR if interest is compounded once each year, monthly, and daily. (<u>Note</u>: The formula will give the EAR as a decimal. You should report your answer as a percent.)

   (c) If you borrow $5000 and make no payments for one year, how much will you owe at the end of a year if interest is compounded monthly? If interest is compounded daily?

   (d) If interest is compounded as often as possible, that is, continuously, then the relationship between APR and EAR is given by

   $$\text{EAR} = e^{\text{APR}} - 1.$$

   Again using an APR of 10%, compare the EAR when the interest is compounded monthly with the EAR when interest is compounded continuously.

2. **An amortization table:** Suppose you borrow P dollars at a monthly interest rate of r (as a decimal) and wish to pay off the loan in t months. Then your monthly payment can be calculated using

$$M = \frac{Pr(1+r)^t}{(1+r)^t - 1} \text{ dollars.}$$

Remember that for monthly compounding you get the monthly rate by dividing the APR by 12. Suppose you borrow \$3500 at 9% APR (meaning that you use $r = 0.09/12$ in the formula above) and pay it back in two years.

(a) What is your monthly payment?

(b) Let's look ahead to the time when the loan is paid off.

    i. What is the total amount you paid to the bank?

    ii. How much of that was interest?

(c) The amount B that you still owe the bank after making k monthly payments can be calculated using the variables r, P, and t. The relationship is given by

$$B = P \times \left(\frac{(1+r)^t - (1+r)^k}{(1+r)^t - 1} \right) \text{ dollars.}$$

    i. How much do you still owe the bank after one year of payments?

    ii. An *amortization table* is a table that shows how much you still owe the bank after each payment. Make an amortization table for this loan.

3. **An amortization table for continuous compounding:** *This is a continuation of Exercise 2.* Suppose you have borrowed P dollars from a lending institution that compounds interest as often as possible, that is, continuously. If the EAR is r as a decimal, and the loan is to be paid off in Y years, then you would calculate your monthly payment using

$$M = \frac{Pr}{12(1 - e^{-rY})} \text{ dollars.}$$

Under these circumstances, the balance B that you owe the bank after k monthly payments is given by

$$B = \frac{P(e^{rY} - e^{\frac{rk}{12}})}{e^{rY} - 1} \text{ dollars.}$$

As we noted in Exercise 1, for continuous compounding, you get the EAR r from the APR a using $e^a - 1$. Suppose we borrow \$3500 at an APR of 9% (meaning that we use $a = 0.09$ in the formula above) and pay off the note in two years.[2]

(a) Calculate your monthly payment and compare your answer with the answer you got in Exercise 2.

---

[2]It is worth pointing out that the formulas we see in this exercise are simpler than the corresponding ones in Exercise 2. Continuous compounding may appear at first sight to be more complicated than monthly compounding, but it is in fact easier to handle. And as you will see when you complete this exercise, in many applications it does not give significantly different answers.

(b) Make an amortization table and compare it with the answer you got in Exercise 2.

4. **Renting motel rooms:** You own a motel with 30 rooms and have a pricing structure that encourages rentals of rooms in groups. One room rents for $85, two rent for $83 each, and in general the group rate per room is found by taking $2 off the base of $85 for each extra room rented.

    (a) How much money do you take in if a family rents two rooms?

    (b) Use a formula to give the rate you charge for each room if you rent n rooms to an organization.

    (c) Find a formula for a function $R = R(n)$ that gives the revenue from renting n rooms to a convention host.

    (d) What is the most money you can make from rental to a single group? How many rooms do you rent?

5. **Inventory:** For retailers who buy from a distributor or manufacturer and sell to the public, a major concern is the cost of maintaining unsold inventory. You must have appropriate stock to do business, but if you order too much at a time, your profits may be eaten up by storage costs. One of the simplest tools for analysis of inventory costs is the *basic order quantity model*. It gives the yearly inventory expense $E = E(c, N, Q, f)$ when taking into account the following inventory and restocking cost factors:

- The *carrying cost c*, which is the cost in dollars per year of keeping a single unsold item in your warehouse.

- The number N of this item that you expect to sell in one year.

- The number Q of items you order at a time.

- The fixed costs f in dollars of processing a restocking order to the manufacturer. (<u>Note:</u> This is not the cost of the order; the price of an item does not play a role here. Rather f is the cost you would incur with any order of any size. It might include the cost of processing the paperwork, fixed costs you pay the manufacturer for each order, shipping charges which do not depend on the size of the order, the cost of counting your inventory, or the cost of cleaning and rearranging your warehouse in preparation for delivery.)

The relationship is given by

$$E = \left(\frac{Q}{2}\right) c + \left(\frac{N}{Q}\right) f \text{ dollars per year.}$$

A new car dealer expects to sell 36 of a particular model car in the next year. It costs $850 per year to keep an unsold car on the lot. Fixed costs associated with preparing, processing, and receiving a single order from Detroit total $230 per order.

(a) Using the information provided, express the yearly inventory expense $E = E(Q)$ as a function of Q, the number of automobiles included in a single order.

(b) What is the yearly inventory expense if 3 cars at a time are ordered?

(c) How many cars at a time should be ordered to make yearly inventory expenses a minimum?

(d) Using the value of Q you found in Part (c), determine how many orders to Detroit will be placed this year.

(e) What is the average rate of increase in yearly inventory expense from the number you found in Part (c) to an order of two cars more?

6. **A population of foxes:** A breeding group of foxes is introduced into a protected area and exhibits logistic population growth. After t years the number of foxes is given by

$$N(t) = \frac{37.5}{0.25 + 0.76^t} \text{ foxes.}$$

(a) How many foxes were introduced into the protected area?

(b) Calculate $N(5)$ and explain the meaning of the number you have calculated.

(c) Explain how the population varies with time. Include in your explanation the average rate of increase over the first ten-year period and the average rate of increase over the second ten-year period.

(d) Find the carrying capacity for foxes in the protected area.

(e) As we saw in the discussion of terminal velocity for a skydiver, the question of when the carrying capacity is reached may lead to an involved discussion. We ask the question differently. When is 99% of carrying capacity reached?

7. **Falling with a parachute:** If an average size man jumps from an airplane with an open parachute, his downward velocity t seconds into the fall is $v(t) = 20(1 - 0.2^t)$ feet per second.

   (a) Find the velocity 2 seconds into the fall.

   (b) Explain how the velocity increases with time. Include in your explanation the average rate of change from the beginning of the fall to the end of the first second and the average rate of change from the fifth to the sixth second of the fall.

   (c) Find the terminal velocity.

   (d) Compare the time it takes to reach 99% of terminal velocity here with the time it took to reach 99% of terminal velocity in Example 2.1. Based on the information we have, which would you expect to reach 99% of terminal velocity first, a feather or a cannonball?

8. **Rolling 4 sixes:** If you roll N dice, then the probability $p = p(N)$ that you will get exactly 4 sixes is given by

$$p = \frac{N(N-1)(N-2)(N-3)}{24} \times \left(\frac{1}{6}\right)^4 \left(\frac{5}{6}\right)^{N-4}.$$

   (a) What is the probability, rounded to three decimal places, of getting exactly 4 sixes if 10 dice are rolled? How many times out of 1000 rolls would you expect this to happen?

   (b) How many dice should be rolled so that the probability of getting exactly 4 sixes is the greatest?

9. **A precocious child and her blocks:** A child has 64 blocks which are 1 inch cubes. She wants to arrange the blocks into a solid rectangle h blocks long and w blocks wide. There is a relationship between h and w which is determined by the restriction that all 64 blocks must go into the rectangle. A rectangle h blocks long and w blocks wide uses a total of $h \times w$ blocks. Thus $hw = 64$. Applying some elementary algebra, we get the relationship we need:

$$w = \frac{64}{h}. \tag{2.3}$$

   (a) Use a formula to express the perimeter P in terms of h and w.

   (b) Using Equation (2.3), find a formula expressing the perimeter P in terms of the height only.

   (c) How should the child arrange the blocks if she wants the perimeter to be the smallest possible?

(d) Do Parts (b) and (c) again, this time assuming the child has 60 blocks rather than 64 blocks. In this situation the relationship between h and w is $w = \dfrac{60}{h}$. (Note: Be careful when you do Part (c). The child will not cut the blocks into pieces!)

10. **Renting paddleboats:** An enterprise rents out paddleboats for all-day use on a lake. The owner knows that he can rent out all 27 of his paddleboats if he charges $1 for each rental. He also knows that he can rent out only 26 if he charges $2 for each rental, and that in general there will be one less paddleboat rental for each extra dollar he charges per rental.

    (a) What would the owner's total revenue be if he charged $3 for each paddleboat rental?

    (b) Use a formula to express the number of rentals as a function of the amount charged for each rental.

    (c) Use a formula to express the total revenue as a function of the amount charged for each rental.

    (d) How much should the owner charge to get the largest total revenue?

11. **Terminal velocity revisited:** In one of the early "Functions and Change" pilot courses at Oklahoma State University, the instructor asked the class to determine when in Example 2.1 terminal velocity would be reached. Three students gave the following three answers:

    Student 1: 58 seconds into the fall.

    Student 2: 147 seconds into the fall.

    Student 3: Never.

    Each student's answer was accompanied by what the instructor judged to be an appropriate supporting argument, and each student received full credit for the problem. What supporting arguments might the students have used to convince the instructor that these three different answers could all be deserving of full credit? (Hint: Consider the formula given in Example 2.1. For Student 1, look at a table of values where the entries are rounded to two decimal places. For Student 2, look at a table of values made by using all the digits beyond the decimal point that the calculator can handle. In this case that was nine. For Student 3, consider what value 0.834^t must have to make $176(1 - 0.834^t)$ equal to 176.)

2 GRAPHS

A graph is a picture of a function, and just as tables do, graphs can show features of a function that are difficult to see by looking at formulas. For many applications the graph is more useful than a table of values. The graphing calculator can generate the graph of a function as easily as it makes tables.

Hand-drawn graphs from formulas

If we have a function given by a formula, there is a standard procedure for generating the graph. We will make a hand-drawn picture of the graph of $f = f(x)$ where $f = x^2 - 1$. The first step is to make a table of values as shown below.

x	$f = x^2 - 1$
-2	3
-1	0
0	-1
1	0
2	3

This table tells us that the points $(-2, 3), (-1, 0), (0, -1), (1, 0), (2, 3)$ lie on the graph. We plot the individual points as shown in Figure 2.15. We complete the graph by joining up the dots with a smooth curve as is shown in Figure 2.16.

Figure 2.15: *Plotting points*

Figure 2.16: *Completing the graph:* $x^2 - 1$

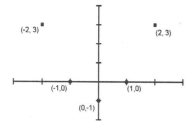

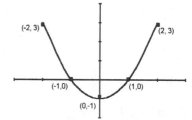

Graphing with the calculator

For more complicated formulas this process is tedious and subject to inaccuracy. But the graphing calculator can make graphs easily. It does exactly what we did above, but it doesn't mind doing all that arithmetic and can accurately plot many more points than we used. For calculator-specific instructions on how to make graphs from formulas see Chapter 2 of the *Keystroke Guide*.

There are two steps involved in using a calculator to get a graph from a formula. We will look at them in the context of making the graph of $f = x^2 - 1$.

Step 1, Entering the function: First we have to tell the calculator which function we want to use. We use the function entry screen to do this exactly as we did in Section 2.1 when we made tables of values. In Figure 2.17 we have cleared old formulas from the function entry screen and entered $\boxed{2.12}$ the new one. As expected, the calculator has chosen its own names for functions and variables, and it is important to record the associations:

$$Y_1 \; = \; f, \text{ corresponding to the vertical axis}$$

$$X \; = \; x, \text{ corresponding to the horizontal axis.}$$

Step 2, Selecting the viewing window: As with any picture, the graph looks different from different points of view. If you are sure that your viewing window is set up to show the graph as you wish to see it, you can skip this step and go directly to the graph $\boxed{2.13}$. If this produces an unsatisfactory view, then it will be necessary to make adjustments in the viewing window. In Figure 2.18 we have shown what we will refer to as the *standard view* $\boxed{2.14}$, which shows the graph in a window extending from -10 to 10 in the horizontal direction and from -10 to 10 in the vertical direction. This standard display is satisfactory for some graphs, and we will use it occasionally in what follows. Below we will discuss what to do when the standard view is unsatisfactory.

Figure 2.17: *Entering* $f = x^2 - 1$

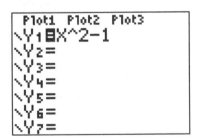

Figure 2.18: *The standard view of the graph*

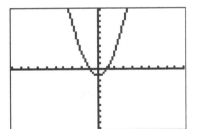

Tracing the graph

Once we have a graph on the screen there are several ways to adjust the view or to get information from it. We will continue working with $f = x^2 - 1$, and to follow the discussion it is important that you have a graph on your screen which matches the one in Figure 2.18. If you are having difficulty getting that picture see Chapter 2 of the *Keystroke Guide*.

Most calculators allow you to *trace* $\boxed{2.15}$ a graph which appears on the screen. This means to put a movable cursor on the screen which follows the graph and is controlled by left and right arrow keys. In Figure 2.19 we have used this feature to add a cursor to the screen and move it along the graph. As the cursor moves, its location is recorded on the screen. In Figure 2.19 the X=2.3404255 prompt at the bottom of the screen shows where we are relative to the horizontal axis, and the Y=4.4775917 prompt shows where we are relative to the vertical axis. This tells us that the cursor is located at the point (2.3404255, 4.4775917). Since this point lies on the graph it also tells us that $f(2.3404255) = 4.4775917$, or, rounded to two decimal places, $f(2.34) = 4.48$.

Figure 2.19: *Tracing the graph*

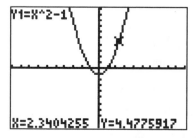

Figure 2.20: *Getting f(3)*

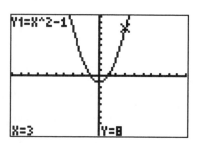

The trace feature cannot directly show all function values. For example, on our calculator we were unable, using the arrow keys, to make the cursor land exactly on X=3. Most calculators allow alternative ways $\boxed{2.16}$ to locate the cursor at X=3. We have used this feature in Figure 2.20, and we read from the prompt at the bottom of the screen that $f(3) = 8$ as expected. It is important that you be able to move the graphing cursor to any point on the graph, and you are encouraged to consult Chapter 2 of the *Keystroke Guide* for information on how to do this.

Choosing a viewing window in practical settings

The key to making a usable graph on the calculator is to make a proper choice of a viewing window. Finding a viewing window that shows a good picture can sometimes be a bit frustrating, but often appropriate settings can be determined from practical considerations.

Consider, for example, a leather craftsman who has produced 25 belts which he intends to sell at an upcoming art fair for $22.75 each. He has invested a total of $300 in leather, buckles, and other accessories for the belts. We want to look at his net profit $p = p(n)$ in dollars as a function of the number n of belts that he sells:

$$\begin{aligned} \text{Net profit} \quad &= \quad \text{Profit from sales} \; - \; \text{Investment} \\ &= \quad \text{Price per item} \; \times \; \text{Number sold} \; - \; \text{Investment} \\ p \quad &= \quad 22.75n - 300 \, . \end{aligned}$$

Let's make a graph of p versus n, that is, a graph which shows net profit as a function of the number of belts sold. First we enter $\boxed{2.17}$ the function as shown in Figure 2.21. The proper variable associations are

$$Y_1 \; = \; p, \text{ net profit on the vertical axis}$$

$$X \; = \; n, \text{ number sold on the horizontal axis.}$$

If we look at the standard $\boxed{2.18}$ view of the function as shown in Figure 2.22 we see no graph at all! We need to choose a different viewing window that will show the graph. To do

Figure 2.21: *Entering a function for net profit on the sale of belts*

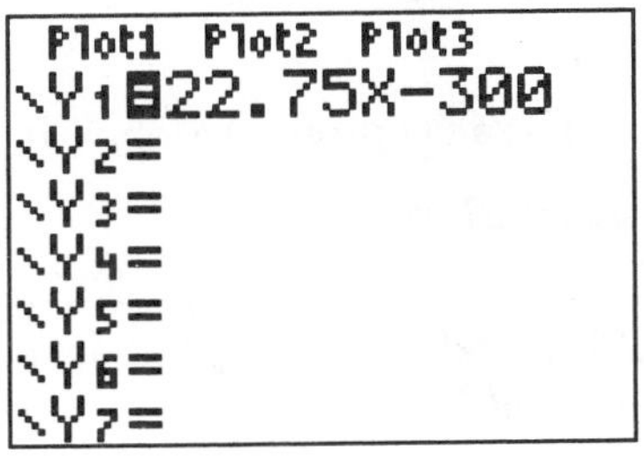

Figure 2.22: *The standard view of the graph is unsatisfactory*

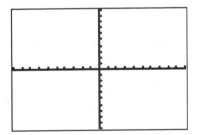

```
Plot1 Plot2 Plot3
\Y1■22.75X-300
\Y2=
\Y3=
\Y4=
\Y5=
\Y6=
\Y7=
```

that, we note first of all that there are 25 belts available for sale. So we are only interested in the function for values of n between 0 and 25. As a consequence, the horizontal axis should extend from 0 to 25 rather than from the standard setting of -10 to 10. But it is not immediately apparent how to choose a vertical span. A good way to handle this is to look at a table of values, which we already know how to make. As shown in Figure 2.23, we choose a starting value of 0 and a table increment of 5. When we look at the table in Figure 2.24 we see that the possible values for net profit range from a low of $-\$300$ for no sales to $\$268.75$ for the sale of all 25 belts. This tells us how to choose the vertical span of the viewing window; we want it to go from -300 to 268.75. Allowing a little extra margin, we show in Figure 2.25 a window setup $\boxed{2.19}$ where the horizontal span goes from 0 to 25 and the vertical span goes from -325 to 300. Now when we graph $\boxed{2.20}$ we get a very nice picture of the function as shown in Figure 2.26. If we now trace the graph, we can get usable information from it. For example, in Figure 2.27 we have put the cursor $\boxed{2.21}$ at X=10, and we see that if 10 belts are sold, the net profit is -72.50 dollars. That is, the craftsman will lose \$72.50. In Figure 2.28 we have put

Figure 2.23: Setting up the table

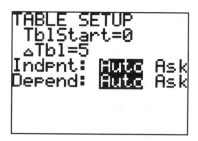

Figure 2.24: Table for belt sales

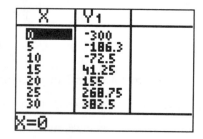

Figure 2.25: Setting up the viewing window

Figure 2.26: Graph of profit versus sales

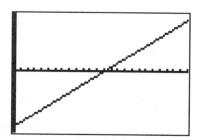

the cursor at X=20, and we see that if 20 belts are sold, there will be a net profit of $155. What is the practical significance of the place where the graph crosses the horizontal axis?

Figure 2.27: $72.50 lost if only 10 belts are sold

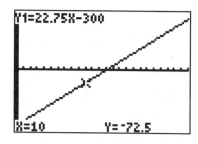

Figure 2.28: A net profit of $155 when 20 belts are sold

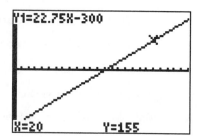

The example of the leather craftsman illustrates that in many situations the horizontal span can be determined from practical considerations. We noted in the example that the reasonable values of the variable n were between 0 and 25. Traditionally this has been called the *domain* of the function, and so the horizontal span of the graphing window is a way of visualizing the domain. The corresponding values taken by the function form what has traditionally been called the *range* of the function. We found this in the example by using a table for the function p to get the vertical span of -325 to 300. Again, the vertical span of the graphing window gives a helpful way of visualizing the range of a function.

EXAMPLE 2.3 *Labor Productivity*

For a manufacturing company with w workers working h hours per day and producing a total of I items, *labor productivity* $P = P(w, h, I)$ is measured by the function

$$P = \frac{I}{wh} \text{ items per worker hour.}$$

1. Suppose a company produces 750 items per day and each worker works 8 hours per day. The company budget allows for no more than 11 workers.

   (a) Find a formula for labor productivity P as a function of the number of workers w and make its graph.

   (b) Use the graph to find labor productivity if there are 7 workers.

   (c) What happens to labor productivity if the number of workers is increased? What happens if the number is decreased?

2. Suppose your company has 5 workers who work 8 hours each day. Company resources and product demand dictate that between 500 and 1000 items per day must be produced. Use a graph to show how labor productivity changes when I increases.

Solution to Part 1 (a): Under the given conditions we have $I = 750$ and $h = 8$. Thus labor productivity is given by the formula $P = \dfrac{750}{8w}$.

To make the graph, we first enter $\boxed{2.22}$ the function as shown in Figure 2.29 and record the variable correspondences:

$$\mathsf{Y_1} \;\; = \;\; P \text{, productivity on the vertical axis}$$

$$\mathsf{X} \;\; = \;\; w \text{, number of workers on the horizontal axis.}$$

Next we need to choose a window for graphing. Since the company budget allows for at most 11 workers, we will set the horizontal span of the window from 1 to 11. To find the vertical span, we made the table of values shown in Figure 2.30 with a starting value of 1 and an increment $\boxed{2.23}$ value of 2. Consulting the table and adding a little margin at the top and bottom, we use a window setup $\boxed{2.24}$ with a horizontal span from 1 to 11 and a vertical span from 0 to 100. This is shown in Figure 2.31. These settings give the graph $\boxed{2.25}$ in Figure 2.32.

Solution to Part 1(b): We want to get $P(7)$ from the graph. To do this, we locate the cursor $\boxed{2.26}$ at X=7. We read from the Y= prompt at the bottom of the screen in Figure 2.32 that using 7 workers gives a labor productivity of about 13.39 items per worker hour.

Figure 2.29: Entering the labor productivity function

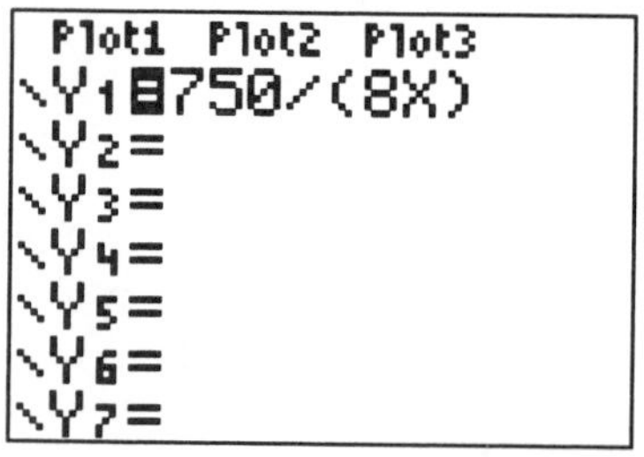

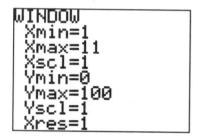

Figure 2.30: A table of values for labor productivity

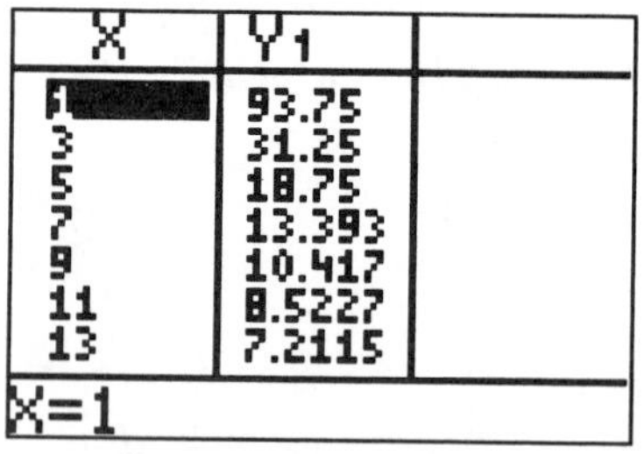

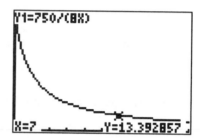

Figure 2.31: Setting up the window

```
WINDOW
 Xmin=1
 Xmax=11
 Xscl=1
 Ymin=0
 Ymax=100
 Yscl=1
 Xres=1
```

Figure 2.32: Graph of labor productivity versus number of workers

Solution to Part 1(c): The graph in Figure 2.32 is decreasing. This shows that as the number of workers increases, labor productivity decreases. This is not a surprise since we are increasing the number of worker hours but holding the number of items produced constant at 750 items. Thus each worker is producing less.

If we decrease the number of workers, we see from Figure 2.32 that labor productivity increases. Once again, this is not surprising since we are decreasing worker hours but holding the number of items produced constant. Thus each worker is producing more.

Solution to Part 2: In this scenario, we are holding the number of workers constant at 5 and the number of hours at 8. We are looking at labor productivity as a function of the number of items produced:

$$P = \frac{I}{5 \times 8} = \frac{I}{40} \text{ items per worker hour.}$$

We enter ⌐2.27⌐ this function and record the variable correspondences:

Y_1 = P, productivity on the vertical axis

X = I, number of items produced on the horizontal axis.

Since the company must produce between 500 and 1000 items per day, we will set the horizontal span of the graphing window from 500 to 1000. To get the vertical span for the window, we made the table of values in Figure 2.33 with a starting value of 500 and a table increment of 100. From this table, we choose a window setup $\boxed{2.28}$ with a horizontal span of 500 to 1000 and a vertical span from 0 to 30. This gives the graph in Figure 2.34. It is interesting to note that if productivity is viewed as a function of the number of items produced, then its graph is a straight line, but if productivity is viewed as a function of the number of worker hours, then the graph is curved.

Figure 2.33: *A table of values for productivity as a function of items produced*

X	Y₁	
500	12.5	
600	15	
700	17.5	
800	20	
900	22.5	
1000	25	
1100	27.5	

X=500

Figure 2.34: *A graph of productivity versus items produced*

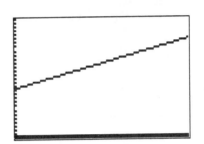

Getting limiting values from graphs

We have seen that tables of values can be helpful in determining limiting values. Graphs can do that as well, and in addition they provide an informative picture of how limiting values may be approached. The height $h = h(t)$ of some plants as a function of time t closely follows a *logistic formula*. Interestingly enough this is the same type of formula that is often used to study population growth. For a certain variety of sunflower[3] growing under ideal conditions, and starting at a time when the plant is already a few centimeters tall, its height may be given by the function

$$h = \frac{13}{0.93^t + 0.05} ,$$

where h is measured in centimeters, and t is measured in days. We want to make a graph of h versus t and see what information we can gain from it. In particular, we would like to be able to describe how the sunflower grows and figure out its maximum height.

First we enter $\boxed{2.29}$ the function as shown in Figure 2.35 and record variable correspondences:

$$Y_1 = h, \text{ height on the vertical axis}$$

$$X = t, \text{ time on the horizontal axis.}$$

[3]This example is adapted from the presentation on page 53 of *Exploring Differential Equations via Graphics and Data* by D. Lomen and D. Lovelock, John Wiley & Sons, 1996.

Next we need to determine a viewing window. In contrast with what happened in Example 2.3, we are not provided with explicit information on the values of t to help us set up the window. But our everyday experience with annual plants can give us the information we need. To get an idea of what is happening over a possible four month growing period, we have in Figure 2.36 made a table of values showing the height of the sunflower from 60 to 120 days in 10 day intervals. We see that in 120 days the sunflower will be just over 259

Figure 2.35: *Entering a logistic growth function*

Figure 2.36: *A table of values for four months of growth*

X	Y1
60	206.83
70	231.23
80	245.23
90	252.64
100	256.38
110	258.24
120	259.14

X=60

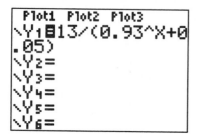

centimeters tall, and if you pan further down the table, you will see that almost no further growth occurs. Adding as usual a bit of extra room, we have in Figure 2.37 plotted the graph in a window where the horizontal span is from 0 to 120 and the vertical span is 0 to 300. Also in this figure, we have traced and moved the cursor toward the right-hand end of the graph where the height is just over 259 centimeters. From the graph and from the table of values, it appears that the maximum height of the sunflower is about 260 centimeters.

Limiting values can be shown to good advantage if the corresponding horizontal line is added to the graph. We added 2.30 $Y_2 = 260$ to the function list, and in Figure 2.38 both graphs are shown. Note the inflection point in the graph of h, where the graph changes from concave up to concave down. This represents the point of most rapid growth. This picture shows a growth model for sunflowers which agrees with our everyday experience. Growth is slow when the plant is very young, but once a healthy plant is established, it grows rapidly. When maturity is reached, growth slows and little if any additional height is attained.

Figure 2.37: Sunflower growth

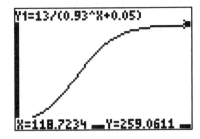

Figure 2.38: Limiting height added

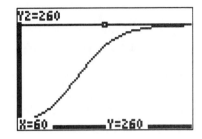

EXAMPLE 2.4   Blood Cholesterol

The amount $C = C(t)$ of cholesterol (in milligrams per deciliter) in the blood of a certain man on an unhealthful diet is given by

$$C = 235 - 105e^{-0.3t} \, ,$$

where t is time measured in months.

1. Make a graph that shows the blood cholesterol level as a function of time if the unhealthful diet is continued.

2. The doctor has issued a warning that this man may experience severe health problems if cholesterol levels in excess of 200 milligrams per deciliter of blood are reached. Is there a danger of exceeding this level?

3. If the unhealthful diet is continued indefinitely, what eventual cholesterol level will be reached?

4. Is the graph of C concave up or concave down? Explain in practical terms what your answer means.

Solution to Part 1: We first enter ⌐2.31⌐ the function and record the variable correspondences:

$$Y_1 \; = \; C, \text{ blood cholesterol on vertical axis}$$
$$X \; = \; t, \text{ time on horizontal axis.}$$

We have very little information beyond the formula itself to help us set up the graphing window. Thus we experiment a little with tables of values. First we make a table of values starting at $t = 0$ with a table increment of 1. This is shown in Figure 2.39, where we see that after six months cholesterol levels are still increasing. If you pan further down the table, you will see the levels continue to increase. If you are patient and pan down far enough, you can see what eventually happens. A quicker way is to return to the table setup and increase the table increment value. In Figure 2.40 we have changed the starting value to 6 and viewed the table in six-month intervals. We see that blood cholesterol has leveled out at 235 milligrams per deciliter by 36 months. As usual

Figure 2.39: *A table for blood cholesterol in one-month intervals*

X	Y1	
0	130	
1	157.21	
2	177.37	
3	192.31	
4	203.37	
5	211.57	
6	217.64	

X=0

Figure 2.40: *A table for blood cholesterol in six-month intervals*

X	Y1	
6	217.64	
12	232.13	
18	234.53	
24	234.92	
30	234.99	
36	235	
42	235	

X=6

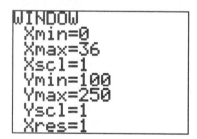

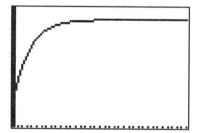

we allow some margin and set up the graphing window [2.32] in Figure 2.41 with a horizontal span from 0 to 36 and a vertical span from 100 to 250. The graph with these settings is in Figure 2.42.

Figure 2.41: *Setting up the graphing window*

Figure 2.42: *Blood cholesterol versus time*

Solution to Part 2: We have already seen from our table of values that a blood cholesterol level of 200 will certainly be exceeded. But we can use the graph to show this in a striking way. In Figure 2.43 we have added [2.33] the horizontal line corresponding to a blood level of 200 milligrams of cholesterol per deciliter. Note that the point where these graphs cross gives the time when the danger level will be exceeded. In Figure 2.43 we have used the trace option and moved the graphing cursor as close as we could to the crossing point, and we see from the prompt at the bottom of the screen that this individual may incur health risks in about three and a half months.

We will return to this problem in Section 4 of this chapter, where we will show how to locate accurately crossing points such as this one.

Solution to Part 3: The table of values shows that the limiting value for blood cholesterol in this case is about 235 milligrams per liter. In Figure 2.44 we have added ⌐2.34⌐ the horizontal line corresponding to this limiting value.

Figure 2.43: *When blood cholesterol reaches 200 milligrams per deciliter*

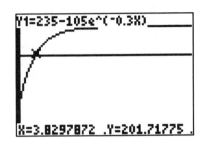

Figure 2.44: *Limiting value for blood cholesterol*

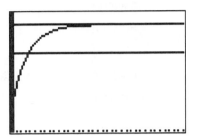

Solution to Part 4: The graph is concave down, so the function is increasing at a decreasing rate. Cholesterol levels increase rapidly at first, but the rate of increase slows near the limiting value.

Exercise Set 2.2

In each of the following exercises you are asked to produce a graph which you should turn in as part of the solution. Ideally you would transfer the graph via a computer link to a printer. If such technology is not available to you, you should provide hand-drawn copies of calculator-generated graphs. Be sure to label your graphs, include identifying names for the horizontal and vertical axes, and indicate the graphing window you use. Note also that the graphing windows you choose may not be the same as we used to make the Odd-Answer Key. Thus you should not expect your graphs to match ours exactly.

1. **Baking a potato:** A potato is placed in a preheated oven to bake. Its temperature $P = P(t)$ is given by

$$P = 400 - 325e^{-\frac{t}{50}} \, ,$$

where P is measured in degrees Fahrenheit and t is the time in minutes since the potato was placed in the oven.

(a) Make a graph of P versus t. (Suggestion: In choosing your graphing window, it is reasonable to look at the potato over no more than a 2 hour period. After that, it will surely be burned to a crisp. You may wish to look at a table of values to select a vertical span.)

(b) What was the initial temperature of the potato?

(c) Did the potato's temperature rise more during the first thirty minutes of baking or the second 30 minutes of baking? What was the average rate of change per minute during the first 30 minutes? What was the average rate of change per minute during the second 30 minutes?

(d) Is this graph concave up or concave down? Explain what that tells you about how the potato heats up, and relate this to Part (c).

(e) The potato will be done when it reaches a temperature of 270 degrees. Approximate the time when the potato will be done.

(f) What is the temperature of the oven? Explain how you got your answer. (Hint: If the potato were left in the oven for a long time, its temperature would match that of the oven. You may wish to make a new graph showing the temperature over a long period of time.)

2. **Ohm's law** says that when electric current is flowing across a resistor, the current i, measured in amperes, can be calculated from the voltage v, measured in volts, and the resistance R, measured in ohms. The relationship is given by

$$i = \frac{v}{R} \text{ amperes.}$$

(a) A resistor in a radio circuit is rated at 4000 ohms.

    i. Find a formula for the current as a function of the voltage.

    ii. Plot the graph of i versus v. Include values of the voltage up to 12 volts.

    iii. What happens to the current when voltage increases?

(b) The lights on your car operate on a 12 volt battery.

    i. Find a formula for the current in your car lights as a function of the resistance.

    ii. Plot the graph of i versus R. We suggest a horizontal span here of 0 to 25.

    iii. What happens to the current when resistance increases?

3. **The economic order quantity model** tells a company how many items at a time to order so that inventory costs will be minimized. The number $Q = Q(N, c, h)$ of items that should be included in a single order depends on the demand N per year for the product, the fixed cost c in dollars associated with placing a single order (not the price of the item), and the carrying cost h in dollars. (This is the cost of keeping an unsold item in stock.) The relationship is given by

$$Q = \sqrt{\frac{2Nc}{h}}.$$

(a) Assume the demand for a certain item is 400 units per year and the carrying cost is \$24 per unit per year. That is, $N = 400$ and $h = 24$.

    i. Find a formula for Q as a function of the fixed ordering cost c and plot its graph. For this particular item, we do not expect the fixed ordering costs ever to exceed \$25.

    ii. Use the graph to find the number of items to order at a time if the fixed ordering cost is \$6 per order.

    iii. How should increasing fixed order cost affect the number of items you order at a time?

(b) Assume the demand for a certain item is 400 units per year and the fixed ordering cost is \$14 per order.

    i. Find a formula for Q as a function of the carrying cost h and make its graph. We do not expect the carrying cost for this particular item ever to exceed \$25.

ii. Use the graph to find the optimal order size if the carrying cost is $15 per unit per year.

iii. How should an increase in carrying cost affect the optimal order size?

iv. What is the average rate of change per dollar in optimal order size if the carrying cost increases from $15 to $18?

v. Is this graph concave up or concave down? Explain what that tells you about how optimal order size depends on carrying costs.

4. **Monthly payment for a home:** If you borrow $120,000 at an EAR of 8% in order to buy a home, and if the lending institution compounds interest continuously, then your monthly payment $M = M(Y)$ depends on the number of years Y you take to pay off the loan. The relationship is given by

$$M = \frac{120000 \times 0.08}{12(1 - e^{-0.08Y})} \text{ dollars.}$$

(a) Make a graph of M versus Y. In choosing a graphing window, you should note that rarely does a home mortgage extend beyond 30 years.

(b) Use the graph to find your monthly payment if you pay off the loan in 20 years.

(c) Use the graph to find your monthly payment if you pay off the loan in 30 years.

(d) From Part (b) to Part (c) of this problem you increased the debt period by 50%. Did it decrease your monthly payment by 50%?

(e) Is the graph concave up or concave down? Explain your answer in practical terms.

(f) Calculate the average decrease per year in your monthly payment from a loan period of 25 to 30 years.

5. **An annuity:** Suppose you are able to find an investment that pays a monthly interest rate of r as a decimal. You want to invest P dollars that will help support your child. If you want your child to be able to withdraw M dollars per month for t months, then the amount you must invest is given by

$$P = M \times \frac{1}{r} \times \left(1 - \frac{1}{(1+r)^t}\right) \text{ dollars.}$$

A fund such as this is known as an *annuity*. For the remainder of this problem we suppose that you have found an investment with a monthly interest rate of 0.01, and you want your child to be able to withdraw $200 from the account each month.

(a) Find a formula for your initial investment P as a function of t, the number of monthly withdrawals you want to provide, and make a graph of P versus t. Be sure your graph shows up through 40 years (480 months).

(b) Use the graph to find out how much you need to invest so that your child can withdraw $200 per month for 4 years.

(c) How much would you have to invest if you wanted your child to be able to withdraw $200 per month for 10 years?

(d) A *perpetuity* is an annuity that allows for withdrawals for an indefinite period. How much money would you need to invest so that your descendants could withdraw $200 per month from the account forever? Be sure to explain how you got your answer.

6. **Alexander's formula:** One interesting problem in the study of dinosaurs is to determine from their tracks how fast they ran. The scientist R. McNeill Alexander developed a formula giving the velocity of any running animal in terms of its stride length and the height of its hip above the ground.[4] The stride length of a dinosaur can be measured from successive prints of the same foot, while the hip height (roughly the leg length) can be estimated based on the size of a footprint, so Alexander's formula gives a way of estimating from dinosaur tracks how fast the dinosaur was running.

If the velocity v is measured in meters per second, and the stride length s and hip height h are measured in meters, then Alexander's formula is

$$v = 0.78 s^{1.67} h^{-1.17} .$$

(For comparison, a length of one meter is 39.37 inches, and a velocity of one meter per second is about 2.2 miles per hour.)

(a) First we study animals with varying stride lengths but all with a hip height of 2 meters (so $h = 2$).

  i. Find a formula for the velocity v as a function of the stride length s.

  ii. Make a graph of v versus s. Include stride lengths from 2 to 10 meters.

  iii. What happens to the velocity as the stride length increases? Explain your answer in practical terms.

  iv. Some dinosaur tracks show a stride length of 3 meters, and a scientist estimates that the hip height of the dinosaur was 2 meters. How fast was the dinosaur running?

---

[4]See his article "Estimates of speeds of dinosaurs," *Nature* **261** (1976), 129–130. See also his book *Animal Mechanics*, 2nd edition, 1983, Blackwell.

(b) Now we study animals with varying hip heights but all with a stride length of 3 meters (so $s = 3$).

    i. Find a formula for the velocity v as a function of the hip height h.

    ii. Make a graph of v versus h. Include hip heights from 0.5 to 3 meters.

    iii. What happens to the velocity as the hip height increases? Explain your answer in practical terms.

7. **Artificial gravity:** To compensate for weightlessness in a space station, artificial gravity can be produced by rotating the station.[5] The number N of rotations per minute needed is a function of two variables: the distance r to the center of rotation, and a, the desired acceleration (or size of artificial gravity). The formula is

$$N = \frac{60}{2\pi} \times \sqrt{\frac{a}{r}}.$$

We measure r in meters and a in meters per second per second.

(a) First we assume that we want to simulate the gravity of Earth, so $a = 9.8$ meters per second per second.

    i. Find a formula for the number N of rotations per minute needed as a function of the distance r to the center of rotation.

    ii. Make a graph of N versus r. Include distances from 10 to 200 meters.

    iii. What happens to the number of rotations per minute needed as the distance increases? Explain your answer in practical terms.

    iv. What number of rotations per minute is necessary to produce Earth gravity if the distance to the center is 150 meters?

(b) Now we assume that the distance to the center is 150 meters (so $r = 150$).

    i. Find a formula for the number N of rotations per minute needed as a function of the desired acceleration a.

    ii. Make a graph of N versus a. Include values of a from 2.45 (one quarter of Earth gravity) to 9.8 meters per second per second.

    iii. What happens to the number of rotations per minute needed as the desired acceleration increases?

---

[5] This exercise is based on *Space Mathematics* by B. Kastner, published by NASA, 1985.

8. **Plant growth:** The amount of growth of plants in an ungrazed pasture is a function of the amount of plant biomass already present and the amount of rainfall.[6] For a pasture in the arid zone of Australia, the formula

$$Y = -55.12 - 0.01535N - 0.00056N^2 + 3.946R$$

gives an approximation of the growth. Here R is the amount of rainfall (in millimeters) over a three-month period, N is the plant biomass (in kilograms per hectare) at the beginning of that period, and Y is the growth (in kilograms per hectare) of the biomass over that period. (For comparison, 100 millimeters is about 3.9 inches, and 100 kilograms per hectare is about 89 pounds per acre.)

For this exercise assume that the amount of plant biomass initially present is 400 kilograms per hectare, so $N = 400$.

   (a) Find a formula for the growth Y as a function of the amount R of rainfall.

   (b) Make a graph of Y versus R. Include values of R from 40 to 160 millimeters.

   (c) What happens to Y as R increases? Explain your answer in practical terms.

   (d) How much growth will there be over a three-month period if initially there are 400 kilograms per hectare of plant biomass and the amount of rainfall is 100 millimeters?

9. **More on plant growth:** *This is a continuation of Exercise 8.* Now we consider the amount of growth Y as a function of the amount of plant biomass N already present.

   (a) First we assume that the rainfall is 100 millimeters, so $R = 100$.

      i. Find a formula for the growth Y as a function of the amount N of plant biomass already present.

      ii. Make a graph of Y versus N. Include biomass levels N from 0 to 800 kilograms per hectare.

      iii. What happens to the amount of growth Y as the amount N of plant biomass already present increases? Explain your answer in practical terms.

---

[6]This exercise and the next are based on the work of G. Robertson, "Plant dynamics." In: G. Caughley, N. Shepherd, and J. Short, eds. *Kangaroos*, 1987, Cambridge University Press, Cambridge, England.

(b) Next we assume that the rainfall is 80 millimeters, so $R = 80$.

    i. With this lower rainfall level, find a formula for the growth Y as a function of the amount N of plant biomass already present.

    ii. Add the graph of Y versus N for the lower rainfall to the graph you found in Part (a).

    iii. According to the graph you just found, what is the effect of lowered rainfall on plant growth? Is your answer consistent with that in Part (c) of Exercise 8?

10. **Viewing Earth:** Astronauts looking at Earth from a spacecraft can see only a portion of the surface.[7] The fraction F of the surface of Earth visible at a height h (in kilometers) above the surface is given by the formula

$$F = \frac{0.5h}{R + h} \, .$$

Here R is the radius of Earth, about 6380 kilometers. (For comparison, 1 kilometer is about 0.62 mile, and the Moon is about 380,000 kilometers from Earth.)

(a) Make a graph of F versus h covering heights up to 100,000 kilometers.

(b) A value of F equal to 0.25 means that 25%, or one-quarter, of Earth's surface is visible. At what height is this fraction visible?

(c) During one flight of a space shuttle, astronauts performed an extravehicular activity at a height of 280 kilometers. What fraction of the surface of Earth is visible at that height?

(d) Is the graph of F concave up or concave down? Explain your answer in practical terms.

(e) Determine the limiting value for F as the height h gets larger. Explain your answer in practical terms.

11. **Magazine circulation:** The circulation C of a certain magazine as a function of time t is given by the formula

$$C = \frac{5.2}{0.1 + 0.3^t} \, .$$

Here C is measured in thousands and t is measured in years since the beginning of 1992, when the magazine was started.

(a) Make a graph of C versus t covering the first 6 years of the magazine's existence.

(b) What was the circulation of the magazine 18 months after it was started?

---

[7] This exercise is based on *Space Mathematics* by B. Kastner, published by NASA, 1985.

(c) Over what time interval is the graph of C concave up? Explain your answer in practical terms.

(d) At what time was the circulation increasing the fastest?

(e) Determine the limiting value for C. Explain your answer in practical terms.

12. **Growth:** The length L (in inches) of a certain fish is given by the formula

$$L = 15 - 19 \times 0.6^t \, ,$$

while its weight W (in pounds) is given by the formula

$$W = (1 - 1.3 \times 0.6^t)^3 \, .$$

Here t is the age of the fish (in years), and both formulas are valid from the age of 1 year.

(a) Make a graph of the length of the fish against its age, covering ages 1 to 8.

(b) To what limiting length does the fish grow? At what age does it reach 90% of this length?

(c) Make a graph of the weight of the fish against its age, covering ages 1 to 8.

(d) To what limiting weight does the fish grow? At what age does it reach 90% of this weight?

(e) Of the graphs you made in Parts (a) and (c), it should be that one has an inflection point while the other is always concave down. Identify which is which, and explain in practical terms what this means. Include in your explanation the approximate location of the inflection point.

13. **Buffalo:** In 1993, the Nature Conservancy introduced buffalo to the Tallgrass Prairie Preserve near Pawhuska, Oklahoma. Biologists have carefully studied the growth of buffalo herds as well as the ability of tallgrass to sustain buffalo populations. Based on their work, a logistic model of growth is used, which estimates the number N of buffalo in the herd by the formula

$$N = \frac{353}{0.18 + 0.58^t} \, .$$

Here t is the number of years since 1993.

(a) Make a graph of N versus t covering the first 20 years of the herd's existence (corresponding to dates up to 2013).

(b) How many buffalo were introduced in 1993?

(c) When will the number of buffalo first exceed 1200?

(d) How many buffalo will there eventually be on the Tallgrass Prairie Preserve?

(e) When is the graph of N, as a function of t, concave up? When concave down? What does this mean in terms of the growth of the buffalo herd?

2.3 SOLVING LINEAR EQUATIONS

Many times we will need to find when two functions are equal or when a function value is zero. To do this we will need to solve an equation. For our purposes, equations come in two types, *linear equations* and *nonlinear equations*. Linear equations are the simplest kind of equation and basically are those that do not involve powers, square roots, or other complications of the variable for which we want to solve.

The basic operations

As you may recall from your elementary algebra course, linear equations can always be solved using two basic operations: subtraction (or addition) and division (or multiplication). The following is presented as a reminder.

Adding to both sides of an equation: You may add (or subtract) the same thing to (or from) both sides of an equation. This can be thought of as moving a term from one side of the equation to the other, provided you change its sign, positive to negative or negative to positive.

Dividing both sides of an equation: You may divide (or multiply) both sides of an equation by any nonzero number. The same thing may be accomplished if you divide (or multiply) each term of an equation by any nonzero number; however, the results may sometimes appear to be different.

We will show how to use these operations to solve $5x + 7 = 18 - 2x$. The steps in solving a linear equation are always the same.

Step 1: Move all terms with the variable to one side of the equation and all terms that do not involve the variable to the other side of the equation, changing the sign of each moved term. This is accomplished by adding (or subtracting) the same thing to (or from) both sides of an equation.

In our case we need to move $-2x$ to the left side of the equation and 7 to the right side. We can do this by adding $2x$ to both sides of the equation and subtracting 7 from both sides. We get

$$\underbrace{5x + 7}_{\text{add } 2x \text{ and subtract } 7} = \underbrace{18 - 2x}_{\text{add } 2x \text{ and subtract } 7}$$

$$5x + 2x = 18 - 7 \;.$$

Step 2: Combine terms. Combining $5x + 2x$ into a single term is an easy task. For example, 5 cars + 2 cars is 7 cars, and it works the same with x's: $5x + 2x = 7x$. We get

$$\underbrace{5x + 2x}_{\text{combine to } 7x} = \underbrace{18 - 7}_{\text{combine to } 11}$$

$$7x = 11 \;.$$

Step 3: Divide out the coefficient of x. In our case, we want to divide both sides of the equation by 7:

$$\underbrace{7x}_{\text{divide by } 7} = \underbrace{11}_{\text{divide by } 7}$$

$$x = \frac{11}{7} = 1.57 \;,$$

rounding to two decimal places.

A word of caution is in order concerning division in equations. Sometimes we may have more than two terms in our equation when we need to divide, and the results will *appear* to be different depending on whether we divide both sides of the equation or whether we divide each term of the equation. For example, suppose in Step 3 above we had $7x = 11 + 8y$, and our goal was to solve for x. We could divide both sides of the equation by 7 to get $x = \dfrac{11 + 8y}{7}$. Alternatively, we could divide each term by 7, but in that case we must remember to divide both 11 and $8y$. The result is $x = \dfrac{11}{7} + \dfrac{8y}{7}$, or (rounding to two decimal places) $x = 1.57 + 1.14y$. Even though the two answers, $x = \dfrac{11 + 8y}{7}$ and $x = 1.57 + 1.14y$, may appear to be different, rules of elementary algebra can be used to show that (allowing for round-off error) they are the same.

EXAMPLE 2.5   *Rental Cars*

A car rental company charges $38.00 in insurance and other fees plus a flat rate of $12.00 per day.

1. Use a formula to express the cost of renting a car as a function of the number of days that you keep it.

2. Your expense account allows you $200 to spend on a rental car. How many days can you keep the car without exceeding your expense account?

Solution to Part 1: We need to choose variable and function names. We let d be the number of days that we keep the car and C the cost (in dollars) of renting the car.

The cost is the initial fee of $38 plus $12 per day:

$$\text{Cost} = \text{Initial fee} + 12 \times \text{Number of days}$$
$$C = 38 + 12d \, .$$

Solution to Part 2: We want to know how many rental days will make the cost be $200. That is, we want to solve the following equation for d:

$$C = 200$$
$$38 + 12d = 200 \, .$$

We solve it using the steps described above:

$$12d = 200 - 38 \text{ (Subtract 38 from both sides.)}$$
$$12d = 162$$

$$d = \frac{162}{12} = 13.5 \text{ (Divide by 12.).}$$

Thus you can rent the car for 13 and a half days without exceeding your expense account. Since the rental company will probably charge you a full day's price for a half day's rental, you may only be able to rent the car for 13 days.

Reversing the roles of variables

Sometimes when we are given formulas expressing one thing in terms of another, we can gain additional information by changing the roles of the variables. In formal mathematical terms, we are finding the *inverse* of a function.

For example, the relationship between the temperature in degrees Fahrenheit F and the temperature in degrees Celsius C is given by the formula

$$F = 1.8C + 32 \, . \tag{2.4}$$

This formula tells us that a room whose temperature is 30 degrees Celsius has a temperature of $F = 1.8 \times 30 + 32 = 86$ degrees Fahrenheit.

But Equation (2.4) does not immediately tell us the temperature in degrees Celsius of a room that is 75 degrees Fahrenheit. To find this we need to rearrange the formula so that it shows C in terms of F. That is, we need to solve Equation (2.4) for C. Since the equation is linear, we know how to do that:

$$
\begin{aligned}
F &= 1.8C + 32 \\
F - 32 &= 1.8C \text{ (Subtract 32 from both sides.)} \\
\frac{F - 32}{1.8} &= C \text{ (Divide by 1.8.).}
\end{aligned}
$$

We can use this formula to find out the temperature in degrees Celsius of a room whose temperature is 75 degrees Fahrenheit:

$$
C = \frac{75 - 32}{1.8} = 23.89 \text{ degrees Celsius.}
$$

In the derivation above, we divided both sides of the equation by 1.8. Alternatively, we could have divided each term by 1.8, giving

$$
C = \frac{F}{1.8} - \frac{32}{1.8} \ .
$$

The two expressions for C in terms of F look different, but they are in fact the same, and either is acceptable.

EXAMPLE 2.6   *A Moving Car*

If a car moves at a constant velocity, then the distance traveled d can be expressed as a function of velocity v and the time traveled t. If we measure d in miles, v in miles per hour, and t in hours, then the relationship is

$$
\begin{aligned}
\text{Distance} &= \text{Velocity} \times \text{Time} \\
d &= vt \ .
\end{aligned}
$$

1. What is the distance traveled if you drive 55 miles per hour for two and a half hours?

2. Use a formula to express v as a function of d and t. Use your function to find velocity in case it takes you three hours to drive 172 miles.

3. Use a formula to express time as a function of velocity and distance traveled. Use your function to find the time it takes to travel 230 miles at a velocity of 40 miles per hour.

Solution to Part 1: We simply plug in the given numbers for v and t:

$$
d = 55 \times 2.5 = 137.5 \text{ miles.}
$$

Solution to Part 2: We need to solve the equation $d = vt$ for v. We divide both sides by t to obtain

$$v = \frac{d}{t} \text{ miles per hour.}$$

Note that this is the familiar formula that says velocity is distance divided by time:

$$\text{Velocity} = \frac{\text{Distance}}{\text{Time}} \, .$$

If we take 3 hours to drive 172 miles, then our velocity is $v = \frac{172}{3} = 57.33$ miles per hour.

Solution to Part 3: This time we want to solve the equation $d = vt$ for t. We divide both sides by v and obtain

$$t = \frac{d}{v}.$$

This says

$$\text{Time} = \frac{\text{Distance}}{\text{Velocity}} \, .$$

If we drive 230 miles at 40 miles per hour, then we take $\frac{230}{40} = 5.75$ hours to complete the trip.

EXAMPLE 2.7 *Car Rentals Again*

Company Alpha charges an initial fee of $28.00, a daily rate of $4.00, and a rate of 29 cents per mile. Company Beta charges an initial fee of $32.00, a daily rate of $6.00, and a rate of 14 cents per mile. You need a rental car for 3 days.

1. Use a formula to express the cost of renting a car from Company Alpha as a function of the number of miles you drive.

2. Use a formula to express the cost of renting a car from Company Beta as a function of the number of miles you drive.

3. For what number of miles driven is the cost of car rental from Company Alpha and Company Beta the same?

4. On the same screen, plot the graphs of the cost of renting from Company Alpha and the cost of renting from Company Beta.

5. How do you decide from which company to rent?

Solution to Part 1: First choose variable and function names: Let m be the number of miles driven and A the cost (in dollars) of renting a car from Company Alpha. The initial fee for Company Alpha is $28, and you pay $12 for the three day rental. In addition, you pay $0.29 for each mile you drive:

$$\begin{aligned}
\text{Cost for Alpha} &= \text{Initial fee} + \text{three day price} + 0.29 \times \text{miles driven} \\
A &= 28 + 12 + 0.29m \\
&= 40 + 0.29m \,.
\end{aligned}$$

Solution to Part 2: Let B be the cost (in dollars) of renting a car from Company Beta. For Company Beta you pay the $32 initial fee, plus $18 for the three day rental, plus $0.14 for each mile you drive:

$$\begin{aligned}
\text{Cost for Beta} &= \text{Initial fee} + \text{three day price} + 0.14 \times \text{miles driven} \\
B &= 32 + 18 + 0.14m \\
&= 50 + 0.14m \,.
\end{aligned}$$

Solution to Part 3: We want to find when the cost for the two companies is the same:

$$\begin{aligned}
\text{Cost for Alpha} &= \text{Cost for Beta} \\
A &= B \\
40 + 0.29m &= 50 + 0.14m \\
0.29m - 0.14m &= 50 - 40 \quad \text{(Subtract 40 and } 0.14m \text{ from both sides.)} \\
0.15m &= 10 \\
m &= \frac{10}{0.15} = 66.67 \quad \text{(Divide by 0.15.)} \,.
\end{aligned}$$

Thus the cost for the two companies will be the same if you drive 66.67 miles.

Solution to Part 4: We need to enter $\boxed{2.35}$ both functions in the calculator. This is shown in Figure 2.45, and we list the appropriate correspondences:

$$\begin{aligned}
Y_1 &= A, \text{ cost for Alpha on vertical axis} \\
Y_2 &= B, \text{ cost for Beta on vertical axis} \\
X &= m, \text{ miles driven on horizontal axis.}
\end{aligned}$$

Before we make the graphs, let's think about how to set up the window. We surely want the picture to show where the cost is the same, and so we want to include 66.67 in the horizontal span. This will show nicely if the horizontal span is from 0 to 100 miles. To choose the vertical span we look at the table of values shown in Figure 2.46. We used a starting value of 0 and a table increment of 20. Note in this table the Y_1 column corresponds to Company Alpha and the Y_2 column corresponds to Company Beta. Allowing a little extra room, we choose a vertical span of 20 to 90.

Figure 2.45: *Entering both functions*

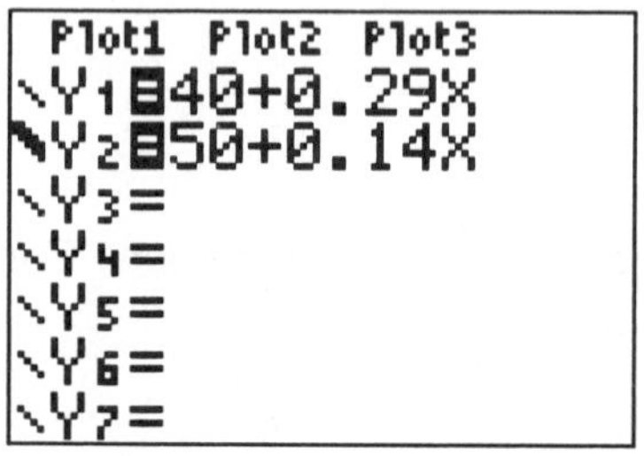

Figure 2.46: *A table of values for car rentals*

X	Y1	Y2
0	40	50
20	45.8	52.8
40	51.6	55.6
60	57.4	58.4
80	63.2	61.2
100	69	64
120	74.8	66.8

X=0

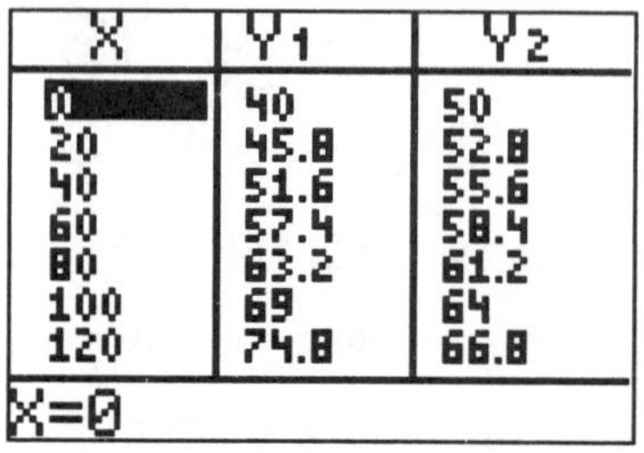

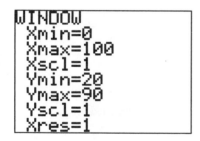

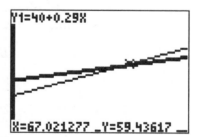

The properly configured window [2.36] is in Figure 2.47. When we graph, both plots appear as in Figure 2.48. In this figure, the thinner line represents Company Alpha, and

Figure 2.47: *Configuring the window*

Figure 2.48: *Thin graph for Company Alpha, thick graph for Company Beta*

the thicker line [2.37] represents Company Beta.

Solution to Part 5. We want to spend only as much money as is necessary. Figure 2.48 shows that the graph of the cost of Alpha is initially on the bottom, indicating that it costs less. But after the graphs cross at about $m = 67$ miles, the graph of the cost of Beta is on the bottom, indicating that it costs less. We found the exact value of this crossing point in Part 3. Thus, if we are going to drive for less than 66.67 miles, we should rent from Company Alpha. If we need to drive further, we should rent from Company Beta.

Exercise Set 2.3

1. **Gas mileage:** The distance d in miles that you can travel without stopping for gas depends on the number of gallons g of gasoline in your tank and the gas mileage m in miles per gallon your car gets. The relationship is

$$d = gm. \tag{2.5}$$

   (a) How far can you drive if you have 12 gallons of gas in your tank and your car gets 24 miles per gallon?

   (b) Solve Equation (2.5) for m.

       i. Explain in everyday terms what this new equation means.

       ii. Use this equation to determine the gas mileage of your car if you can drive 335 miles on a full 13 gallon tank of gas.

   (c) A Detroit engineer wants to be sure that the car she is designing can go 425 miles on a full tank of gas, and she must design a gas tank to ensure that. She does not yet know what gas mileage this new model car will get, and so she decides to make a graph of the size of the gas tank as a function of gas mileage.

       i. Solve Equation (2.5) for g using 425 for the distance.

       ii. Make the graph that the engineer made. Is it a straight line?

2. **Sales strategy:** A small business is considering hiring a new sales representative (sales rep) to market its product in a nearby city. Two pay scales are under consideration.

   Pay scale 1: Pay the sales rep a base yearly salary of $10,000 plus 8% commission on total sales.

   Pay scale 2: Pay the sales rep a base yearly salary of $13,000 plus 6% of total sales.

   (a) For each of the pay scales above, use a formula to express the total yearly earnings for the sales rep as a function of total yearly sales. Be sure to identify clearly what the letters you use mean.

   (b) What amount of total yearly sales would result in the same total yearly earnings for the sales rep no matter which of the two pay scales above is used?

   (c) On the same screen, plot the graphs of the functions you made in Part (a). Copy the picture onto the paper you turn in. Be sure to label the horizontal and vertical axes, and be sure your picture includes the number you found in Part (b).

(d) If you were a sales rep negotiating for the new position, under what conditions would you prefer pay scale 1? Under what conditions would you prefer pay scale 2?

3. **Net profit:** Suppose you pay rent of R dollars per month to rent space for the production of dolls. You pay c dollars in material and labor to make each doll, which you then sell for d dollars.

   (a) If you produce n dolls per month, use a formula to express your net profit p per month as a function of R, c, d, and n. (Suggestion: First make a formula using the words *rent*, *cost of a doll*, *selling price*, and *number of dolls*. Then replace the words by appropriate letters.)

   (b) What is your net profit per month if the rent is $1280 per month, it costs $2 to make each doll, which you sell for $6.85, and you produce 826 dolls per month?

   (c) Solve the equation you got in Part (a) for d.

   (d) Your accountant tells you that you need to make a net profit of $4000 per month. Your rent is $1200 per month, it costs $2 to make each doll, and your production line can only make 700 of them in a month. Under these conditions, what price do you need to get for each doll?

4. **Ohm's law** says that when electric current is flowing across a resistor, then the voltage v, measured in volts, is the product of the current i, measured in amperes, and the resistance R, measured in ohms. That is, $v = iR$.

   (a) What is the voltage if the current is 20 amperes and the resistance is 15 ohms?

   (b) Find a formula expressing resistance as a function of current and voltage. Use your function to find the resistance if the current is 15 amperes and the voltage is 12 volts.

   (c) Find a formula expressing current as a function of voltage and resistance. Use your function to find the current if the voltage is 6 volts and the resistance is 8 ohms.

5. **Temperature conversion:** In everyday experience the measures of temperature most often used are Fahrenheit F and Celsius C. Recall that the relationship is given by

$$F = 1.8C + 32 .$$

Physicists and chemists often use the Kelvin temperature scale.[8] You can get *kelvins* K

---

[8]With this temperature scale, physicists traditionally do not use the term *degrees*. Rather, if the temperature is 100 on the Kelvin scale, they say that it is "100 kelvins." The temperature scale is named after Lord Kelvin, and its name is capitalized; but the unit of temperature, the kelvin, is not.

from degrees Celsius using

$$K = C + 273.15 \, .$$

(a) What is the temperature in kelvins of a room that is 30 degrees Celsius?

(b) Find a formula expressing the temperature F in degrees Fahrenheit as a function of the temperature K in kelvins.

(c) What is the temperature in degrees Fahrenheit of an object that is 310 kelvins?

6. **The ideal gas law:** A *mole* of a chemical compound is a fixed number,[9] like a dozen, of molecules (or atoms in the case of an element) of that compound. A mole of water, for example, is about 18 grams, or just over a half an ounce in your kitchen. Chemists often use the mole as the measure of the amount of a chemical compound.

A mole of carbon dioxide has a fixed mass, but the volume V that it occupies depends on pressure p and temperature T; greater pressure tends to compress the gas into a smaller volume, whereas increasing temperature tends to make the gas expand into a larger volume. If we measure the pressure in atmospheres (1 atm is the pressure exerted by the atmosphere at sea level), the temperature in kelvins, and the volume in liters, then the relationship is given by the *ideal gas law*:

$$pV = 0.082T \, .$$

(a) Solve the ideal gas law for the volume V.

(b) What is the volume of one mole of carbon dioxide under 3 atm of pressure at a temperature of 300 kelvins?

(c) Solve the ideal gas law for pressure.

(d) What is the pressure on one mole of carbon dioxide if it occupies a volume of 0.4 liter at a temperature of 350 kelvins?

(e) Solve the ideal gas law for temperature.

(f) At what temperature will one mole of carbon dioxide occupy a volume of 2 liters under a pressure of 0.3 atm?

---

[9]You may be interested to know that this number is 6.02217×10^{23}. It may appear to be a strange number, but, as it develops, this is a natural unit of measure for chemists to use.

7. **Running ants:** A scientist observed that the speed S at which certain ants ran was a function of T, the ambient temperature.[10] He discovered the formula

$$S = 0.2T - 2.7 \, ,$$

where S is measured in centimeters per second and T is in degrees Celsius.

   (a) Using functional notation, express the speed of the ants if the ambient temperature is 30 degrees Celsius, and calculate that speed using the formula above.

   (b) Solve the formula above for T to express the ambient temperature T as a function of the speed S at which the ants run.

   (c) If the ants are running at a speed of 3 centimeters per second, what is the ambient temperature?

8. **Growth in weight and height:** Between the ages of 7 and 11 years, the weight w (in pounds) of a certain girl is given by the formula

$$w = 8t \, .$$

Here t represents her age in years.

   (a) Use a formula to express the age t of the girl as a function of her weight w.

   (b) At what age does she attain a weight of 68 pounds?

   (c) The height h (in inches) of this girl during the same period is given by the formula

$$h = 1.8t + 40 \, .$$

     i. Use your answer to Part (b) to determine how tall she is when she weighs 68 pounds.

     ii. Use a formula to express the height h of the girl as a function of her weight w.

     iii. Answer the question in Part (i) again, this time using your answer to Part (ii).

---

[10]See the study by H. Shapley, "Note on the thermokinetics of Dolichoderine ants," *Proc. Nat. Acad. Sci.* **10** (1924), 436–439.

9. **Plant growth:** The amount of growth of plants in an ungrazed pasture is a function of the amount of plant biomass already present and the amount of rainfall.[11] For a pasture in the arid zone of Australia, the formula

$$Y = -55.12 - 0.01535N - 0.00056N^2 + 3.946R \qquad (2.6)$$

gives an approximation of the growth. Here R is the amount of rainfall (in millimeters) over a three month period, N is the plant biomass (in kilograms per hectare) at the beginning of that period, and Y is the growth (in kilograms per hectare) of the biomass over that period. (For comparison, 100 millimeters is about 3.9 inches, and 100 kilograms per hectare is about 89 pounds per acre.)

(a) Solve Equation (2.6) for R.

(b) Ecologists are interested in the relationship between the amount of rainfall and the initial plant biomass if there is to be no plant growth over the period. Put $Y = 0$ in the equation you found in Part (a), and so get a formula for R in terms of N which describes this relationship.

(c) Use the formula you found in Part (b) to make a graph of R versus N (again with $Y = 0$). Include values of N from 0 to 800 kilograms per hectare.

This graph is called the *isocline for zero growth*. It shows the amount of rainfall needed over the three month period just to maintain a given initial plant biomass.

(d) With regard to the isocline for zero growth you found in Part (c), what happens to R as N increases? Explain your answer in practical terms.

(e) How much rainfall is needed just to maintain the initial plant biomass if that biomass is 400 kilograms per hectare?

(f) A point below the zero isocline graph corresponds to having less rainfall than is needed to sustain the given initial plant biomass, and in this situation the plants would die back. A point above the zero isocline graph corresponds to having more rainfall than is needed to sustain the given initial plant biomass, and in this situation the plants would grow. If the initial plant biomass is 500 kilograms per hectare and there are 40 millimeters of rain, what will happen to the plant biomass over the period?

---

[11] This exercise is based on the work of G. Robertson, "Plant dynamics." In: G. Caughley, N. Shepherd, and J. Short, eds. *Kangaroos*, 1987, Cambridge University Press, Cambridge, England.

10. **Collision:** In reconstructing an automobile accident, investigators study the *total momentum*, both before and after the accident, of the vehicles involved. The total momentum of two vehicles moving in the same direction is found by multiplying the weight of each vehicle by its speed and then adding the results.[12]   For example, if one vehicle weighs 3000 pounds and is traveling at 35 miles per hour, and another weighs 2500 pounds and is traveling at 45 miles per hour in the same direction, then the total momentum is $3000 \times 35 + 2500 \times 45 = 217,500$.

In this exercise we study a collision in which a vehicle weighing 3000 pounds ran into the rear of a vehicle weighing 2000 pounds.

(a) After the collision, the larger vehicle was traveling at 30 miles per hour, and the smaller vehicle was traveling at 45 miles per hour. Find the total momentum of the vehicles after the collision.

(b) The smaller vehicle was traveling at 30 miles per hour before the collision, but the speed V (in miles per hour) of the larger vehicle before the collision is unknown. Find a formula expressing the total momentum of the vehicles before the collision as a function of V.

(c) The *principle of conservation of momentum* states that the total momentum before the collision equals the total momentum after the collision. Using this principle with Parts (a) and (b), determine at what speed the larger vehicle was traveling before the collision.

---

[12]See *Accident Reconstruction* by J. C. Collins, 1979, Charles C. Thomas Publisher, Springfield, Il. In physics the mass is used instead of the weight, but in this exercise we ignore the distinction.

2.4 SOLVING NONLINEAR EQUATIONS

Nonlinear equations, those which involve powers, square roots, or other complicating factors, are in general much more difficult to solve than are linear equations. In fact, most nonlinear equations cannot be solved by simple hand calculation. There are some notable exceptions. You may, for example, recall from elementary algebra that the *quadratic formula* can be used to solve *second degree* equations. We will nonetheless treat all nonlinear equations the same and solve them using the graphing calculator.

Let's look, for example, at an aluminum bar which must be heated before it can be properly worked. The bar is placed in a preheated oven and its temperature $A = A(t)$, t minutes later, is given by

$$A = 800 - 730e^{-0.06t} \text{ degrees Fahrenheit.}$$

The aluminum bar will be ready for bending and shaping when it reaches a temperature of 600 degrees. How long should the bar remain in the oven?

We want to know when the temperature A is 600 degrees. That is, we want to solve the equation

$$800 - 730e^{-0.06t} = 600$$

for t. This equation is not a linear equation like the ones we solved in Section 2.3. Rather, it is *nonlinear*, and it cannot be solved using the methods of Section 2.3. We will show two methods for solving equations of this type using the graphing calculator.

The crossing graphs method

We illustrate the crossing graphs method using the equation

$$800 - 730e^{-0.06t} = 600.$$

For this method we will make two graphs, the graph of the left-hand side of the equation, $800 - 730e^{-0.06t}$, and the graph of the right-hand side of the equation, the function with a constant value of 600. Then we will find out where they are the same, that is, where the graphs cross.

The first step is to enter $\boxed{2.38}$ both functions as shown in Figure 2.49. We record the appropriate correspondences:

$$Y_1 \;=\; A, \text{ temperature on vertical axis}$$

$$Y_2 \;=\; 600, \text{ target temperature}$$

$$X \;=\; t, \text{ minutes on horizontal axis.}$$

To get a rough idea of when the temperature will reach 600 degrees, we look at the table of values for $800 - 730e^{-0.06t}$ in Figure 2.50. We used a starting value of 0 and an increment of 5 minutes. We see that the temperature will reach 600 degrees between 20 and 25 minutes after the bar is placed in the oven.

Figure 2.49: Entering left-hand and right-hand sides of the equation for the crossing graphs method

Figure 2.50: A table of values for temperature

Based on the table in Figure 2.50 we use a window setup $\boxed{2.39}$ with a horizontal span from 0 to 30 minutes and a vertical span from 0 to 700 degrees. The graphs are in Figure 2.51. We want the point where the graphs cross, or intersect. You can get reasonably close to the answer if you trace one graph and move the cursor as close as you can to the crossing point, but your calculator can locate the crossing point quite accurately using only a few keystrokes $\boxed{2.40}$. The result is shown in Figure 2.52, and we see from the prompt at the bottom of the screen that the temperature will be 600 degrees at the time 21.58 minutes after the bar is placed in the oven. Consult the *Keystroke Guide* to find out the exact keystrokes required to execute

Figure 2.51: Graphs of temperature and the target temperature

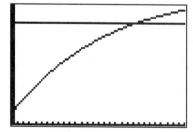

Figure 2.52: When the bar reaches 600 degrees

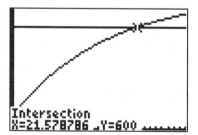

the crossing graphs method on your calculator.

The single graph method

There is a slightly different method for solving nonlinear equations, the *single graph method*, which finds where a function equals 0. To illustrate it, we consider the account for a health care plan a firm offers its employees. When the firm instituted the plan, it pledged that for five years there would be no increase in the insurance premium, but since then health expenses have risen dramatically. The plan manager estimates that the plan's account balance B (in millions of dollars), as a function of the number of years t since the plan was instituted, is given by the formula

$$B = 102 + 12t - 100e^{0.1t} .$$

When will the plan's account run out of money? Will this happen before the five year period is over? To answer these questions we need to know when the balance B is 0, so we need to solve the equation

$$102 + 12t - 100e^{0.1t} = 0$$

for t. This is not a linear equation, and we will use the calculator to solve it.

We want to see where $102 + 12t - 100e^{0.1t}$ is zero, and this is where the graph crosses the horizontal axis. Such points are knows as *zeros* or *roots*. We enter $\boxed{2.41}$ the function as shown in Figure 2.53. As in the crossing graphs method, we first look at a table of values. We are interested in the period between 0 and 5 years after the plan was instituted, and from the table in Figure 2.54 we see that the account will run out of money before the end of the five year period.

Figure 2.53: *Entering the function for the single graph method*	Figure 2.54: *Getting an estimate with a table of values*

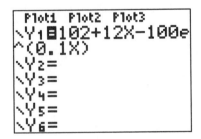

Now we set up the graphing window. We want to see the graph near the place where it crosses the horizontal axis, so, based on the table, we use a window setup $\boxed{2.42}$ with a horizontal span from 0 to 5 and a vertical span from -5 to 6. The graph is in Figure 2.55. As in the crossing graphs method, we can get a pretty good estimate of where the graph crosses the horizontal axis by tracing. But once again, your calculator can find $\boxed{2.43}$ the point quickly and accurately, as shown in Figure 2.56. Consult the *Keystroke Guide* for specific instructions regarding your calculator. We see from the prompt at the bottom of the screen in Figure 2.56

that the account will run out of money 4.26 years, or about 4 years and 3 months, after the plan is instituted.

___

Figure 2.55: *Picture for the single graph method*

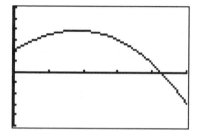

___

Figure 2.56: *Finding the root*

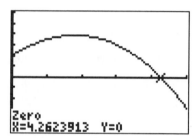

The single graph method applies directly when we want to find where a single function is equal to zero. Since more general equations can be put in this form, the single graph method can be applied in those situations as well. See Example 2.9 below for an illustration of this. The crossing graphs method is widely applicable and is often easier to use.

We should note that whenever the calculator executes a routine to solve an equation, it is seldom able to find the value it is looking for exactly. Rather, it provides an approximation which is normally so accurate that we need not be concerned. But the presentation of approximate rather than exact answers is not a failing of the calculator. It is rather a mathematical necessity reflected in how the calculator functions.

___

EXAMPLE 2.8   *The George Deer Reserve Again*

For this example we look again at the logistic population growth formula introduced in Section 2.1 for deer on the George Reserve. The number N of deer expected to be present on the reserve after t years has been determined by ecologists to be

$$N = \frac{6.21}{0.035 + 0.45^t} \text{ deer.}$$

1. Plot the graph of N versus t and explain how the deer population increases with time.

2. For planning purposes, the wildlife manager for the reserve needs to know when to expect there to be 85 deer on the reserve. The answer to that is the solution of an equation. Which equation?

3. Solve the equation you found in Part 2.

Solution to Part 1: We enter [2.44] the population function as shown in Figure 2.57 and record the proper variable associations:

$$\begin{aligned} \mathsf{Y_1} &= N, \text{ population on vertical axis} \\ \mathsf{X} &= t, \text{ years on horizontal axis.} \end{aligned}$$

To determine a good window size, we look at the table of values in Figure 2.58. This table leads us to choose the window setup in Figure 2.59. Now when we graph, we see in Figure 2.60 the classic S-shaped curve which is characteristic of logistic population growth. This graph shows that the deer population increases rapidly for the first few years, but as population size nears the *carrying capacity* of the reserve, the rate of increase slows, and the population levels out at around 177 deer.

Figure 2.57: *Entering a logistic function for population growth*

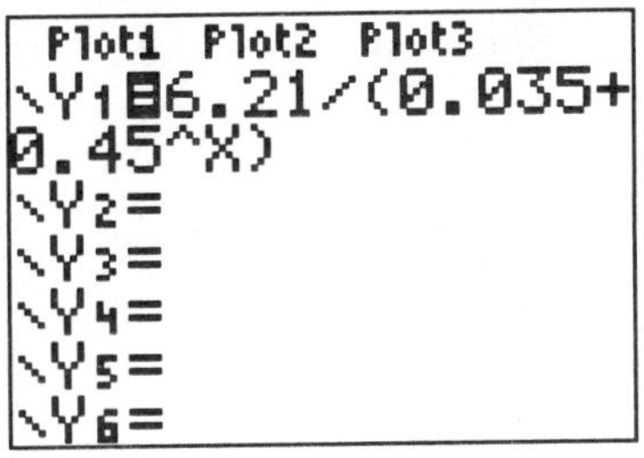

Figure 2.58: *Searching for a good window size*

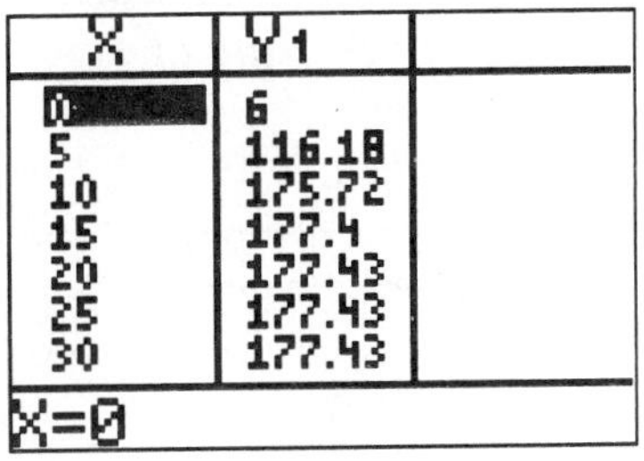

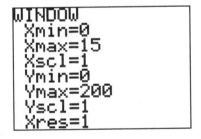

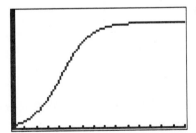

Figure 2.59: *Selecting the window size*

Figure 2.60: *The graph of deer population versus time*

Solution to Part 2: We want to know when the deer population will reach 85. We write the equation first in words and then replace the words by the appropriate letters:

$$\begin{aligned} \text{Deer population} &= 85 \\ N &= 85 \\ \frac{6.21}{0.035 + 0.45^t} &= 85. \end{aligned}$$

The time when the deer population will reach 85 is the value of t that makes this equation true. That is, we need to solve this equation for t.

Solution to Part 3: To solve this equation we will use the crossing graphs method. Since we have already entered the left-hand side of the equation, we need only enter $\boxed{2.45}$ the right-hand side of the equation, 85. Now graph to see the picture in Figure 2.61. We use the calculator to find $\boxed{2.46}$ the crossing point shown in Figure 2.62. We see that 85 deer can be expected to be present in about 4.09 years, or about 4 years and one month.

Figure 2.61: *Adding the line at height 85*

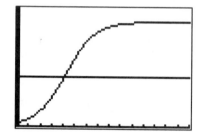

Figure 2.62: *Solution using the crossing graphs method*

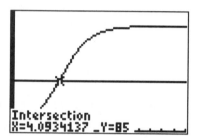

We should note that this equation is difficult to solve by hand calculation. That is why we do it with the calculator. We should also note that acquiring the skills necessary to solve such an equation is a significant step forward in your mathematical development.

EXAMPLE 2.9 *A Floating Ball*

According to *Archimedes' law*, the weight of water that is displaced by a floating ball is equal to the weight of the ball. A certain wooden ball of diameter 4 feet weighs 436 pounds. If the ball is allowed to float in pure water, Archimedes' law can be used to show that d feet of its diameter will be below the surface of the water, where d is a solution of the equation

$$62.4\pi d^2 \left(2 - \frac{d}{3}\right) = 436 .$$

(The left-hand side of this equation represents the weight of the displaced water, and the right-hand side is the weight of the ball.) How much of this ball's diameter is below the surface of the water?

Solution: This time we use the single graph method. As it stands, the equation does not involve the zero of a function, so the first step is to subtract 436 from both sides of the equation:

$$62.4\pi d^2\left(2-\frac{d}{3}\right)-436=0\,.$$

Now it is clear that we are looking for a zero of the function

$$62.4\pi d^2\left(2-\frac{d}{3}\right)-436\,.$$

The next step is to enter [2.47] the function into the calculator and record the appropriate correspondences:

Y1 = weight of displaced water less weight of ball on vertical axis

X = d, diameter below surface on horizontal axis.

It is important to remember that the diameter of the ball is 4 feet. Thus we are interested in viewing the graph only on the span from $d = 0$ to $d = 4$. (See Exercise 2 at the end of this section.) To get the proper vertical span of the viewing screen, we consult the table in Figure 2.63. Anticipating the need to leave a little extra space at the bottom for the Intersection prompt, we set the viewing window as shown in Figure 2.64. Now when we graph, we get the picture in Figure 2.65. We use the calculator to find [2.48] where the graph crosses the horizontal axis as shown in Figure 2.66. We read the answer $d = 1.18$ feet from the X= display in Figure 2.66.

Figure 2.63: *A table for setting vertical screen span*

X	Y1	
0	-436	
1	-109.3	
2	609.52	
3	1328.3	
4	1655	
5	1197.6	
6	-436	

X=4

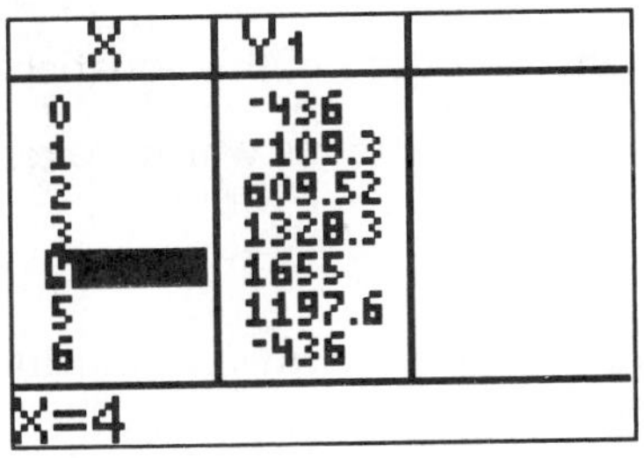

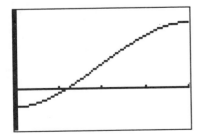

Figure 2.64: *The properly configured* WINDOW *menu*

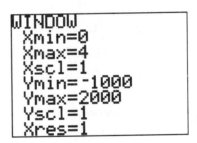

Figure 2.65: *Preparing to solve using the single graph method*

Figure 2.66: *The part of a floating ball that is below the water line*

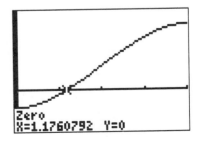

Exercise Set 2.4

1. **A population of foxes:** A breeding group of foxes is introduced into a protected area, and the population growth follows a logistic pattern. After t years the population of foxes is given by

$$N = \frac{37.5}{0.25 + 0.76^t} \text{ foxes.}$$

   (a) How many foxes were introduced into the protected area?

   (b) Make a graph of N versus t and explain in words how the population of foxes increases with time.

   (c) When will the fox population reach 100 individuals?

2. **Revisiting the floating ball:** In Example 2.9 we solved the equation

$$62.4\pi d^2 \left(2 - \frac{d}{3}\right) - 436 = 0$$

   on the interval from $d = 0$ to $d = 4$. If you graph $62.4\pi d^2 \left(2 - \frac{d}{3}\right) - 436$ on the span from $d = -2$ to $d = 7$ you will see that there are two other solutions of this equation. Find these solutions. Is there a physical interpretation of these solutions which makes sense?

3. **The skydiver again:** When a skydiver jumps from an airplane, her downward velocity (in feet per second) before she opens her parachute is given by $v = 176(1 - 0.834^t)$, where t is the number of seconds that have elapsed since she jumped from the airplane. We found earlier that the terminal velocity for the skydiver is 176 feet per second. How long does it take to reach 90% of terminal velocity?

4. **Falling with a parachute:** If an average size man with a parachute jumps from an airplane, he will fall $12.5(0.2^t - 1) + 20t$ feet in t seconds. How long will it take him to fall 140 feet?

5. **A cup of coffee:** The temperature C of a fresh cup of coffee t minutes after it is poured is given by

$$C = 125e^{-0.03t} + 75 \text{ degrees Fahrenheit.}$$

(a) Make a graph of C versus t.

(b) The coffee is cool enough to drink when its temperature is 150 degrees. When will the coffee be cool enough to drink?

(c) What is the temperature of the coffee in the pot? (<u>Note</u>: We are assuming the coffee pot is being kept hot and is the same temperature as the cup of coffee when it was poured.)

(d) What is the temperature in the room where you are drinking the coffee? (<u>Hint</u>: If the coffee is left to cool a long time, it will reach room temperature.)

6. **Van der Waals equation:** In Exercise Set 2.3 we discussed the *ideal gas law*, which shows the relationship among volume V, pressure p, and temperature T for a fixed amount (one mole) of a gas. But chemists believe that in many situations, the *van der Waals equation* gives more accurate results. If we measure temperature T in kelvins, volume V in liters, and pressure p in atmospheres (1 atm is the pressure exerted by the atmosphere at sea level), then the relationship for carbon dioxide is given by

$$p = \frac{0.082T}{V - 0.043} - \frac{3.592}{V^2} \text{ atm.}$$

What volume does this equation predict for one mole of carbon dioxide at 500 kelvins and 100 atm? (<u>Suggestion</u>: Consider volumes ranging from 0.1 to 1 liter.)

7. **Radioactive decay:** The *half-life* of a radioactive substance is the time H that it takes for half of the substance to change form through radioactive decay. This number does not depend on the amount with which you start. For example carbon 14 is known to have a half-life of $H = 5770$ years. Thus if you begin with one gram of carbon 14, then 5770 years later you will have one-half gram of carbon 14. Or if you begin with 30 grams of carbon 14, then after 5770 years there will be 15 grams left. In general radioactive substances decay according to the formula

$$A = A_0 \times 0.5^{\frac{t}{H}} ,$$

where H is the half-life, t is the elapsed time, A_0 is the amount you start with (the amount when $t = 0$), and A is the amount left at time t.

(a) Uranium 228 has a half-life H of 9.3 minutes. Thus the decay function for this isotope of uranium is

$$A = A_0 \times 0.5^{\frac{t}{9.3}} ,$$

where t is measured in minutes. Suppose we start with 8 grams of U_{228}.

  i. How much U_{228} is left after 2 minutes?

  ii. How long will you have to wait until there are only 3 grams left?

(b) Uranium 235 is the isotope of uranium that can be used to make nuclear bombs. It has a half-life of 713 million years. Suppose we start with 5 grams of U_{235}.

  i. How much U_{235} is left after 200 million years?

  ii. How long will you have to wait until there are only 3 grams left?

8. **Radiocarbon dating:** *This is a continuation of Exercise 7.* We rewrite the decay formula as

$$\frac{A}{A_0} = 0.5^{\frac{t}{H}} \ .$$

This formula shows the fraction of the original amount that is present as a function of time.

Carbon 14 has a half-life of 5.77 thousand years. It is thought that in recent geological time (the last few million years or so) the amount of C_{14} in the atmosphere has remained constant. As a consequence all organisms that take in air (trees, people, etc.) maintain the same level of C_{14} so long as they are alive. When a living organism dies, it no longer takes in C_{14}, and the amount present at death begins to decay according to the formula above. This can be used to date some archaeological objects.

Suppose the amount of carbon 14 in charcoal from an ancient campfire is $\frac{1}{3}$ of the amount in a modern, living tree. In terms of the formula above, this means $\frac{A}{A_0} = \frac{1}{3}$. When did the tree that was used to make the campfire die? Be sure to explain how you get your answer.

9. **Monthly payment on a loan:** If you borrow $5000 at an EAR of r (as a decimal) from a lending institution which compounds interest continuously, and if you wish to pay off the note in 3 years, then your monthly payment M can be calculated using

$$M = \frac{5000r}{12(1 - e^{-3r})} \text{ dollars.}$$

Your budget will allow a payment of $150 per month, and you are shopping for an interest rate that will give a payment of this size. What interest rate do you need to find?

10. **Grazing kangaroos:** The amount of vegetation eaten in a day by a grazing animal is a function of the amount V of food available (measured as biomass, in units such as pounds per acre).[13] This relationship is called the *functional response*. If there is little vegetation available, the daily intake will be small, since the animal will have difficulty finding and eating the food. As the food biomass increases, so does the daily intake. Clearly, though, there is a limit to the amount the animal will eat, regardless of the amount of food available. This maximum amount eaten is the *satiation level*.

(a) For the western grey kangaroo of Australia the functional response is

$$G = 2.5 - 4.8e^{-0.004V} ,$$

where $G = G(V)$ is the daily intake (measured in pounds) and V is the vegetation biomass (measured in pounds per acre).

  i. Draw a graph of G against V. Include vegetation biomass levels up to 1000 pounds per acre.

  ii. Is the graph you found in Part (i) concave up or concave down? Explain in practical terms what your answer means about how this kangaroo feeds.

  iii. There is a *minimal* vegetation biomass level below which the western grey kangaroo will eat nothing. (Another way of expressing this is to say that the animal cannot reduce the biomass below this level.) Find this minimal level.

  iv. Find the satiation level for the western grey kangaroo.

(b) For the red kangaroo of Australia the functional response is

$$R = 1.9 - 1.9e^{-0.033V} ,$$

where R is the daily intake (measured in pounds) and V is the vegetation biomass (measured in pounds per acre).

  i. Add the graph of R against V to the graph of G you drew in Part (a).

  ii. A simple measure of the grazing efficiency of an animal involves the minimal vegetation biomass level described above: the lower the minimal level for an animal, the more efficient it is at grazing. Which is more efficient at grazing, the western grey kangaroo or the red kangaroo?

---

[13]This exercise and the next are based on the work of J. Short, "Factors affecting food intake of rangelands herbivores." In: G. Caughley, N. Shepherd, and J. Short, eds. *Kangaroos*, 1987, Cambridge University Press, Cambridge, England.

11. **Grazing rabbits and sheep:** *This is a continuation of Exercise 10.* In addition to the kangaroos, major grazing mammals of Australia are merino sheep and rabbits. For sheep the functional response is

$$S = 2.8 - 2.8e^{-0.01V},$$

while for rabbits it is

$$H = 0.2 - 0.2e^{-0.008V}.$$

Here S and H are the daily intake (measured in pounds), and V is the vegetation biomass (measured in pounds per acre).

(a) Find the satiation level for sheep and that for rabbits.

(b) One concern in the management of rangelands is whether the various species of grazing animals are forced to compete for food. It is thought that competition will not be a problem if the vegetation biomass level provides at least 90% of the satiation level for each species. Find the biomass level which guarantees that competition between sheep and rabbits will not be a problem.

12. **Growth rate:** The per capita growth rate r (on an annual basis) of a population of grazing animals is a function of V, the amount of vegetation available. A positive value of r means that the population is growing, while a negative value of r means that the population is declining. For the red kangaroo of Australia, the relationship has been given[14] as

$$r = 0.4 - 2e^{-0.008V}.$$

Here V is the vegetation biomass, measured in pounds per acre.

(a) Draw a graph of r versus V. Include vegetation biomass levels up to 1000 pounds per acre.

(b) The population size will be stable if the per capita growth rate is zero. At what vegetation level will the population size be stable?

---

[14] This formula is based on the work of P. Bayliss and G. Caughley. In: G. Caughley, *Ibid.*

2.5 OPTIMIZATION

One key feature of a graph is its *zeros*, the places where it crosses or touches the horizontal axis. We learned how to find these in Section 2.4. Other important features of a graph include its *maxima* and *minima*. The need to locate these may occur, for example, if you want to find the number of items you can produce that will yield a maximum profit, or if you wish to lay a pipeline so that the cost is a minimum. Graphs make it easy to find maxima and minima. Thus if we have a function given by a formula, we can *optimize* it (find the optimal values) by first getting the graph.

Optimizing at peaks and valleys

In many instances optimal values for a function are located at peaks or valleys of its graph. For example, if a cannon is placed at the origin (X=0, Y=0) and elevated at an angle of 45 degrees, then when it is fired, the cannonball will follow the graph of

$$y = x - g\left(\frac{x}{m}\right)^2 .$$

(This simple model ignores air resistance.) Here x represents the number of feet downrange, y is the height in feet, and m is the *muzzle velocity* (in feet per second) of the cannonball. The constant g is the acceleration due to gravity near the surface of the earth, about 32 feet per second per second. So, if the cannon is fired with a muzzle velocity of 250 feet per second, then it will follow the graph of

$$y = x - 32\left(\frac{x}{250}\right)^2 .$$

Let's use the graph to analyze the flight of the cannonball. The first step is to enter [2.49] the function and record the appropriate correspondences:

$\mathsf{Y_1}$ = y, height on vertical axis

X = x, distance downrange on horizontal axis.

We need to look at a table of values to help us choose a window. Since we expect the cannonball to travel several hundreds of feet, we made the table [2.50] in Figure 2.67 with a starting value of 0 and a table increment of 500. Allowing a little extra room, we set the window as in Figure 2.68.

Now graphing produces the cannonball path shown in Figure 2.69. In this picture the horizontal axis shows feet downrange, and the vertical axis shows height above the ground. Thus the horizontal axis is ground level. In Figure 2.69 we have located the place where the

Figure 2.67: *A table of values for a cannonball flight*

X	Y₁	
0	0	
500	372	
1000	488	
1500	348	
2000	-48	
2500	-700	
3000	-1608	

X=0

Figure 2.68: *Window settings from the table*

```
WINDOW
 Xmin=0
 Xmax=2500
 Xscl=1
 Ymin=-100
 Ymax=600
 Yscl=1
 Xres=1
```

graph crosses the horizontal axis $\boxed{2.51}$, that is, where the cannonball strikes the ground, and we see that this will happen about 1953 feet downrange. (That is just over a third of a mile.)

Let's find the maximum height of the cannonball, which is at the top of the graph in Figure 2.69. We can get a reasonable estimate for the maximum if we trace the graph and move the cursor as close to the peak as possible. But most graphing calculators have the ability to find maximum values $\boxed{2.52}$ such as this in a fashion similar to the way they find zeros or where graphs cross. You should consult the *Keystroke Guide* for the exact keystrokes needed to accomplish this. In Figure 2.70 we have used this feature to locate the peak of the graph, and we see from the prompt at the bottom of the screen that the cannonball reaches a maximum height of 488.28 feet at 976.56 feet downrange.

Figure 2.69: *Where the cannonball lands*

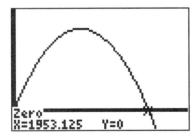

Zero
X=1953.125   Y=0

Figure 2.70: *The peak of the cannonball's flight*

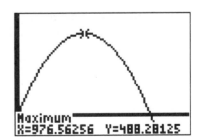

Maximum
X=976.56256   Y=488.28125

EXAMPLE 2.10 *Growth of Forest Stands*

In forestry management it is important to know the *growth* and the *yield* of a forest stand.[15] The growth G is the amount by which the volume of wood will increase in a unit of time, and the yield Y is the total volume of wood. In a stand of age A, a forest manager has determined that the growth $G = G(A)$ is given by the formula

$$G = 32A^{-2}e^{10-32A^{-1}} ,$$

and that the yield $Y = Y(A)$ is given by the formula

$$Y = e^{10-32A^{-1}} .$$

Here G is measured in cubic feet per acre per year, Y in cubic feet per acre, and A in years.

1. Draw a graph of growth as a function of age which includes ages up to 60 years.

2. At what age is growth maximized?

3. Draw a graph of yield as a function of age. What is the physical meaning of the point on this graph corresponding to your answer to Part 2?

4. To meet market demand, loggers are considering harvesting a relatively young stand of trees. This area was initially clear-cut[16] and left barren. The forest was replanted, with plans to get a new harvest within the next 14 years. At what time in this 14 year period will growth be maximized?

Solution to Part 1: The first step is to enter ⌊2.53⌋ the growth function and record the appropriate correspondences:

$$Y_1 = G, \text{ growth on vertical axis}$$
$$X = A, \text{ age on horizontal axis.}$$

We need to look at a table of values to help us set the window size. We made the table ⌊2.54⌋ in Figure 2.71 using a starting value of 0 and a table increment of 10. We were told specifically that our graph should include ages up to 60, and so we made the graph in Figure 2.72 using a window ⌊2.55⌋ with a horizontal span from $A = 0$ to $A = 60$ and a vertical span from $G = 0$ to $G = 500$.

---

[15]See Thomas E. Avery and Harold E. Burkhart, *Forest Measurements*, 4th edition, 1994, McGraw-Hill, New York.
[16]Clear-cutting is the practice of harvesting by cutting *all* the trees in an area. The result shows as large swaths of land that are almost barren.

Figure 2.71: A table of values for growth

X	Y_1	
0	ERROR	
10	287.31	
20	355.77	
30	269.53	
40	197.94	
50	148.66	
60	114.86	

X=0

Figure 2.72: A graph of forest growth versus age

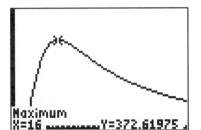

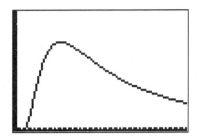

Solution to Part 2: Growth is maximized at the peak of the graph in Figure 2.72. We have used the calculator [2.56] to locate this point in Figure 2.73. We see from the prompt at the bottom of the screen that a maximum growth of 372.62 cubic feet per acre per year occurs when the stand is 16 years old.

Solution to Part 3: Now we want to enter [2.57] the yield function

$$Y = e^{10-32A^{-1}} \text{ cubic feet per acre.}$$

In Figure 2.74 we see in the Y_2 column values for the yield. It would be good to make graphs of both functions on the same screen, but the relative sizes of function values for growth and yield make this impractical. (Try it to see what happens.) Thus we turn off [2.58] the growth function and graph the yield in a window with horizontal span from 0 to 60 and vertical span from 0 to 13,000. The result is in Figure 2.75.

Figure 2.73: The age of maximum growth

Figure 2.74: A table of values for yield

X	Y_1	Y_2
0	ERROR	ERROR
10	287.31	897.85
20	355.77	4447.1
30	269.53	7580.5
40	197.94	9897.1
50	148.66	11614
60	114.86	12922

X=0

In Figure 2.75 we have put the cursor at the age $A = 16$, corresponding to maximum growth that we found in Part 2. We see that if we harvest at the time of maximum growth, we will get a yield of $Y = 2980.96$ cubic feet per acre. Examination of the graph of the yield function shows that it is the steepest at age $A = 16$, which is where the inflection point occurs. The physical meaning of this is that maximum growth corresponds to the fastest increase of yield.

Solution to Part 4: For this part of the problem, we want to look at growth again. Thus we go to the function entry screen and turn Y off and G on $\boxed{2.59}$. We reset the window so that the horizontal span is from 0 to 14 and the vertical span is from 0 to 500. The resulting graph is in Figure 2.76. We see that the graph of growth will be increasing over this span, and so the maximum growth will occur at the right-hand endpoint. That is, we get a maximum growth of $G = 365.73$ cubic feet per acre per year if we wait to the end of the 14 year period to harvest. Do you think it is appropriate to proceed with harvesting at this time?

Figure 2.75: Yield at maximum growth

Figure 2.76: Maximum growth for a young stand of trees

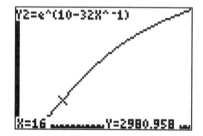

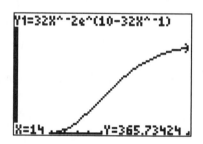

Optimizing at endpoints

Let's look back at the graph of the flight of the cannonball in Figure 2.69. If we adjust the window $\boxed{2.60}$ we see the view in Figure 2.77. This shows that the graph of $y = x - 32 \left(\dfrac{x}{250}\right)^2$ extends far below the horizontal axis and to the left of the vertical axis. This indicates, among other things, that the cannonball would travel underground, which is nonsense. In fact, the function $y = x - 32 \left(\dfrac{x}{250}\right)^2$ represents the path of the cannonball only during the period when it is flying through the air. This is from the place where the cannonball was fired ($x = 0$) to the location $x = 1953.125$ feet downrange where it strikes the ground. Beyond these limits, the graph in Figure 2.77 is not a representation of the physical situation. A more accurate picture would show the graph starting at $x = 0$ and ending where the cannonball lands at $x = 1953.125$. That is, an accurate graphical model would have endpoints, and if we wanted to find the *minimum* height of the cannonball, we would not find it at a peak or valley of the graph. Rather, we would find it at ground level: at the endpoints of the graph.

A similar situation occurs in Part 4 of Example 2.10, where we are finding the maximum growth for a clear-cut forest. Because of market demand, loggers want to harvest the stand within 14 years from planting, and under these conditions, growth is maximized at the end of the 14 year period. This is at the right endpoint of the graph in Figure 2.76.

For the cannonball, we don't need a graph to tell us that the minimum height will be at ground level, but there are many physical situations where the optimal value is not so obvious but occurs at endpoints of the graph. Let's look, for example, at a simple geometry problem. Suppose we have 100 yards of fence from which we wish to construct two pens. We will use part of the fence to make a square pen and the rest to make a circular pen as illustrated in Figure 2.78.

Figure 2.77: The formula represents the cannonball path only on a restricted range

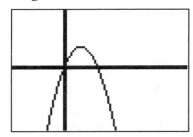

Figure 2.78: Cutting fence to make two pens

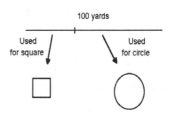

If we use s yards of the 100 yard stretch of fence for the square, then the total area $A = A(s)$ enclosed by the two pens turns out to be

$$A = \left(\frac{s}{4}\right)^2 + \frac{(100 - s)^2}{4\pi} \text{ square yards.}$$

We want to investigate how much fence to use for the square, s yards, and how much to use for the circle, $100 - s$ yards, to optimize the total area A. The first step is to enter [2.61] the function and list the variable correspondences:

$$\mathsf{Y_1} \quad = \quad A, \text{ total area on vertical axis}$$

$$\mathsf{X} \quad = \quad s, \text{ yards used for square on horizontal axis.}$$

Bearing in mind that we have only 100 yards of fence, so that s is between 0 and 100, we made the table of values in Figure 2.79 with a starting value of 0 and an increment of 20. This led us to choose a window with a horizontal span from 0 to 100 and a vertical span from 0 to 1000, which gives the graph in Figure 2.80. In Figure 2.80 we have located the minimum value [2.62] for area. We see from the prompt at the bottom of the screen that we enclose the minimum amount of area, $A = 350$ square yards, if we use about $s = 56$ yards of fence to make the square. That leaves $100 - 56 = 44$ yards of fence to make the circle.

Now, let's find how we should use the fence to enclose the maximum area. In Figure 2.80, we found the minimum at a *valley* of the graph, but there is no *peak* which corresponds to the maximum. The critical factor here is that we have only 100 yards of fence to work with,

Figure 2.79: A table of values for area

X	Y_1
0	795.77
20	534.3
40	386.48
60	352.32
80	431.83
100	625
120	931.83

X=0

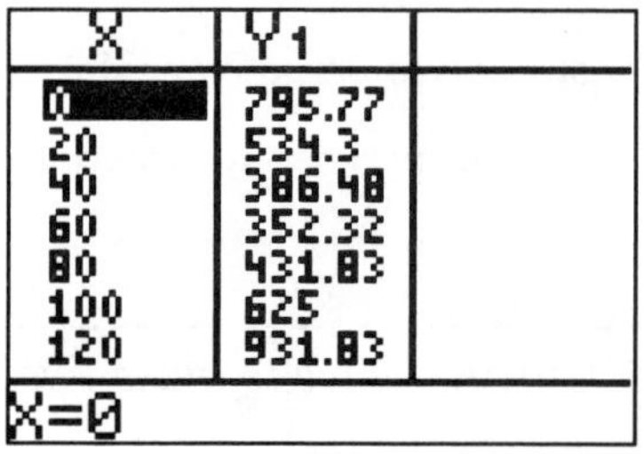

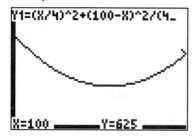

Figure 2.80: The minimum area enclosed by a square and a circle

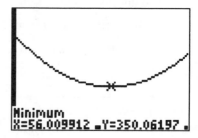

and so s is between 0 and 100. The graph shows clearly that the maximum occurs at an endpoint.

We have traced the graph to locate the cursor at the right-hand endpoint in Figure 2.81 and at the left-hand endpoint in Figure 2.82. We see in Figure 2.82 that we get the maximum

Figure 2.81: The right-hand endpoint: the area when all the fence is used for the square

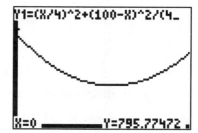

Figure 2.82: The maximum area at the left-hand endpoint, when all the fence is used for the circle

area of $A = 795.77$ square yards if we use $s = 0$ yards for the square. That is, all the fence goes to make the circle.

Exercise Set 2.5

1. **The cannon at a different angle:** Suppose a cannon is placed at the origin and elevated at an angle of 60 degrees. If the cannonball is fired with a muzzle velocity of 0.15 miles per second, it will follow the graph of $y = x\sqrt{3} - \dfrac{160x^2}{297}$, where distances are measured in miles.

    (a) Make a graph that shows the path of the cannonball.

    (b) How far downrange does the cannonball travel? Explain how you got your answer.

    (c) What is the maximum height of the cannonball, and how far downrange does that occur?

2. **Enclosing a field:** You have 16 miles of fence which you will use to enclose a rectangular field.

    (a) Draw a picture to show that you can arrange the 16 miles of fence into a rectangle of width 3 miles and length 5 miles. What is the area of this rectangle?

    (b) Draw a picture to show that you can arrange the 16 miles of fence into a rectangle of width 2 miles and length 6 miles. What is the area of this rectangle?

    (c) The first two parts of this exercise are designed to show you that you can get different areas for rectangles of the same perimeter, 16 miles. In general, if you arrange the 16 miles of fence into a rectangle of width w miles, then it will enclose an area of $A = w(8 - w)$ square miles.

        i. Make a graph of area enclosed as a function of w, and explain what the graph is showing.

        ii. What width w should you use to enclose the most area?

        iii. What is the length of the maximum-area rectangle that you have made, and what kind of figure do you have?

3. **An aluminum can:** The cost of making a can is determined by how much aluminum A (in square inches) is required to make it. This in turn depends on the radius r and the height h (both measured in inches) of the can. If the can must hold 15 cubic inches, then we can express both h and A in terms of the radius r:

$$h = \frac{15}{\pi r^2}$$
$$A = 2\pi r^2 + \frac{30}{r}.$$

(a) What is the height and how much aluminum is needed to make the can if the radius is 1 inch? (This is a tall, thin can.)

(b) What is the height and how much aluminum is needed to make the can if the radius is 5 inches? (This is a short, fat can.)

(c) The first two parts of this problem are designed to illustrate that for an aluminum can, different surface areas can enclose the same volume of 15 cubic inches.

    i. Make a graph of A versus r and explain what the graph is showing.

    ii. What radius should you use to make the can using the least amount of aluminum?

    iii. What is the height of the can that uses the least amount of aluminum?

4. **Laying phone cable:** City A lies on the north bank of a river that is one mile wide. You need to run a phone cable from City A to City B, which lies on the opposite bank five miles down the river. You will lay L miles of the cable along the north shore of the river, and from the end of that stretch of cable you will lay W miles of cable running under water directly toward City B. (See Figure 2.83.) It costs $300 per mile to run cable on land but $500 per mile to lay it under water. Under these conditions, the length of cable you run under water, and the total cost C of the project (in dollars), can be expressed in terms of L:

$$W = \sqrt{1 + (5 - L)^2}$$
$$C = 300L + 500\sqrt{1 + (5 - L)^2}.$$

(a) Find the amount of cable that runs under water and the total cost of the project if you make L one mile long. Draw and properly label a picture that shows this cable plan.

(b) Find the amount of cable that runs under water and the total cost of the project if you make L three miles long. Draw and properly label a picture that shows this cable plan.

Figure 2.83: *Laying phone cable*

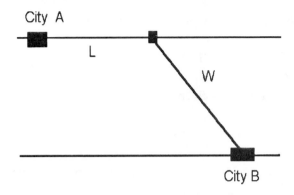

(c) Make a graph of the cost C as a function of L, and explain what the graph is showing.

(d) What value of L gives the least cost for the project?

(e) Find the value of W corresponding to your answer in Part (d) and draw a picture showing the least-cost project.

(f) Potential legal disputes about easements have caused the cost of laying cable on land to increase \$700 per mile. With this change, the new function for cost is

$$C = 700L + 500\sqrt{1 + (5 - L)^2}\,.$$

How does this affect your plans for the least-cost project?

5. **Mean annual growth:** *This is a continuation of Example 2.10.* Forest managers study the *mean annual growth* $M = M(A)$, defined as the yield at age A divided by A.

(a) Use the formula for Y in Example 2.10 to find a formula for M as a function of A.

(b) Add the graph of M as a function of A to the graph of G.

(c) The *rotation age* may be determined as the age of maximum mean annual growth. Determine the rotation age using the graph of M.

(d) From your two graphs in Part (b) can you suggest a different way to find the rotation age?

6. **Growth of fish biomass:** An important model for commercial fisheries is that of Beverton and Holt.[17] It begins with the study of a single *cohort* of fish; that is, all the fish in the study are born at the same time. For a cohort of the North Sea plaice (a type of flatfish), the number $N = N(t)$ of fish in the population is given by

$$N = 1000e^{-0.1t} ,$$

and the weight $w = w(t)$ of each fish is given by

$$w = 6.32(1 - 0.93e^{-0.095t})^3 .$$

Here w is measured in pounds and t in years. The variable t measures the so-called recruitment age, which we refer to simply as the age.

The *biomass* $B = B(t)$ of the fish cohort is defined to be the total weight of the cohort, so it is obtained by multiplying the population size by the weight of a fish.

(a) If a plaice weighs 3 pounds, how old is it?

(b) Use the formulas for N and w given above to find a formula for $B = B(t)$, and then make a graph of B against t. (Include ages through 20 years.)

(c) At what age is the biomass the largest?

(d) In practice, fish below a certain size can't be caught, so the biomass function becomes relevant only at a certain age.

   i. Suppose we want to harvest the plaice population at the largest biomass possible, but a plaice has to weigh 3 pounds before we can catch it. At what age should we harvest?

   ii. Work Part i under the assumption that we can catch plaice weighing at least 2 pounds.

---

[17]See R. J. H. Beverton and S. J. Holt, *On the Dynamics of Exploited Fish Populations*, Fishery Investigations, Series 2, Volume 19, 1957, Ministry of Agriculture, Fisheries and Food, London.

7. **Spawner-recruit model:** In fish management it is important to know the relationship between the abundance of the *spawners* (also called the parent stock) and the abundance of the *recruits*, that is, those hatchlings surviving to maturity.[18] According to the *Ricker model*, the number of recruits R as a function of the number of spawners P has the form

$$R = APe^{-BP}$$

for some positive constants A and B. This model describes well a phenomenon observed in some fisheries: A large spawning group can actually lead to a small group of recruits.[19]

In a study of the sockeye salmon, it was determined that $A = 4$ and $B = 0.7$. Here we measure P and R in thousands of salmon.

(a) Make a graph of R against P for the sockeye salmon. (Assume there are at most 3000 spawners.)

(b) Find the maximum number of salmon recruits possible.

(c) If the number of recruits R is greater than the number of spawners P, then the difference $R - P$ of the recruits can be removed by fishing, and next season there will once again be P spawners surviving to renew the cycle. What value of P gives the maximum value of $R - P$, the number of fish available for removal by fishing?

8. **Rate of growth:** The rate of growth G in the weight of a fish is a function of the weight w of the fish. For the North Sea cod the relationship is given by

$$G = 2.1w^{\frac{2}{3}} - 0.6w .$$

Here w is measured in pounds and G in pounds per year. The maximum size for a North Sea cod is about 40 pounds.

(a) Make a graph of G against w.

(b) Find the greatest rate of growth among all cod weighing at least 5 pounds.

(c) Find the greatest rate of growth among all cod weighing at least 25 pounds.

---

[18]See W. E. Ricker, "Stock and recruitment," *J. Fish. Res. Board Can.* **11** (1954), 559–623.

[19]Biological mechanisms which contribute to this phenomenon are suspected to include competition and cannibalism of the young.

9. **Health plan:** The manager of an employee health plan for a firm has studied the balance B (in millions of dollars) in the plan account as a function of t, the number of years since the plan was instituted. He has determined that the account balance is given by the formula $B = 60 + 7t - 50e^{0.1t}$.

   (a) Make a graph of B versus t over the first 7 years of the plan.

   (b) At what time is the account balance at its maximum?

   (c) What is the smallest value of the account balance over the first 7 years of the plan?

10. **Size of high schools:** The farm population has declined dramatically in the years since World War II, and with that decline, rural school districts have been faced with consolidation so as to be economically efficient. One researcher studied data from the early 1960's on expenditures for high schools ranging from 150 to 2400 in enrollment.[20] He considered the cost per pupil as a function of the number of pupils enrolled in the high school, and he found the approximate formula $C = 743 - 0.402n + 0.00012n^2$, where n is the number of pupils enrolled and C is the cost (in dollars) per pupil.

    (a) Make a graph of C versus n.

    (b) What enrollment size gives a minimum per-pupil cost?

    (c) If a high school had an enrollment of 1200, how much in per-pupil cost would be saved by increasing enrollment to the optimal size found in Part (b)?

11. **Radioactive decay:** Radium 223 is a radioactive substance which itself is a product of the radioactive decay of thorium 227. For one experiment the amount A of radium 223 present, as a function of the time t since the experiment began, is given by the formula $A = 3(e^{-0.038t} - e^{-0.059t})$, where A is measured in grams and t in days.

    (a) Make a graph of A versus t covering the first 60 days of the experiment.

    (b) What was the largest amount of radium 223 present over the first 60 days of the experiment?

    (c) What was the largest amount of radium 223 present over the first 10 days of the experiment?

    (d) What was the smallest amount of radium 223 present over the first 60 days of the experiment?

---

[20]This exercise is based on the study by J. Riew, "Economies of scale in high school operation," *Review of Economics and Statistics* **48** (August, 1966), 280–287. In our presentation of his results we have suppressed variables such as teacher salaries.

2.6 CHAPTER SUMMARY

Each type of function presentation has it advantages and disadvantages, and one of the most effective methods of mathematical analysis is to take a function presented in one way and obtain its presentation in another form. It is particularly tedious to get a graph or table of values from a function given by a complicated formula, but doing so can also be very illuminating. The graphing calculator takes the tedium out of this procedure by producing graphs and tables on demand.

Tables and Trends

It is often difficult to determine limiting values or maximum and minimum values from a formula, but as was seen in Chapter 1, this is easily done from a table of values. The calculator produces such tables on demand and in this regard becomes an important problem-solving tool. For example, the probability of getting exactly 3 sixes when you roll N dice is given by the formula

$$P = \frac{N(N-1)(N-2)}{750} \left(\frac{5}{6}\right)^N .$$

It is by no means clear from this formula how many dice should be rolled to make the probability of achieving exactly 3 sixes the greatest. A calculator can be used to produce quickly the following partial table of values.

N	P
15	0.23626
16	0.24231
17	0.2452
18	0.2452
19	0.24264
20	0.23789
21	0.23128

The table shows clearly that the maximum probability is achieved by rolling 17 or 18 dice.

Graphs

As with tables, the graphing calculator can quickly produce graphs from formulas, and effective analysis can proceed. Graphs viewed from different perspectives can have quite different appearances. Effective graphical analysis is dependent upon one's ability to view the graph in an appropriate window. This is done through an intelligent choice of *horizontal span* and *vertical span*. When the function is a model for some physical phenomenon, choosing the horizontal span, or *domain*, may be a matter of common sense. The vertical span, or *range*, can then be chosen by looking at a table of values.

As an example, suppose a leather craftsman has produced 25 belts which he intends to sell at an upcoming art fair for $22.75 each. He has invested a total of $300 in materials. The net profit p, in dollars, from the sale of n belts is given by

$$p = 22.75n - 300.$$

In order to graph this function, we need to choose a horizontal and a vertical span. Since he has only 25 belts, he will sell somewhere between 0 and 25 belts. This is the horizontal span, or domain, for our function. To get an appropriate vertical span, we look at the table of values shown in Figure 2.84. This shows that we should choose a vertical span of about -325 to 300. The resulting graph, which is now available for further analysis, is in Figure 2.85.

Figure 2.84: A table of values for net profit on belt sales

X	Y1
0	-300
5	-186.5
10	-73
15	40.5
20	154
25	267.5
30	381

X=0

Figure 2.85: A graph of net profit on belt sales

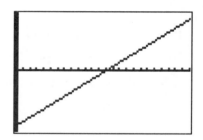

Solving Linear Equations

Many times the analysis required to understand a phenomenon involves solving an equation. For many equations, this can be a difficult process, but for *linear equations*, solutions are easy. Linear equations are those that do not involve powers, roots, or other complications of the variable; they are the simplest of all equations. Any linear equation can be solved using two basic rules.

Moving terms across the equal sign: You may add (or subtract) the same thing to (or from) both sides of the equation. This can be thought of as moving a term from one side of the equation to the other, provided you change its sign.

Dividing each term of an equation: You may divide (or multiply) both sides of an equation by any nonzero number.

As an example, we show how to solve $5x + 7 = 18 - 2x$ using these two rules.

$$5x + 7 = 18 - 2x$$

$$5x = 18 - 7 - 2x \qquad \text{(Move 7 across and change its sign.)}$$

$$5x + 2x = 11 \qquad \text{(Move } 2x \text{ across and change its sign.)}$$

$$7x = 11$$

$$x = \frac{11}{7} = 1.57 \qquad \text{(Divide each term by 7.)}$$

Solving Nonlinear Equations

Equations that are not linear may be difficult or impossible to solve exactly by hand. But the graphing calculator can provide approximate (yet highly accurate) solutions to virtually any equation quickly and easily.

The most useful method for solving equations with a calculator is referred to in this text as the *crossing graphs method*. We will illustrate it with an example. Suppose the temperature, after t minutes, of an aluminum bar being heated in an oven is given by

$$A = 800 - 730e^{-0.06t} \text{ degrees Fahrenheit.}$$

We want to know when the temperature will reach 600 degrees. In terms of an equation, that means we need to solve

$$800 - 730e^{-0.06t} = 600$$

for t. This is done by entering the functions $800 - 730e^{-0.06t}$ and 600 into the calculator. We want to know when these functions are the same, that is, where the graphs cross. We see in Figure 2.86 that this occurs 21.58 minutes after the bar is put in the oven.

Figure 2.86: Finding when the temperature is 600 degrees

Figure 2.87: The maximum height of a cannonball

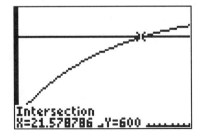

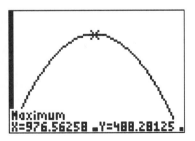

Optimization

A strategy similar to that used to solve nonlinear equations may be used to optimize functions given by formulas. We use the calculator to make the graph and then look for peaks or valleys. To illustrate the method, we look for the maximum height of a cannonball following the graph of

$$x - 32\left(\frac{x}{250}\right)^2,$$

where x is the distance downrange, measured in feet. We graph the function as shown in Figure 2.87. That figure shows that the cannonball reaches its maximum height of 488.28 feet when it is 976.56 feet downrange.

CHAPTER 3 *Straight Lines and Linear Functions*

Straight lines in the form of city streets, directions, boundaries, and many other things are among the most obvious mathematical objects that we experience in daily life. Historically mathematicians made extensive studies of the *geometry* of straight lines several centuries before they were associated with a *linear formula*. Today it is understood that it is the combination of pictures and formulas that make lines so useful and so easy to handle.

3.1 THE GEOMETRY OF LINES

Characterizations of straight lines

A familiar characterization of straight lines is that they are determined by two points. For example, to describe a straight ramp it is only necessary to give the locations of the ends of the ramp. In Figure 3.1 we have depicted a ramp with one end atop a 4-foot-high retaining wall and the other on the ground 10 feet away. It is often convenient to represent such lines on coordinate axes. If we choose the horizontal axis to be ground level and let the vertical axis follow the retaining wall, we get the picture in Figure 3.2. The ends of the ramp in Figure 3.1

Figure 3.1: A ramp on a retaining wall

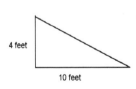

Figure 3.2: Representing the ramp line on coordinate axes

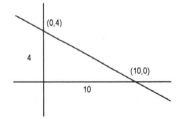

match the points in Figure 3.2 where the line crosses the horizontal and vertical axes. These crossing points are known as the *horizontal and vertical intercepts*. (Sometimes they are called the *x-intercept* and the *y-intercept*.) Thus in Figure 3.2 the horizontal intercept is 10 and the vertical intercept is 4.

To illustrate these ideas further, in Figure 3.3 we have drawn a roof line which is 8 feet high at the outside wall and 8.5 feet high 1 foot toward the interior of the structure. If we want to represent the roof line on coordinate axes, it is natural to let the horizontal axis correspond to ground level and let the vertical axis follow the outside wall. We have done this in Figure 3.4. We note that the vertical intercept of this line is 8, but in this case it is not immediately apparent what the horizontal intercept is or what its physical significance might be. We will return to this in Example 3.1.

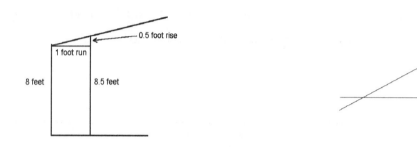

Figure 3.3: *The roof of a building*   **Figure 3.4:** *Representing the roof line*

Another way of characterizing a straight line is to say that it rises or falls at the same rate everywhere on the line. (Curved graphs rise or fall at different rates at different places.) If we stand at the outside wall of the house depicted in Figure 3.3 and move one foot toward its interior, the roof rises from 8 feet to 8.5 feet. That is, in 1 horizontal foot the roof rises by 0.5 vertical foot. Since the roof line is straight, we know that if we move one more foot toward the interior, the roof will rise by that same amount, 0.5 foot, and the roof at that point will be 9 feet high. Each 1 foot horizontal movement toward the interior results in a 0.5 foot vertical rise in the roof line. In this setting, horizontal change is often referred to as the *run* and vertical change as the *rise*, and we would say that the roof line rises 0.5 foot for each 1 foot run.

EXAMPLE 3.1   *Further Examination of the Roof Line*

In Figure 3.5 we have extended the roof to its peak 14 horizontal feet toward the interior of the structure.

1. How high is the roof line at its peak?

2. In Figure 3.6 locate the horizontal intercept and explain its physical significance.

Solution to Part 1: We know that the roof rises 0.5 foot for each 1 foot run. To get to the peak of the roof, we need to make a 14 foot run. That will result in 14 half-foot rises. Thus the roof rises by $14 \times 0.5 = 7$ feet. Since the roof is 8 feet high at the outside wall, that makes it $8 + 7 = 15$ feet high at its peak. This is illustrated in Figure 3.5.

Solution to Part 2: In Figure 3.6, movement 1 foot to the right results in a 0.5 foot rise in the line. Thus a 1 foot movement to the left would give a fall of 0.5 foot. To get down to the horizontal axis, we need to drop a total of 8 units. Since we do that in 0.5 foot steps, we need to move 16 units to the left of the vertical axis to get to the place in Figure 3.6 where the line crosses the horizontal axis. Thus the horizontal intercept is −16 (negative, since it's to the left of zero). This is where an extended roof line would meet the ground. It might, for example, be important if we wanted to build an "A-frame" structure. The minus sign here means that an extended roof line would meet the ground 16 feet in the exterior direction from the wall.

Figure 3.5: *Extending the roof line to its peak*

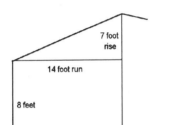

Figure 3.6: *Finding the horizontal intercept*

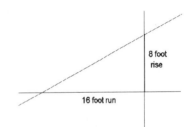

The slope of a line

The number 0.5 associated with the roof line in Example 3.1 is the *rate of change* in height with respect to horizontal distance. For straight lines, this is universally termed the *slope* and is commonly denoted by the letter m. The slope is one of the most important and useful features of a line. It tells the vertical change along the line when there is a horizontal change of 1 unit. In other words, a line of slope m exhibits m units of rise for each unit of run. Another way to express this is: The rise is proportional to the run, and the constant of proportionality is called the slope. Horizontal change is sometimes denoted by Δx and vertical change by Δy. For a line of slope m, the following three equations say exactly the same thing using these different terms:

$$\text{rise} \;=\; m \times \text{run} \tag{3.1}$$

$$\text{Vertical change} \;=\; m \times \text{Horizontal change} \tag{3.2}$$

$$\Delta y \;=\; m \times \Delta x. \tag{3.3}$$

Figure 3.7 shows lines of slopes 0, 1, and 2. Notice that the horizontal line has slope $m = 0$. This makes sense because a horizontal change results in no vertical change at all. The line with slope $m = 1$ rises 1 unit for each unit of run. Thus it slants upward at a 45 degree angle. The line of slope $m = 2$ is much steeper, showing a rise of 2 units for each unit of run. In general, larger positive slopes mean steeper lines that point upward from left to right.

If a line has negative slope, then a run of 1 unit will result in a *drop* rather than a rise. The line in Figure 3.8 with $m = -1$ drops 1 unit for each unit of run. Similarly, the line with $m = -2$ drops 2 units for each unit of run. In general, large negative slopes mean steep lines that point downward.

Figure 3.7: Some lines with positive slope

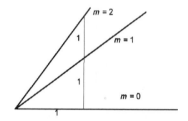

Figure 3.8: Some lines with negative slope

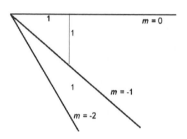

Getting slope from points

Equations (3.1), (3.2), and (3.3) show us how to use the slope of a line to determine vertical change. These are all linear equations, and so we can use division to solve each of them for m. This will give us equations that tell how to find the slope m of a line. Once again, all three equations say the same thing using different notations:

$$\text{Rate of change} = \text{slope} = m \quad = \quad \frac{\text{rise}}{\text{run}} \tag{3.4}$$

$$\text{Rate of change} = \text{slope} = m \quad = \quad \frac{\text{Vertical change}}{\text{Horizontal change}} \tag{3.5}$$

$$\text{Rate of change} = \text{slope} = m \quad = \quad \frac{\Delta y}{\Delta x}. \tag{3.6}$$

KEY IDEA 3.1: THE SLOPE OF A LINE

- The slope, or rate of change, m, of a line shows how steeply it is increasing or decreasing. It tells the vertical change along the line when there is a horizontal change of 1 unit.

  - If m is positive, the line is rising from left to right. Larger positive values of m mean steeper lines.
  - If $m = 0$, the line is horizontal.
  - If m is negative, the line is falling from left to right. Negative values of m which are larger in size correspond to lines that fall more steeply.

- The slope m of a line can be calculated using

$$m = \frac{\text{Vertical change}}{\text{Horizontal change}}.$$

- The slope m can be used to calculate vertical change:

$$\text{Vertical change} = m \times \text{Horizontal change}.$$

EXAMPLE 3.2   A Circus Tent

The outside wall of the circus tent depicted in Figure 3.9 is 10 feet high. Five feet toward the center pole, the tent is 12 feet high. The center pole of the tent is 60 feet from the outside wall.

1. What is the slope of the line that follows the roof of the circus tent?

2. Use the slope to find the height of the tent at its center pole.

3. A rope is attached to the roof of the tent at the outside wall and anchors the tent to a stake in the ground. The rope follows the line of the tent roof as shown in Figure 3.10. How far away from the tent wall is the anchoring stake?

Figure 3.9: A circus tent

Figure 3.10: An anchor rope

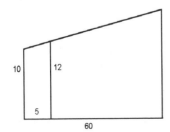

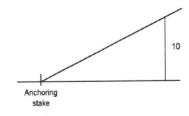

Solution to Part 1: To calculate the slope we use Equation (3.4). At the outside wall, the circus tent is 10 feet high. If we run 5 feet toward the center pole, the height increases to 12 feet; that is a rise of 2 feet. Thus

$$m = \frac{\text{rise}}{\text{run}} = \frac{2}{5} = 0.4 \text{ feet per foot.}$$

Solution to Part 2: To find the height at the center pole, we use Equation (3.1) and the value of the slope we found in Part 1. From the outside wall to the center post is a run of 60 feet, so

$$\text{rise} = m \times \text{run} = 0.4 \times 60 = 24 \text{ feet.}$$

Thus from the outside wall to the center pole, the height increases by 24 feet. That makes the height at the center pole $10 + 24 = 34$ feet.

Solution to Part 3: Since the slope of the line is 0.4, if we stand at the anchoring stake and move toward the tent, each horizontal foot we move results in a 0.4 foot rise in the rope. We need the total rise to be 10 feet:

$$
\begin{aligned}
\text{rise} &= m \times \text{run} \\
10 &= 0.4 \times \text{run} \\
\frac{10}{0.4} &= \text{run} \\
25 &= \text{run} .
\end{aligned}
$$

Thus the anchoring stake is located 25 feet outside the wall of the tent. If the vertical axis corresponds to the outside wall of the tent, then the horizontal intercept of the roof line is -25.

Exercise Set 3.1

1. **A line with given intercepts:** On coordinate axes, draw a line with vertical intercept 4 and horizontal intercept 3. Do you expect its slope to be positive or negative? Calculate the slope.

2. **A line with given intercept and slope:** On coordinate axes, draw a line with vertical intercept 3 and slope 1. What is its horizontal intercept?

3. **Lines with the same slope:** On the same coordinate axes, draw two lines, each of slope 2. The first line has vertical intercept 1 and the second has vertical intercept 3. Do the lines cross? In general what can you say about lines with the same slope?

4. **Where lines with different slopes meet:** On the same coordinate axes, draw one line with vertical intercept 2 and slope 3 and another with vertical intercept 4 and slope 1. Do these lines cross? If so, do they cross to the right or left of the vertical axis? In general, if one line has its vertical intercept below the vertical intercept of another, what conditions on the slope will insure that the lines cross to the right of the vertical axis?

5. **A ramp to a building:** The base of a ramp sits on the ground. Its slope is 0.4, and it extends to the top of the front steps of a building 15 horizontal feet away.

   (a) How high is the ramp 1 horizontal foot toward the building from the base of the ramp?

   (b) How high is the top of the steps relative to the ground?

6. **A wheelchair service ramp:** The *Americans with Disabilities Act* (or ADA) requires, among other things, that wheelchair service ramps have a slope not exceeding $\frac{1}{12}$.

   (a) Suppose the front steps of a building are 2 feet high. You want to make a ramp conforming to ADA standards that reaches from the ground to the top of the steps. How far away from the building is the base of the ramp?

   (b) Another way to give specifications on a ramp is to give allowable inches of rise per foot of run. In these terms, how many inches of rise does the ADA requirement allow in one foot of run?

7. **A cathedral ceiling:** A cathedral ceiling shown in Figure 3.11 is 8 feet high at the west wall of a room. As you go from the west wall toward the east wall, the ceiling slants upward. Three feet from the west wall, the ceiling is 10.5 feet high.

Figure 3.11: A cathedral ceiling

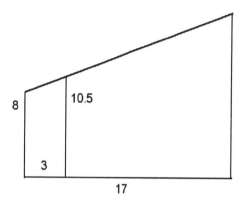

(a) What is the slope of the ceiling?

(b) The width of the room (the distance from the west wall to the east wall) is 17 feet. How high is the ceiling at the east wall?

(c) You want to install a light in the ceiling as far away from the west wall as possible. You intend to change the bulb, when required, by standing at the top of your small stepladder. If you stand on your stepladder, you can reach 12 feet high. How far from the west wall should you install the light?

8. **The Mississippi River:** For purposes of this exercise, we will think of the Mississippi River as a straight line beginning at its headwaters, Lake Itasca, Minnesota, at an elevation of 1475 feet above sea level, and sloping downward to the Gulf of Mexico 2340 miles to the south.

(a) Think of the southern direction as pointing to the right along the horizontal axis. What is the slope of the line representing the Mississippi River? Be sure to say what units you are using.

(b) Memphis, Tennessee, sits on the Mississippi River 1982 miles south of Lake Itasca. What is the elevation of the river as it passes Memphis?

(c) How many miles south of Lake Itasca would you find the elevation of the Mississippi to be 200 feet?

9. **A road up a mountain:** You are driving on a straight road which is inclined upward toward the peak of a mountain. You pass a sign that says *elevation 4130 feet*. Three horizontal miles further you pass a sign that says *elevation 4960 feet*.

    (a) Thinking of the direction you are driving as the positive direction, what is the slope of the road? (<u>Note</u>: Units are important here.)

    (b) What is the elevation of the road 5 horizontal miles from the first sign?

    (c) You know that the peak of the mountain is 10,300 feet above sea level. How far in horizontal distance is the first sign from the peak of the mountain?

10. **An underground water source:** An underground aquifer near Seiling, Oklahoma, sits on an impermeable layer of limestone. West of Seiling the limestone layer is thought to slope downward in a straight line. In order to map the limestone layer, hydrologists started at Seiling heading west and drilled *sample wells*. Two miles west of Seiling the limestone layer was found at a depth of 220 feet. Three miles west of Seiling the limestone layer was found at a depth of 270 feet.

    (a) What would you expect the depth of the limestone layer to be 5 miles west of Seiling?

    (b) You want to drill a well down to the limestone layer as far west from Seiling as you can. Your budget will allow you to drill a well that is 290 feet deep. How far west of Seiling can you go to drill the well?

    (c) Four miles west of Seiling someone drilled a well and found the limestone layer at 273 feet. Were the hydrologists right in saying that the limestone layer slopes downward in a straight line west of Seiling? Explain your reasoning.

3.2 LINEAR FUNCTIONS

Constant rates of change

We now turn to some special types of functions, *linear functions*, which are intimately related to straight lines. A linear function is one whose *rate of change* or *slope* is always the same. Let's look at an example to show what we mean. Suppose the CEO of a company wants to have a dinner catered for employees. He finds that he must pay a dining hall rental fee of $275, and an additional $28 for each meal served. Then the cost $C = C(n)$ (in dollars) is a function of the number n of meals served. If an unexpected guest arrives for dinner, then the CEO will have to pay an additional $28. This is the rate of change in C, and it is always the same no matter how many dinner guests are already seated. In more formal terms, when the variable n increases by 1, the function C changes by 28. This means that C is a linear function of n with slope or rate of change 28.

The rate of change or slope of a linear function can be used in ways that remind us of how slope is used for straight lines. Suppose, for example, that the caterer, having anticipated that 35 people would attend the dinner, sent the CEO a bill for $1255. But 48 people actually attended the dinner. What should the total price of the dinner be under these circumstances? Since 13 additional people attended the dinner, and the price per meal is $28, we can calculate the additional charge over $1255:

$$\text{Additional charge} = 28 \times \text{Additional people} = 28 \times 13 = 364 \text{ dollars.}$$

Thus the total cost for 48 people is $1255 + 364 = 1619$ dollars.

In general the slope or rate of change of a linear function is the amount the function changes when the variable increases by 1 unit. Furthermore, we can use the slope m just as we did above to calculate the change in function value corresponding to a given change in the variable. We state the relationship in two equivalent ways:

$$\text{Change in function} \quad = \quad m \times \text{Change in variable}$$

$$\text{Additional function value} \quad = \quad m \times \text{Additional variable value.}$$

We can use division to restate these so that they show how to calculate m from changes in function value corresponding to changes in the variable. Once again, the two equations below say exactly the same thing using different words:

$$m \quad = \quad \frac{\text{Change in function}}{\text{Change in variable}}$$

$$m \quad = \quad \frac{\text{Additional function value}}{\text{Additional variable value}}.$$

KEY IDEA 3.2: LINEAR FUNCTIONS

A function is linear if it has a constant slope or rate of change. This slope is then the amount the function changes when the variable increases by 1 unit.

Suppose $y = y(x)$ is a linear function of x. Then:

- The slope m, or rate of change in y with respect to x, can be calculated from the change in y corresponding to a given change in x:

$$m = \frac{\text{Change in function}}{\text{Change in variable}} = \frac{\text{Change in } y}{\text{Change in } x} \ .$$

- The slope m can be used to calculate the change in y resulting from a given change in x:

$$\text{Change in function} = m \times \text{Change in variable} \ .$$

Or, using letters:

$$\text{Change in } y = m \times \text{Change in } x \ .$$

EXAMPLE 3.3 Oklahoma Income Tax

The amount of income tax $T = T(I)$ (in dollars) owed to the State of Oklahoma is a linear function of the adjusted gross income I (in dollars), at least over a suitably restricted range of incomes. According to the 1994 Oklahoma Income Tax tables, a single Oklahoma resident taxpayer with an adjusted gross income of \$15,000 owes \$780 in Oklahoma income tax. In functional notation this is $T(15,000) = 780$. If the adjusted gross income is \$15,500, then the tables show a tax liability of \$825.

1. Calculate the rate of change in T with respect to I and explain in practical terms what it means.

2. How much does the taxpayer owe if the adjusted gross income is \$15,350?

Solution to Part 1: Since we are thinking of the tax T as a function of the variable I representing income, we calculate the slope as follows:

$$m = \frac{\text{Change in } T}{\text{Change in } I} = \frac{\text{Change in tax}}{\text{Change in income}} = \frac{825 - 780}{15,500 - 15,000} = \frac{45}{500} = 0.09.$$

This means that for each additional dollar earned, the taxpayer can expect to pay 9 cents in Oklahoma income tax. In economics this is known as the *marginal tax rate*, because it is the rate at which new money that you earn is taxed. It is worth noting that the marginal tax rate normally does not appear directly in tax tables but must be calculated as we did here. It can provide crucial information for financial planning.

Solution to Part 2: An income of $15,350 is an additional $350 income over the $15,000 level. Now we can use the marginal tax rate m calculated in Part 1 to get the additional tax:

$$\text{Additional tax} = m \times \text{Additional income} = 0.09 \times 350 = 31.50.$$

Thus we would owe the tax on $15,000, that is, $780, plus an additional tax of $31.50, for a total tax liability of $811.50.

Linear functions and straight lines

If we look more carefully at the catered dinner, we can write a formula for the total cost $C = C(n)$ when there are n dinner guests:

$$
\begin{aligned}
\text{Cost} &= \text{Cost of food} + \text{Rent} \\
\text{Cost} &= \text{Price per meal} \times \text{Number of guests} + \text{Rent} \\
C &= 28n + 275.
\end{aligned}
$$

If we graph this function as we did in Figure 3.12 we see the fundamental relationship between linear functions and straight lines: The graph of a linear function is a straight line. (We used a horizontal span of -1 to 3 and a vertical span of 250 to 400.)

But there is still more to find out. In Figure 3.12 we placed the cursor at X=0, and we see that the vertical intercept of the graph is 275. That is the rental fee for the dining hall, or more to the point, it is the value of the linear function C when the variable n is zero. This value $C(0)$ is often referred to as the *initial value* of the function. It is true in general that the vertical intercept of the graph corresponds to the initial value of the linear function and is denoted b.

In Figure 3.13 we have put the cursor at X=1, and we see that a horizontal change of one unit causes the graph to rise from 275 to 303. That means the slope of the straight line graph is $303 - 275 = 28$, and that is the same as the rate of change of the linear function. Once again, this is true in general: The slope of a linear function is the same as the slope of its graph.

Finally we note that the horizontal intercept of the graph (not shown in Figure 3.12) is the solution of the equation $C(n) = 0$. The form of C and the relationships we have observed here between linear functions and straight-line graphs are typical.

***Figure* 3.12:** *The graph of the linear function* $C = 28n + 275$ *is a straight line*

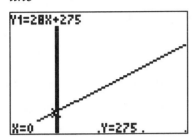

***Figure* 3.13:** *The slope of the graph is the same as the rate of change of the function*

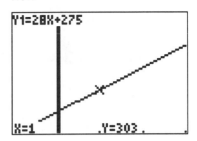

KEY IDEA 3.3: THE RELATIONSHIPS BETWEEN LINEAR FUNCTIONS AND STRAIGHT LINES

1. The formula for a linear function is

$$y = \text{slope} \times x + \text{initial value}$$
$$y = mx + b.$$

2. The graph of a linear function is a straight line.

3. The slope of a linear function is the same as the slope of its graph.

4. The vertical intercept of the graph corresponds to the initial value b of the linear function, that is, the value of the function when the variable is 0.

5. The horizontal intercept of the graph corresponds to the value of the variable when the linear function is zero. This is the solution of the equation

$$\text{Linear function} = 0.$$

EXAMPLE 3.4   *Selling Jewelry at an Art Fair*

Suppose you pay \$192 to rent a booth for selling necklaces at an art fair. The necklaces sell for \$32 each.

1. Explain why the function that shows your net income (revenue from sales minus rental fee) as a function of the number of necklaces sold is a linear function.

2. Write a formula for this function.

3. Use functional notation to show your net income if you sell 25 necklaces, and then calculate the value.

4. Make the graph of the net income function.

5. Identify the vertical intercept and slope of the graph and explain in practical terms what they mean.

6. Find the horizontal intercept and explain its meaning in practical terms.

Solution to Part 1: We choose variable and function names: Let n be the number of necklaces sold and $P = P(n)$ the net income (in dollars). Each time n increases by 1, that is, when one additional necklace is sold, the value of P, the net income, increases by the same amount, $32. Thus the rate of change for P is always the same, and hence it is a linear function.

Solution to Part 2: We pay $192 to rent the booth and take in $32 for each necklace sold:

$$\text{Net income} = \text{Income from sales} - \text{Rent}$$
$$\text{Net income} = \text{Price} \times \text{Number sold} - \text{Rent}$$
$$P = 32n - 192 \text{ dollars.}$$

Solution to Part 3: If 25 necklaces are sold, then in functional notation the net income is $P(25)$. To calculate that, we put 25 into the formula in place of n:

$$P(25) = 32 \times 25 - 192 = 608 \text{ dollars .}$$

Solution to Part 4: First we enter $\boxed{3.1}$ the function and record appropriate correspondences:

$$Y_1 = P, \text{ net income on vertical axis}$$
$$X = n, \text{ necklaces sold on horizontal axis.}$$

To choose a window size, we made the table in Figure 3.14 using a starting value of 0 and an increment of 5. This led us to choose a horizontal span of $n = 0$ to $n = 30$ and a vertical span of $P = -200$ to $P = 800$. This makes the graph in Figure 3.15, and we note that, as expected, the graph of our linear function P is a straight line.

Solution to Part 5: We know in general that the vertical intercept of the graph of $y = mx + b$ is b and the slope is m. So for $P = 32n - 192$, the vertical intercept is -192, and the slope is 32. The vertical intercept -192 is the net income if no necklaces are sold. That is, you lost $192 because you had to pay rent for the booth but sold no necklaces. The slope 32 is the price of each necklace. We should note that economists refer to the slope 32 here as the *marginal income*. It shows the additional income taken when one additional item is sold.

Figure 3.14: A table for setting the window

X	Y₁	
0	-192	
5	-32	
10	128	
15	288	
20	448	
25	608	
30	768	

X=0

Figure 3.15: Net income for selling necklaces

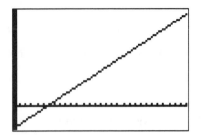

Solution to Part 6: The horizontal intercept occurs where the graph crosses the horizontal axis. That is where $P(n) = 0$. We will show how to find that by direct calculation:

$$
\begin{aligned}
P(n) &= 0 \\
32n - 192 &= 0 \\
32n &= 192 \ \ \text{(Add 192 to both sides.)} \\
n &= \frac{192}{32} = 6 \ \ \text{(Divide by 32.)}.
\end{aligned}
$$

Thus the horizontal intercept occurs at $n = 6$ necklaces. That is where $P(n) = 0$, and in this case it is the "break-even" point. You need to sell 6 necklaces to avoid losing money on this venture.

We found the horizontal intercept here by hand calculation, but we should note that it could also be found with the calculator by looking at the graph or a table.

Linear equations from data

There are a number of ways in which information about a linear function may be given, but the following three situations are quite common. We will illustrate each case using variations on Example 3.4.

Getting a linear equation if you know the slope and initial value: In Example 3.4, we were effectively told the slope, 32, of the linear function P and its initial value, -192. As we have already noted, this allows us immediately to write a formula:

$$
\begin{aligned}
P &= \text{slope} \times n + \text{initial value} \\
P &= 32n - 192 \,.
\end{aligned}
$$

This works in general. If we know the slope m of a linear function and its initial value b, we can immediately write down the formula as $y = mx + b$.

Getting a linear equation if you know the slope and one data point: Suppose that we were given the information for Example 3.4 in a different way: We are told that the price for each necklace is $32 and that if we sell $n = 8$ necklaces we will have a net income of $P = 64$ dollars. Now we know the slope, 32, but we don't know the initial value. We can get it by solving an equation. The information tells us that the formula for P is

$$P = 32n + \text{initial value.} \tag{3.7}$$

We also know that when $n = 8$ then $P = 64$. We put these values into Equation (3.7) and proceed to solve for the initial value:

$$64 = 32 \times 8 + \text{initial value}$$
$$64 = 256 + \text{initial value}$$
$$-192 = \text{initial value}.$$

Now we have the slope and initial value, and we can write the formula for P as $P = 32n - 192$. This works in general. If you know the slope and one data point for a linear function, you can solve an equation as we did above to find the initial value and then write down the formula.

Getting a linear equation from two data points: Suppose the information in Example 3.4 were given as two data points. For example, suppose that we make a net income $P = 64$ dollars when we sell $n = 8$ necklaces and a net income of $P = 160$ dollars if $n = 11$ necklaces are sold. In this case, we know neither the slope nor the initial value. Since the slope is so important, we get it first using a familiar formula:

$$\text{slope} = \frac{\text{Change in function}}{\text{Change in variable}} = \frac{160 - 64}{11 - 8} = 32 \text{ dollars per necklace.}$$

Now we know the slope, 32, and the data point $P = 64$ when $n = 8$. With this information we can proceed exactly as we did above to get the initial value and then the formula for P.

We also know that $P = 160$ when $n = 11$. What would happen if we used this data point with the slope rather than $n = 8$ and $P = 64$? Would the equation we find be different? Let's see:

$$P = \text{slope} \times n + \text{initial value}$$
$$P = 32n + \text{initial value}$$
$$160 = 32 \times 11 + \text{initial value}$$
$$160 = 352 + \text{initial value}$$
$$-192 = \text{initial value.}$$

We got the same initial value as we did before, and hence we will get the same formula for P, namely $P = 32n - 192$.

In general if you are given two data points and need to find the linear function that they determine, you proceed in two steps. First use the formula

$$\text{slope} = \frac{\text{Change in function}}{\text{Change in variable}}$$

to compute the slope. Next you can use the slope you found and put *either* of the given data points into the equation

$$y = \text{slope} \times x + \text{initial value}$$

to solve for the initial value.

KEY IDEA 3.4: HOW TO GET THE EQUATION OF A LINEAR FUNCTION

1. **If you know the slope and the initial value**, use

$$y = \text{slope} \times x + \text{initial value}.$$

2. **If you know the slope and one data point**, put the given slope and data point into the equation
$$y = \text{slope} \times x + \text{initial value}$$
and solve it for the initial value. You can now get the formula as in Part 1 above.

3. **If you know two data points**, first use the formula

$$\text{slope} = \frac{\text{Change in function}}{\text{Change in variable}}$$

to get the slope. Now use either of the data points and proceed as in Part 2 above.

EXAMPLE 3.5 *Changing Celsius to Fahrenheit*

The temperature $F = F(C)$ in Fahrenheit is a linear function of the temperature C in Celsius. A lab assistant placed a Fahrenheit thermometer beside a Celsius thermometer and observed the following. When the Celsius thermometer reads 30 degrees ($C = 30$), the Fahrenheit thermometer reads 86 degrees ($F = 86$). When the Celsius thermometer reads 40 degrees, the Fahrenheit thermometer reads 104 degrees.

1. Use a formula to express F as a linear function of C.

2. At sea level, water boils at 212 degrees Fahrenheit. What temperature in degrees Celsius makes water boil?

3. Explain in practical terms what the slope means in this setting.

Solution to Part 1: Since we know two points, the first step is to calculate the slope. When C changes from 30 to 40, F changes from 86 to 104. Since we are thinking of F as the function and C as the variable, we use

$$\text{slope} = \frac{\text{Change in function}}{\text{Change in variable}} = \frac{\text{Change in } F}{\text{Change in } C} = \frac{104 - 86}{40 - 30} = 1.8$$

Now we use the slope $m = 1.8$ and the data point, $F = 86$ when $C = 30$, to make an equation which we solve for the initial value:

$$
\begin{aligned}
F &= \text{slope} \times C + \text{initial value} \\
86 &= 1.8 \times 30 + \text{initial value} \\
86 &= 54 + \text{initial value} \\
32 &= \text{initial value} .
\end{aligned}
$$

Now we know the slope, 1.8, and the initial value, 32, so $F = 1.8C + 32$. It is worth noting the significance of the initial value we found. The temperature 32 degrees Fahrenheit is the freezing point for water, and this corresponds to a temperature of 0 degrees Celsius. That is, on the Celsius scale, a temperature of 0 is the freezing point for water.

Solution to Part 2: We want to know the value of C when $F = 212$. Thus we put in 212 for F in the equation we found in Part 1 and then solve for C:

$$
\begin{aligned}
F &= 1.8C + 32 \\
212 &= 1.8C + 32 \\
180 &= 1.8C \\
\frac{180}{1.8} &= C \\
100 &= C .
\end{aligned}
$$

Thus, water boils at 100 degrees Celsius.

Solution to Part 3: As we found in Part 1, the slope is 1.8. Remember that the slope tells how much F changes when C changes by 1. Thus, a change of 1 degree Celsius results in a change of 1.8 degrees Fahrenheit.

Exercise Set 3.2

1. **Getting Celsius from Fahrenheit:** Water freezes at 0 degrees Celsius, which is the same as 32 degrees Fahrenheit. Also water boils at 100 degrees Celsius, which is the same as 212 degrees Fahrenheit.

    (a) Use the freezing and boiling points of water to find a formula expressing Celsius temperature C as a linear function of the Fahrenheit temperature F.

    (b) What is the slope of the function you found in Part (a)? Explain its meaning in practical terms.

    (c) In Example 3.5 we showed that $F = 1.8C + 32$. Solve this equation for C and compare the answer with that obtained in Part (a).

2. **A trip to a science fair:** An elementary school is taking a busload of children to a science fair. It costs $130.00 to drive the bus to the fair and back, and the school pays each student's $2.00 admission fee.

    (a) Use a formula to express the total cost C (in dollars) of the science fair trip as a linear function of the number n of children who make the trip.

    (b) Identify the slope and initial value of C and explain in practical terms what they mean.

    (c) Solve the equation $C(n) = 146$ for n. Explain what the answer you get represents.

3. **Digitized pictures on a disk drive:** The hard disk drive on a computer holds 2 gigabytes of information. That is 2000 megabytes. The formatting information, operating system, and applications software take up 781 megabytes of disk space. The operator wants to store on his computer a collection of digitized pictures, each of which requires 2.3 megabytes of storage space.

    (a) Explain why the total amount of storage space used on the disk drive is a *linear* function of the number of pictures that are stored on the drive.

    (b) Find a formula to express the total amount of storage space used on the disk drive as a linear function of the number of pictures that are stored on the drive. (Be sure to identify what the letters you use mean.) Explain in practical terms what the slope of this function is.

(c) After putting a number of pictures on the disk drive, the operator executes a *directory* command, and at the end of the list the computer displays the message *598,000,000 bytes free*. This message means that there are 598 megabytes of storage space left on the computer. How many pictures are stored on the disk drive? How many additional pictures can be added before the drive is filled?

4. **Real estate sales:** A real estate agency has fixed monthly costs associated with rent, staff salaries, utilities, and supplies. It earns its money by taking a percentage commission on total real estate sales. During the month of July the agency had total sales of $832,000 and showed a net income (after paying fixed costs) of $15,704. In August total sales were $326,000 with a net income of only $523.

   (a) Use a formula to express net income as a linear function of total sales. Be sure to identify what the letters you use mean.

   (b) Plot the graph of net income and identify the slope and vertical intercept.

   (c) What are the real estate agency's fixed monthly costs?

   (d) What percentage commission does the agency take on the sale of a home?

   (e) Find the horizontal intercept and explain what this number means to the real estate agency.

5. **Currency conversion:** The number P of British pounds you can get from a bank is a linear function of the number D of American dollars you pay. An American tourist arriving at Heathrow airport in England went to a banking window at the airport and gave the teller 83 American dollars. She received 31 British pounds in exchange. In this exercise assume there is no service charge for exchanging currency.

   (a) What is the rate of change, or slope, of P with respect to D? Explain in practical terms what this number means. (Note: You need two values to calculate a slope, but you were given only one. If you think about it, you know one other value. How many British pounds can you get for zero American dollars?)

   (b) A few days later the American tourist went to a bank in Plymouth and exchanged 130 American dollars for British pounds. How many pounds did she receive?

   (c) Upon returning to the airport, she found that she still had £12.32 in British currency in her purse. In preparation for the trip home, she exchanged that for American dollars. How much money in American dollars did she get?

6. **Growth in height:** Between the ages of 7 and 11 years, a certain boy grows taller by 2 inches each year. At age 9 he is 48 inches tall.

(a) Explain why, during this period, the height of the boy is a *linear* function of his age, and identify the slope of this function.

(b) Use a formula to express the height of the boy as a linear function of his age during this period. Be sure to identify what the letters you use mean.

(c) What is the initial value of the function you found in Part (b)?

(d) Studying a graph of the boy's height as a function of his age from birth to age 7 reveals that the graph is increasing and concave down. Does that indicate that his actual height (or length) at birth was larger or smaller than your answer to Part (c)? Be sure to explain your reasoning.

7. **Adult male height and weight:** Here is a rule of thumb relating weight to height among adult males: If a man is one inch taller than another, then we expect him to be heavier by 3.5 pounds.

(a) Explain why, according to this rule of thumb, among typical adult males the weight is a *linear* function of the height. Identify the slope of this function.

(b) A related rule of thumb is that a typical man who is 70 inches tall weighs 160 pounds. Based on these two rules of thumb, use a formula to express the trend giving weight as a linear function of height. (Be sure to identify what the letters you use mean.)

(c) If a man weighs 152 pounds, how tall would you expect him to be?

(d) An atypical man is 75 inches tall and weighs 170 pounds. Compared to the trend formula you found in Part (b), is he heavy or light for his height?

8. **Budget constraints:** Your family likes to eat fruit, but due to a budget constraint you spend only $5 each week on fruit. Your two choices are apples and grapes. Apples cost $0.50 per pound, and grapes cost $1 per pound. Let a denote the number pounds of apples you buy and g the number of pounds of grapes. Because of your budget constraint it is possible to express g as a linear function of the variable a. To find the linear formula we need to find its slope and initial value.

(a) If you buy one more pound of apples, how much less money do you have to spend on grapes? Then how many fewer pounds of grapes can you buy?

(b) Use your answer to Part (a) to find the slope of g as a linear function of a. (Hint: Recall that the slope is the change in the function which results from increasing the variable by 1. Should the slope of g be positive or negative?)

(c) To find the initial value of g, determine how many pounds of grapes you can buy if you buy no apples.

(d) Use your answers to Parts (b) and (c) to find a formula for g as a linear function of a.

9. **More on budget constraints:** *This is a continuation of Exercise 8.* Another way to find a linear formula giving the number of pounds g of grapes you can buy in terms of a, the number pounds of apples, is to write the budget constraint as an equation and solve it for g.

   (a) If you buy 5 pounds of apples, then of course you spend $0.50 \times 5 = 2.50$ dollars on apples. In general, if you buy a pounds of apples, how much money do you spend on apples? Your answer should be a formula involving a.

   (b) Write a formula involving g to express how much money you spend on grapes if you buy g pounds of grapes.

   (c) Use your answers to Parts (a) and (b) to write an equation expressing the budget constraint of \$5 in terms of a and g.

   (d) Solve the equation you found in Part (c) for g, and so find a formula expressing g as a linear function of a. Compare this with your answer to Part (d) of Exercise 8.

10. **Sleeping longer:** A certain man observed that each night he was sleeping 15 minutes longer than he had the night before, and he used this observation to predict the day of his death.[1] If he made his observation right after sleeping 8 hours, how long would it be until he slept 24 hours (and so would never again wake)?

---

[1] This is a legend about the mathematician De Moivre, who died in 1754.

3.3 MODELING DATA WITH LINEAR FUNCTIONS

Information about physical or social phenomena is frequently obtained by gathering data or sampling. For example, collecting data is the basic tool census takers use to get information about the population of the United States. Once data is gathered, an important key to further analysis is to produce a *mathematical model* describing the data. A model is a function that represents the data either exactly or approximately and that incorporates patterns in the data. In many cases, such a model takes the form of a linear function.

Testing data for linearity

Let's look at a hypothetical example to explain what we mean. One of the most important events in the development of modern physics was Galileo's description of how objects fall. In about 1590 he conducted experiments where he dropped objects and attempted to measure their downward velocities $V = V(t)$ as they fell. Here we measure velocity V in feet per second, and time t as the number of seconds after release. If Galileo had been able to nullify air resistance, and if he had been able to measure velocity without any experimental error at all, he might have recorded the following table of values.

t = seconds	0	1	2	3	4	5
V = feet per second	0	32	64	96	128	160

What conclusions can be drawn from this table of data? In this case, the key is to look at how velocity changes as the rock falls. From $t = 0$ to $t = 1$, the velocity changes from 0 feet per second to 32 feet per second. That is a change of $32 - 0 = 32$ feet per second. From $t = 1$ to $t = 2$, the velocity changes from 32 feet per second to 64 feet per second, a change in velocity of $64 - 32 = 32$ feet per second. Thus the change in velocity is the same, 32 feet per second, during the first and second seconds of the fall. If we continue this, we obtain the following table.

Change in t	From 0 to 1	From 1 to 2	From 2 to 3	From 3 to 4	From 4 to 5
Change in V	32	32	32	32	32

We see that the rate of change in velocity, the *acceleration*, is always the same, 32 feet per second per second. We know this is characteristic of linear functions; their rate of change is always the same. Thus it is reasonable to describe this data with a linear formula. That is, we want a formula of the form $V = mt + b$, where m is the slope and b is the initial value. Since V changes by 32 when t changes by 1, the slope of the linear function is $m = 32$ feet per second per second. We also know that $b = 0$ since the initial velocity is zero. We conclude that the velocity of the rock t seconds after it is dropped is given by

$$V = 32t + 0 = 32t \, .$$

Linear models

This linear function serves as a *mathematical model* for the experimentally gathered data, and it gives us more information than is apparent from the data table alone. For example, we can use the linear formula to calculate the velocity of the rock 10.8 seconds after it is released, information that is not given by the table:

$$V(10.8) = 10.8 \times 32 = 345.6 \text{ feet per second.}$$

The key physical observation that Galileo made was that falling objects have constant acceleration, 32 feet per second per second. He did additional experiments to show that if air resistance is ignored, then this acceleration does not depend on the weight or size of the object. The same table of values would result, and the same acceleration would be calculated, whether the experiment were done with a pebble or with a cannonball. According to tradition, Galileo conducted some of his experiments in public, dropping objects from the top of the leaning tower of Pisa. His observations got him into serious trouble with the authorities because they conflicted with the accepted premise of Aristotle that heavier objects would fall faster than lighter ones.

The feature of this data table that allowed us to construct a linear model was that the change in the function, velocity, was always the same. If a data table (with evenly spaced values for the variable) does not exhibit this behavior, then the data cannot be modeled with a linear function, and some more complicated model may be sought. For example, if instead of measuring velocity, we had measured the distance $D = D(t)$ the rock fell, we would have gotten the following data.

$t = $ seconds	0	1	2	3	4	5
$D = $ feet	0	16	64	144	256	400

The corresponding table of differences is

Change in t	From 0 to 1	From 1 to 2	From 2 to 3	From 3 to 4	From 4 to 5
Change in D	16	48	80	112	144

We see that the change in D is not constant, and so D is not a linear function. A deeper analysis can show that $D(t) = 16t^2$.

KEY IDEA 3.5: WHEN DATA IS LINEAR

A data table for $y = y(x)$ with evenly spaced values for x can be modeled with a linear function if it shows a constant change in y.

If the change in y is not constant, then the data cannot be modeled exactly by a linear function.

EXAMPLE 3.6 Sampling Voter Registration

In this hypothetical experiment, a political analyst compiled data on the number of registered voters in Payne County, Oklahoma, each year from 1990 through 1995. For each of the following possible data tables which the analyst might have gotten, determine if the data can be modeled with a linear function. If so, find such a formula and predict the number of registered voters in Payne County in the presidential election year 2000.

1. Hypothetical Data Table 1:

Date	1990	1991	1992	1993	1994	1995
Registered voters	28,321	28,542	29,466	30,381	30,397	31,144

2. Hypothetical Data Table 2:

Date	1990	1991	1992	1993	1994	1995
Registered voters	28,321	28,783	29,245	29,707	30,169	30,631

Solution for Data Table 1: First we choose variable and function names: Let d be the number of years since 1990 and N the number of registered voters. The table for N as a function of d is then

d	0	1	2	3	4	5
N	28,321	28,542	29,466	30,381	30,397	31,144

To determine if the data can be modeled by a linear function, we need to look at the change in the number of registered voters from year to year.

Change in d	0 to 1	1 to 2	2 to 3	3 to 4	4 to 5
Change in N	221	924	915	16	747

This table shows clearly that the change in registered voters is not constant from year to year, and we conclude that it is not appropriate to model this data with a linear function.

Solution for Data Table 2: Again, we let d be the number of years since 1990 and N the number of registered voters. The table for N as a function of d is then

d	0	1	2	3	4	5
N	28,321	28,783	29,245	29,707	30,169	30,631

We construct a table of differences using Data Table 2.

Change in d	0 to 1	1 to 2	2 to 3	3 to 4	4 to 5
Change in N	462	462	462	462	462

Here we see that the number of registered voters increases by 462 each year, and so we know that we can model the data with a linear function. To determine the formula, we need to know the slope and the initial value. Since 462 is the change in N corresponding to a change in d of 1, the slope of our linear function is 462 registered voters per year. The initial value is given by the first entry in the table for N: $N(0) = 28,321$. Thus the formula is $N = 462d + 28,321$. To predict the number of registered voters in the year 2000, we use this formula, putting in 10 for d:

$$N(10) = 462 \times 10 + 28,321 = 32,941 .$$

We can get this same prediction using only the slope rather than the formula. The change in d from 5 (the year 1995) to 10 (the year 2000) is 5 years:

$$\begin{aligned} \text{Change in } N &= m \times \text{Change in } d \\ \text{Change in } N &= 462 \times 5 \\ \text{Change in } N &= 2310 . \end{aligned}$$

Thus from $d = 5$ (the year 1995) to $d = 10$ (the year 2000), we expect an increase of 2310 registered voters. That gives $30,631 + 2310 = 32,941$ registered voters in the year 2000, the same answer we got using the formula.

We note that the choice of the variable d as the number of years since 1990 greatly simplifies finding the linear formula, since the initial value can then be found as the first entry in the table. We could choose the variable to be the actual date. This would not affect the slope, but we would have to calculate the initial value using the slope and one of the data points in the original table. The resulting initial value would be different from the one we found above, since it would represent the number of registered voters present in the year 0 A.D.! That is of course a nonsensical figure, and for this reason, as well as for the purpose of simplifying the calculation, it is better to choose the variable as we did in the example. This is an important point to remember for subsequent examples and exercises in which the variable is time.

We also note that Data Table 2 in Example 3.6 is not very realistic. It almost never happens that statistics of this sort turn out to be exactly linear. But many times such data can be closely approximated by a linear function. We will see how to deal with that in the next section.

Graphing discrete data

Calculating differences can always tell you if data is linear, but many times it is advantageous to view such data graphically. Most graphing calculators will allow you to enter data tables and to display them graphically. In Figure 3.16 we have entered ⬚3.2 the data for Data Table 1 from Example 3.6. It is graphed ⬚3.3 in Figure 3.17. This picture clearly verifies our earlier contention that this is not linear data; the points do not fall on a straight line.

Figure 3.16: *Voter registration data entered in a calculator table*

Figure 3.17: *A view of nonlinear data*

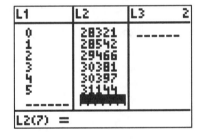

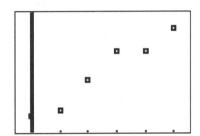

Figure 3.17 contrasts nicely with the plot of linear data from Data Table 2 shown in Figure 3.18. We see that the points do indeed fall on a straight line, clearly showing the linear nature of the data. Finally, we can add ⬚3.4 the graph of the linear function $N = 462d + 28{,}321$ that we found in Part 2 of Example 3.6 to the screen as shown in Figure 3.19.

Figure 3.18: *Linear data points from Data Table 2*

Figure 3.19: *Adding the linear model*

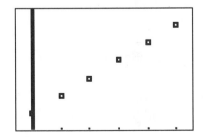

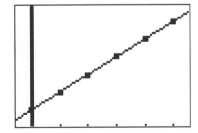

Figure 3.18 and Figure 3.19 illustrate clearly what it means to model linear data. The data is displayed in Figure 3.18. The model is the line in Figure 3.19 which passes through the data points but also fills in the gaps and extends beyond the data points.

EXAMPLE 3.7 *Newton's Second Law of Motion*

Newton's second law of motion shows how force on an object, measured in *newtons*,[2] relates to acceleration of the object, measured in meters per second per second. The following experiment might be conducted in order to discover Newton's second law. Objects of various masses, measured in kilograms, were given an acceleration of 5 meters per second, and the associated forces were measured and recorded in the table below.

Mass	1	1.3	1.6	1.9	2.2
Force	5	6.5	8	9.5	11

1. Check differences to show that this is linear data.

2. Construct a linear model for the data.

3. What force does your model show for an object of mass 1.43 kilograms which is accelerating at 5 meters per second?

4. Make a graph showing the data and overlay it with the graph of the linear model you made in Part 2.

5. Newton's second law of motion says that

$$\text{Force} = \text{Mass} \times \text{Acceleration} .$$

Do the results of our hypothetical experiment provide support for the validity of the second law? What additional experiments might be appropriate to provide further verification?

Solution to Part 1: First we choose variable and function names: Let m be the mass and F the force. Now we make a table of differences.

Change in m	1 to 1.3	1.3 to 1.6	1.6 to 1.9	1.9 to 2.2
Change in F	1.5	1.5	1.5	1.5

Since the change in F is always the same, 1.5 newtons, we conclude that the data is linear.

Solution to Part 2: Part 1 shows that it is appropriate to model the data with a linear function. The first step is to get the slope, but we must be careful. For Data Table 2 of Example 3.6 the common difference in N, the number of registered voters, and the slope were the same. This will occur exactly when the data for the variable is given in steps of 1 unit,

---

[2]One newton is about a quarter of a pound.

as was the case for the variable d in Example 3.6. Here the change in variable is not 1; rather, it is 0.3. Now we get the slope with a familiar formula:

$$\text{slope} = \frac{\text{Change in function}}{\text{Change in variable}} = \frac{\text{Change in } F}{\text{Change in } m} = \frac{1.5}{0.3} = 5 \text{ newtons per kilogram.}$$

(For future reference, it is worth noting that the slope 5 turned out to be exactly the same as the acceleration. This is not an accident!) The next step is to get the initial value b. We cannot find this directly from the table. Instead, we use the slope together with any of the data points we wish. We will make the calculation with the first data point, $F(1) = 5$:

$$
\begin{aligned}
F &= 5m + b \\
5 &= 5 \times 1 + b \\
0 &= b.
\end{aligned}
$$

Thus the initial value b is 0, and we arrive at the linear model $F = 5m$.

In this case we could also have found the initial value by reasoning as follows: If the object has a mass of zero kilograms, the force will surely be 0, so the initial value is 0. Although this method gives the correct answer here, in general we should be cautious about going outside of the table to get data points, since the linear model may be appropriate only over the range of values in the table. See Exercise 9.

Solution to Part 3: To get the force for a mass of 1.43 kilograms, we use this in place of the mass m in our linear model:

$$F = 5m = 5 \times 1.43 = 7.15 \text{ newtons.}$$

Solution to Part 4: The first step is to enter $\boxed{3.5}$ the data as shown in Figure 3.20. Once the data is entered, we can plot $\boxed{3.6}$ it to get Figure 3.21. As expected, the points in Figure

---

Figure 3.20: *Data for force versus mass entered*

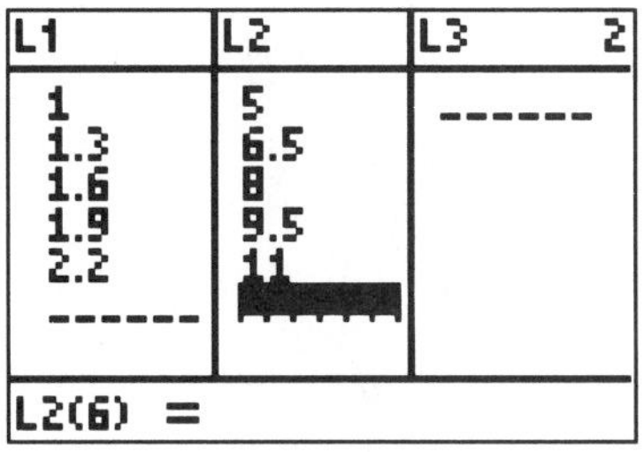

Figure 3.21: *A plot of data for force versus mass*

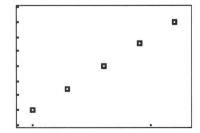

3.21 fall on a straight line, giving striking visual verification to our conclusion in Part 1

that it is appropriate to make a linear model for the data. To overlay the data with the model we made, we enter the function $\boxed{3.7}$ as shown in Figure 3.22 and then make the graph $\boxed{3.8}$. We see in Figure 3.23 that our model overlays the data exactly. This

---

Figure 3.22: Entering the linear model

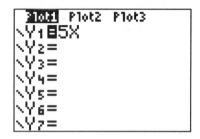

Figure 3.23: The linear model overlaying the data

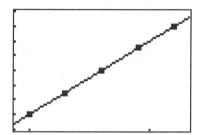

---

provides a good way for us to check our work. If the line had missed some of the points, we would know that we had made an error somewhere.

Solution to Part 5: Our model verifies Newton's second law for objects with an acceleration of 5 meters per second per second. There are many other experiments which might be appropriate to give further evidence. One important experiment would be to take an object of fixed mass and measure what happens to the force when acceleration is varied. See Exercise 6 below.

Exercise Set 3.3

Note: If no variable and function names are given in an exercise, you are of course expected to choose them and give the appropriate units. In your choice, when the variable is time, you should keep in mind the discussion following Example 3.6. Also, some of the data tables in this exercise set have been altered to allow for exact rather than approximate linear modeling.

1. **Auto parts production:** The following table shows the value, in billions of dollars, of auto parts produced in the United States on the given date.

Date	1988	1989	1990	1991	1992
Value	4.9	7.1	9.3	11.5	13.7

   (a) By calculating differences, show that this data can be modeled using a linear function.

   (b) Plot the data points.

   (c) Find a linear formula that models this data.

   (d) Add the graph of the function you found in Part (c) to your graph of data points.

   (e) Use the formula you found in Part (c) to predict the value of auto parts made in the U.S. in 1996.

2. **Tuition at American private universities:** The following table shows the average yearly tuition and required fees (in dollars) charged by American private universities in the school year beginning in the given year.

Date	1990	1991	1992	1993	1994
Average tuition	$11,400	$12,221	$13,042	$13,863	$14,684

   (a) Show that this data can be modeled by a linear function, and find its formula.

   (b) Plot the data points and add the graph of the linear formula you found in Part (a).

   (c) What prediction does this formula give for average tuition and fees at American private universities for the academic year beginning in 1997?

3. **Tuition at American public universities:** *This is a continuation of Exercise 2.* The following table shows the average yearly in-state tuition and required fees (in dollars) charged by American public universities in the school year beginning in the given year.

Date	1990	1991	1992	1993	1994
Average tuition	$2159	$2375	$2591	$2807	$3023

(a) Show that this data can be modeled by a linear function, and find its formula.

(b) What is the slope for the linear function modeling tuition and required fees for public universities?

(c) What is the slope of the linear function modeling tuition and required fees for private universities? (<u>Note</u>: See Exercise 2 above.)

(d) Explain what the information in Parts (b) and (c) tells you about the rate of increase in tuition in public versus private institutions.

(e) Which shows the larger percentage increase from 1993 to 1994?

4. **Dropping rocks on Mars:** The behavior of objects falling near the Earth's surface depends on the mass of the Earth. On Mars, a much smaller planet than the Earth, things are different. If Galileo had performed his experiment on Mars, he would have obtained the following table of data.

t = seconds	0	1	2	3	4	5
V = feet per second	0	12.16	24.32	36.48	48.64	60.8

(a) Show that this data can be modeled by a linear function and find a formula for the function.

(b) What would be the velocity of a rock dropped near the surface of Mars 10 seconds after release?

(c) Galileo found that the acceleration due to gravity of an object falling near the Earth's surface was 32 feet per second per second. Physicists normally denote this number by the letter g. If Galileo had lived on Mars, what value would he have found for g?

5. **The Kelvin temperature scale:** Physicists and chemists often use the *Kelvin* temperature scale. In order to determine the relation between the Fahrenheit and Kelvin temperature scale, a lab assistant put Fahrenheit and Kelvin thermometers side by side and
took readings at various temperatures. The following data was recorded.

K = kelvins	200	220	240	260	280	300
F = degrees Fahrenheit	-99.67	-63.67	-27.67	8.33	44.33	80.33

(a) Show that the temperature F in degrees Fahrenheit is a linear function of the temperature K in kelvins.

(b) What is the slope of this linear function? (<u>Note</u>: Be sure to take into account that
the table lists kelvins in jumps of 20 rather than in jumps of 1.)

(c) Find a formula for the linear function.

(d) Normal body temperature is 98.6 degrees Fahrenheit. What is that temperature in
kelvins?

(e) If temperature increases by one kelvin, by how many degrees Fahrenheit does it
increase? If temperature increases by one degree Fahrenheit, by how many kelvins
does it increase?

(f) The temperature of 0 kelvins is known as *absolute zero*. It is not quite accurate to
say that all molecular motion ceases at absolute zero, but at that temperature the
system has its minimum possible total energy. It is thought that absolute zero cannot be attained experimentally, though temperatures lower than 0.0000001 kelvin
have been attained. Find the temperature in degrees Fahrenheit of absolute zero.

6. **Further verification of Newton's second law:** This exercise represents a hypothetical
implementation of the experiment suggested in the solution of Part 5 of Example 3.7. A
mass of 15 kilograms was subjected to varying accelerations, and the resulting force was
measured. In the following table, accelerations are in meters per second per second, and
forces are in newtons.

Acceleration	8	11	14	17	20
Force	120	165	210	255	300

(a) Construct a table of differences and explain how it shows that this data is linear.

(b) Find a linear model for the data.

(c) Explain in practical terms what the slope of this linear model is.

(d) Explain how this experiment provides further evidence for Newton's second law
of motion.

7. **Market supply:** The following table shows the quantity S of wheat (in billions of bushels) that wheat suppliers are willing to produce in a year and offer for sale at a price P (in dollars per bushel).

S = quantity of wheat	1.0	1.5	2.0	2.5
P = price	$1.35	$2.40	$3.45	$4.50

In economics it is standard to plot S on the horizontal axis and P on the vertical axis, so we will think of S as a variable and P as a function of S.

(a) Show that this data can be modeled by a linear function, and find its formula.

(b) Make a graph of the linear formula you found in Part (a). This is called the *market supply curve.*

(c) Explain why the market supply curve should be increasing. (<u>Hint</u>: Think about what should happen when the price increases.)

(d) How much wheat would suppliers be willing to produce in a year and offer for sale at a price of $3.90 per bushel?

8. **Market demand:** *This is a continuation of Exercise 7.* The following table shows the quantity D of wheat (in billions of bushels) that wheat consumers are willing to purchase in a year at a price P (in dollars per bushel).

D = quantity of wheat	1.0	1.5	2.0	2.5
P = price	$2.05	$1.75	$1.45	$1.15

In economics it is standard to plot D on the horizontal axis and P on the vertical axis, so we will think of D as a variable and P as a function of D.

(a) Show that this data can be modeled by a linear function, and find its formula.

(b) Add the graph of the linear formula you found in Part (a), called the *market demand curve*, to your graph of the market supply curve from Exercise 7.

(c) Explain why the market demand curve should be decreasing.

(d) The *equilibrium price* is the price determined by the intersection of the market demand curve and the market supply curve. Find the equilibrium price determined by your graph in Part (b).

9. **Sports car:** An automotive engineer is testing how quickly a newly designed sports car accelerates from rest. He has collected data giving the velocity (in miles per hour) as a function of the time (in seconds) since the car was at rest. Here is a table giving a portion of his data:

Time	2.0	2.5	3.0	3.5
Velocity	27.9	33.8	39.7	45.6

(a) By calculating differences, show that the data in this table can be modeled by a linear function.

(b) What is the slope for the linear function modeling velocity as a function of time? Explain in practical terms the meaning of the slope.

(c) Use the data in the table to find the formula for velocity as a linear function of time valid over the time period covered in the table.

(d) What would your formula from Part (c) give for the velocity of the car at time 0? What does this say about the validity of the linear formula over the initial segment of the experiment? Explain your answer in practical terms.

(e) Assume that the linear formula you found in Part (c) is valid from 2 seconds though 5 seconds. The marketing department wants to know from the engineer how to complete the following statement: "This car goes from 0 to 60 mph in _____ seconds." How should they fill in the blank?

10. **High school graduates:** The following table shows the number (in millions) graduating from high school in the United States in the given year.[3]

Year	1985	1987	1989	1991
Number (in millions) graduating	2.83	2.65	2.47	2.29

(a) By calculating differences, show that this data can be modeled using a linear function.

(b) What is the slope for the linear function modeling high school graduations as a function of time? Explain in practical terms the meaning of the slope.

(c) Find a formula for a linear function that models this data.

(d) How many do you think graduated from high school in 1994?

---

[3]The table is adapted from the *1995 Statistical Abstract of the United States.*

3.4 LINEAR REGRESSION

In real life, rarely is information gathered which perfectly fits any simple formula. In cases such as government spending, there are many factors influencing the budget, including the political make-up of the legislature. In the case of scientific experiments, variations may be due to *experimental error*, the inability of the data gatherer to obtain exact measurements; there may also be elements of chance involved. Under these circumstances, it may be necessary to obtain an approximate rather than an exact mathematical model.

The regression line

To illustrate this idea, let's look at total federal education expenditures (in billions of dollars) by the United States as reported by the *1995 Statistical Abstract of the United States* and recorded in the table below.

Date	1985	1986	1987	1988	1989
Expenditures in billions	27	29	30.4	30.9	32.1

To study this data we assign variable and function names: Let t be the number of years since 1985 and E the expenditures (in billions of dollars). The table for E as a function of t is then

t	0	1	2	3	4
E	27	29	30.4	30.9	32.1

It is difficult to discern, by looking at the table, the pattern of spending on public education. We can check to see if this data can be modeled by a linear function by making a table of changes.

Change in t	From 0 to 1	From 1 to 2	From 2 to 3	From 3 to 4
Change in E	2	1.4	0.5	1.2

The change in E is not constant, and so we know that this data cannot be modeled exactly by a linear function, but let's explore further. If we plot the data points as we have done in Figure 3.24, we see that they *almost* fall on a straight line. To show this better, we have added, in Figure 3.25, a straight line that passes *close* to all the data points. This line is known as the *least squares fit* or *linear regression* line, and it appears that the data follows this line fairly closely. Thus while we cannot model the data *exactly* with a linear function, it seems reasonable to use the line in Figure 3.25 as an *approximate* model for federal education spending.

There are many ways of approximating data that is nearly linear, but the regression line is the one that is most often used. It is the line that gives the least possible total of the squares of the vertical distances from the line to the data points. This is why the regression line is sometimes referred to as the least-squares fit. The mathematical concepts needed to derive

Figure 3.24: *Data points that are almost on a straight line*

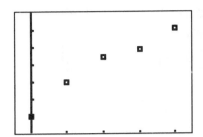

Figure 3.25: *A line that almost fits the data*

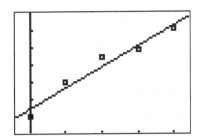

the formula for regression lines are beyond the scope of this course. Furthermore, while it is possible to calculate a formula for the regression line by hand, the procedure is cumbersome. Instead, this calculation is almost always done with a calculator or computer, and that is how we will do it here. You should refer to Chapter 4 of the *Keystroke Guide* to see the exact keystrokes needed to execute this important procedure.

With the data properly entered $\boxed{3.9}$ in the calculator as shown in Figure 3.26, we can get $\boxed{3.10}$ the regression line information shown in Figure 3.27. As usual, the calculator chooses its own letter[4] names, and we need to record the appropriate correspondences:

$$
\begin{aligned}
\mathsf{X} &= & t, \text{ the variable} \\
\mathsf{Y} &= & E, \text{ the function name} \\
\mathsf{a} = 1.21 &= & \text{slope of the regression line} \\
\mathsf{b} = 27.46 &= & \text{vertical intercept of the regression line.}
\end{aligned}
$$

We use this information to make the regression line model for E as a function of t:

$$E = 1.21t + 27.46 . \tag{3.8}$$

This is the line we added $\boxed{3.11}$ to the data plot in Figure 3.24 to get the picture in Figure 3.25.

It is important to remember that even though we have written an equality, Equation (3.8) in fact is a model which only approximates the relationship between t and E given by the data table. For example, the initial value according to the linear model in Equation (3.8) is 27.46, while the entry in the table for $t = 0$ is $E = 27$. We will use the equals sign as above, but you should be aware that in this setting many would prefer to replace it by an approximation symbol, $\approx$, or to use different letters for the regression equation. If you do use different letters for function and variable, it is crucial that you state clearly what they represent.

---

[4]An additional number, often denoted r, is shown at the bottom of the screen display for some calculators. This is a statistical measure of how closely the line fits the data, and we will not make use of it.

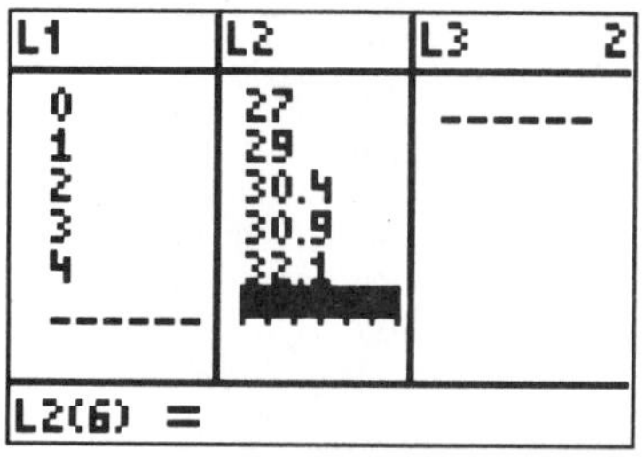

Figure 3.26: Data for federal education spending

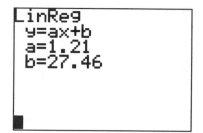

Figure 3.27: The regression line parameters

Uses of the regression line: slope and trends

The most useful feature of the regression line is its slope, and, in many cases, it provides the key to understanding data. For education spending, the slope, 1.21 billion dollars per year, of the regression line tells us that during the period from 1985 to 1989, education spending grew by about 1.21 billion dollars per year. It tells how the data is changing.

A plot which shows the regression line with the data can be useful in analyzing trends. For example, in Figure 3.25 the third data point (corresponding to 1987) lies above the regression line, while the fourth data point (corresponding to 1988) lies below it. Certainly spending was higher in 1988 than in 1987, but it could be argued, based on the position of the data points relative to the line, that in 1987 spending was ahead of the trend, while in 1988 it was slightly behind the trend.

It is tempting to use the regression line to predict the future. We will show how to do this and then discuss the pitfalls in such practice. What level of spending does the regression line in Equation (3.8) predict for 1990? Since t is the number of years since 1985, we want to get the value of E when $t = 5$. This information is not shown in the table, and so we approximate the value of $E(5)$ using the regression line with $t = 5$:

$$\text{Projected spending in 1990} = 1.21 \times 5 + 27.46 = 33.51 \text{ billion dollars.}$$

Consulting the source for our data, the *1995 Statistical Abstract of the United States*, we find that federal education expenditures in 1990 were in fact $E = 33.4$ billion dollars. In this case, the projection given by the regression line was pretty accurate.

Let's try to use the regression line to go the other way, that is, to try to estimate money spent before 1985. In particular, we want to determine the level of federal education expenditures in 1970. That is 15 years before 1985, and so we want the value of E when $t = -15$.

We use the regression line with $t = -15$:

$$\text{Estimated spending in 1970} = 1.21 \times (-15) + 27.46 = 9.31 \text{ billion dollars.}$$

Once again, we refer to the *1995 Statistical Abstract of the United States* and see that the actual federal expenditure for education in 1970 was 26.6 billion dollars. In this case, the value given by the regression line is a very bad estimate for the real value.

The regression line is a powerful tool for analysis of certain kinds of data, but caution in how it is used is essential. We saw above that the regression line gave a good estimate for 1990 expenditures but a very bad one for 1970. In general, using the regression line to extrapolate beyond the limits of the data is risky, and the risk increases dramatically for long-range extrapolations. What the regression line really shows is the *linear trend* established by the data. Thus, for our 1990 projection, it would be appropriate to say that "If the trend established in the late 80's had persisted, federal education spending in 1990 would have been about 33.51 billion dollars." A check of the 1990 data showed that this trend did indeed persist into 1990. An appropriate statement of our 1970 analysis might be that "If the trend established in the late 80's had been valid since 1970, then federal expenditures in 1970 would have been about 9.31 billion dollars." Since the actual expenditure was much more, we can draw the important conclusion that sometime between 1970 and the late 80's, the nature of federal education expenditures changed. Armed with this information, we might proceed by gathering more data to see exactly how it changed and then conduct historical, political, and economic investigations into why it changed.

It is common in newspaper and magazine articles to use data gathered today to make predictions about what might be expected in the future, and in many cases this is done with the regression line. When such data is handled by professionals who have sophisticated tools and insights available to help them, valuable information can be gained. But sometimes predictions based on gathered data are made by people who know a great deal less about handling data than you will by the time you complete this course. As a citizen it is important that you be able to bring your own insight into such analyses. It is risky to use the regression line to make projections unless you have good reason to believe that the data you are looking at is nearly linear and that the linear nature of the data will continue into the future. On the other hand, the regression line can show clearly the trend of almost linear data and can appropriately be used to determine if trends persist. If information is available which indicates that linear trends are persisting, then the regression line can be used to make forecasts.

As an example, suppose a legislator who took office in 1989 ran on a platform calling for a "significant" increase in federal education spending in 1990 and 1991. Does the record show that the legislator was able to fulfill his campaign promise? If so, then we should be able to detect a "significant" increase in 1990 and 1991 over the trend established from 1985 through 1989. We have already seen that the trend of the late 80's leads us to project an expenditure of about 33.51 billion dollars. A similar analysis gives projected spending in 1991:

Projected spending in 1991 = $1.21 \times 6 + 27.46 = 34.72$ billion dollars.

Checking the record, we find that actual federal spending, 33.4 billion in 1990 and 34.4 billion in 1991, was slightly less than the trend of the late 80's would suggest. That is what we might expect to happen if no additional impetus for funding increase were applied. It is easy to conceive of this legislator taking credit for a 1.3 billion dollar increase in 1990 and a 1 billion dollar increase in 1991. We leave it to the reader to decide whether a re-election vote would be merited.

EXAMPLE 3.8 *Military Expenditures*

The following table shows the amount $M = M(t)$ of money in billions of dollars spent by the United States on national defense.[5] In the time row, $t = 0$ corresponds to 1985, $t = 1$ refers to 1986, and so on.

t = years since 1985	0	1	2	3	4
M = billions of dollars	252.7	263.3	282	290.4	303.6

1. Plot the data points. Does it appear that it is appropriate to approximate this data with a straight line?

2. Find the equation of the regression line for M as a function of t, and add the graph of this line to your data plot.

3. Explain in practical terms the meaning of the slope of the regression line model we found in Part 2.

4. Compare the slope of the regression line for defense spending with the slope of the regression line for spending on education that we made earlier. What conclusions do you draw from this comparison?

5. Use the regression equation to estimate military spending by the U.S. in 1990 and 1995. The actual military expenditures in 1990 and 1995 were $299.3 billion and $271.6 billion, respectively. Did the trend established in the late 80's persist until 1990? Did it persist through 1995?

---

[5]From the *1995 Statistical Abstract of the United States.*

Solution to Part 1: First we enter [3.12] the data as shown in Figure 3.28. Next we plot [3.13] the data as shown in Figure 3.29. Note that the horizontal axis corresponds to years since 1985, and the vertical axis corresponds to billions of dollars of military expenditures. From Figure 3.29, it is clear that the data points do not lie exactly on a line, but they do

Figure 3.28: *National defense spending data*

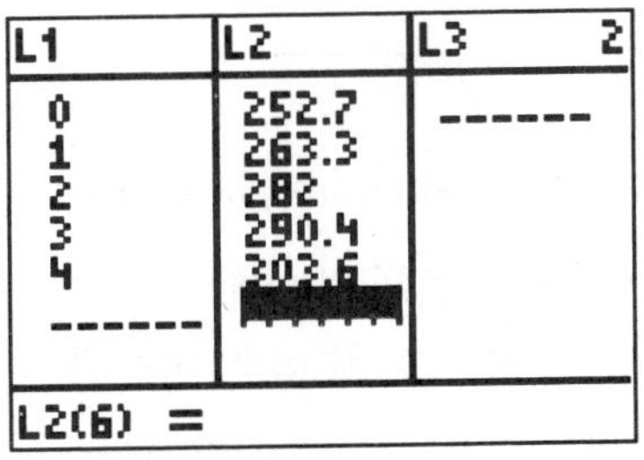

Figure 3.29: *A plot of national defense spending*

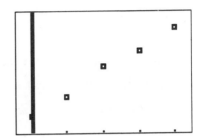

almost line up, and it is not unreasonable to model this data with a straight line.

Solution to Part 2: We use the calculator to get [3.14] the regression line parameters shown in Figure 3.30. We record the appropriate associations:

$$X \;=\; t, \text{ the variable, years since 1985}$$
$$Y \;=\; M, \text{ the function name, billions spent}$$
$$a = 12.89 \;=\; \text{slope of the regression line}$$
$$b = 252.62 \;=\; \text{vertical intercept of the regression line.}$$

Thus the regression line model we want is $M = 12.89t + 252.62$. We emphasize once more that even though we have written an equality, this does not establish an exact relationship between t and M. Rather, it is the approximation provided by the regression model. We add [3.15] this equation to the function list and get the graph in Figure 3.31. It is important to note here that the regression line we found passes near the data points. If it didn't, then we would look for a mistake in our work; if we found that no error had been made, we would conclude that it might not be appropriate to model the data using the regression line.

Solution to Part 3: The slope of the regression line shows the rate at which military expenditures were growing during the late 80's. We can conclude that in this period, defense spending was growing at a rate of about 12.89 billion dollars per year.

Figure 3.30: *Regression line parameters for defense spending*

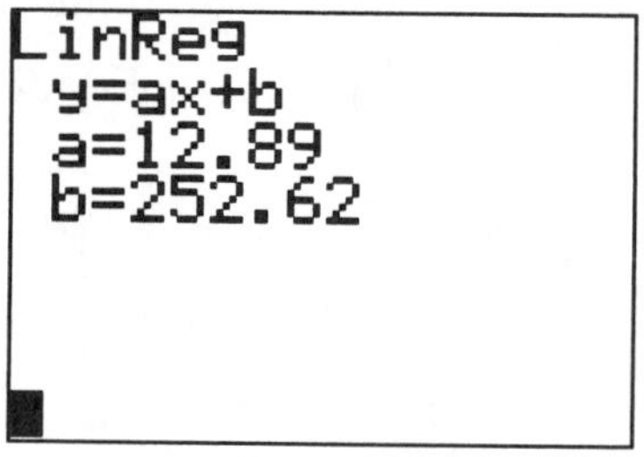

Figure 3.31: *The regression line added to the data plot*

```
LinReg
 y=ax+b
 a=12.89
 b=252.62
```

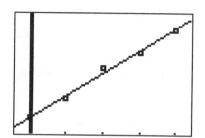

Solution to Part 4: In Part 3 we noted that during the late 80's defense spending was growing at a rate of 12.89 billion dollars per year. Earlier we found that the slope of the regression line for federal education spending was 1.21 billion dollars per year. Thus not only was military spending much greater during the late 80's; it was growing over ten times as fast as federal education spending.

There are other comparisons which may be made that do not show such a sharp distinction. You may note, for example, that the percentage increases are nearly the same.

Solution to Part 5: The year 1990 corresponds to $t = 5$, and 1995 corresponds to $t = 10$. We use these values in the regression equation to make the projections:

$$\text{Projected 1990 spending} = 12.89 \times 5 + 252.62 = 317.07 \text{ billion dollars}$$

$$\text{Projected 1995 spending} = 12.89 \times 10 + 252.62 = 381.52 \text{ billion dollars.}$$

Since the actual military expenditures in 1990 and 1995 were \$299.3 billion and \$271.6 billion, we see that the trend of the late 80's did not persist even until 1990. By 1995 the departure from the trend is dramatic.

We emphasize that the failure of the regression model to predict accurately spending in 1990 and 1991 is not a shortcoming of the model. Rather it provides a convincing argument that something happened to change the trend of military spending in the late 80's. It shows in a striking way the influence of the end of the Cold War on military spending.

Exercise Set 3.4

1. **Is a linear model appropriate?** The number in thousands of bacteria in a petri dish is given by the table below. Time is measured in hours.

The growth of bacteria

Time in hours since experiment began	0	1	2	3	4	5	6
Number of bacteria in thousands	1.2	2.4	4.8	9.6	19.2	38.4	76.8

The table below shows enrollment[6] (in millions of people) in public colleges in the United States during the years from 1986 through 1992.

Enrollment in public colleges

Date	1986	1987	1988	1989	1990	1991	1992
Enrollment in millions	9.8	10	10.3	10.3	10.7	11.1	11.3

(a) Plot the data points for number of bacteria. Does it look reasonable to approximate this data with a straight line?

(b) Plot the data points for college enrollment. Does it look reasonable to approximate this data with a straight line?

2. **College enrollment:** *This is a continuation of Exercise 1.* We use the data in the college enrollment table there.

(a) Find the equation of the regression line model for college enrollment as a function of time and add its graph to the data plot made in Exercise 1. (Round the regression line parameters to two decimal places.)

(b) Explain the meaning of the slope of the line you found in Part (a).

(c) Enrollment in American public colleges in 1983 was 8.6 million. Does it appear that the trend established in the late 80's and early 90's was valid as early as 1983?

---

[6]Taken from the *1995 Statistical Abstract of the United States.*

3. **Tourism:** The number in millions of tourists who visited the United States from other countries is given in the following table.

Date	1987	1988	1989	1990	1991
Millions of tourists	29.7	34.2	36.6	39.8	42.7

(a) Plot the data.

(b) Find the equation of the regression line and add its graph to your data plot.

(c) Explain in practical terms the meaning of the slope.

(d) How many tourists do you think visited the U.S. in 1985?

4. **Cable TV:** The following table gives the percentage of American homes with cable TV.

Date	1985	1986	1987	1988
Percent of homes with cable	46.2	48.1	50.5	53.8

(a) Plot the data points.

(b) Find the equation of the regression line and add its graph to the plotted data.

(c) In 1990, 58.6% of American homes had cable TV. If you had been a marketing strategist in 1988 with only the data in the table above available, what would have been your prediction for the percentage of American homes with cable TV in 1990?

5. **The effect of sampling error on linear regression:** A stream that feeds a lake is flooding, and during this flooding period the depth of water in the lake is increasing. The actual depth of the water at a certain point in the lake is given by the linear function $D = 0.8t + 52$ feet, where t is measured in hours since the flooding began. A hydrologist does not have this function available and is trying to determine experimentally how the water level is rising. She sits in a boat and each half hour drops a weighted line into the water to measure the depth to the bottom. The motion of the boat and the waves at the surface make exact measurement impossible. Her compiled data is given in the table below.

t = hours since flooding began	0	0.5	1	1.5	2
D = measured depth in feet	51.9	52.5	52.9	53.3	53.7

(a) Plot the data points.

(b) Find the equation of the regression line for D as a function of t, and explain in practical terms the meaning of the slope.

(c) Add the graph of the regression line to the plot of the data points.

(d) Add the graph of the depth function $D = 0.8t + 52$ to the picture. Does it appear that the hydrologist was able to use her data to make a close approximation of the depth function?

(e) What was the actual depth of the water at $t = 3$ hours?

(f) What prediction would the hydrologist's regression line give for the depth of the water at $t = 3$?

6. **Gross national product:** The United States gross national product in trillions of dollars is given in the table below.

Date	1985	1986	1987	1988	1989
Gross national product	4.02	4.23	4.52	4.88	5.20

(a) Find the equation of the regression line, and explain in practical terms the meaning of its slope. (Round regression line parameters to two decimal places.)

(b) Plot the data points and the regression line.

(c) In 1989 a prominent economist predicted that by 1993 the gross national product would reach 6.6 trillion dollars. Does your information from Part (a) support that conclusion? If not, when would you predict a gross national product of 6.6 trillion dollars?

7. **Japanese auto sales in the U.S.:** For 1992 through 1996, the table below shows the total U.S. sales (in millions) of Japanese automobiles (excluding light trucks).[7]

Date	1992	1993	1994	1995	1996
Japanese cars sold	2.48	2.38	2.39	2.32	2.29

(a) Plot the data points. Does it appear appropriate to approximate this data with a linear function?

(b) Get the equation of the regression line (rounding parameters to three decimal places), and explain in practical terms the meaning of the slope. In particular, comment on the meaning of the sign of the slope.

(c) Add the graph of the regression line to the data plot in Part (a). In your opinion, does this picture make the use of the regression line here appear to be more or less appropriate?

(d) The U.S. Department of Commerce, International Trade Administration, forecasts that there will be 2.11 million Japanese cars sold in the United States in 1997 and 2.04 million in 1998. How does the forecast obtained from the regression line compare with these figures?

---

[7]From the 1995 *U.S. Industrial Outlook*.

8. **Running speed versus length:** The following table gives the length L (in inches) of an animal and its maximum speed R (in feet per second) when it runs.[8] (For comparison, 10 feet per second is about 6.8 miles per hour.)

Animal	Length L	Speed R
Deermouse	3.5	8.2
Chipmunk	6.3	15.7
Desert crested lizard	9.4	24.0
Grey squirrel	9.8	24.9
Red fox	24	65.6
Cheetah	47	95.1

(a) Based on this table, is it generally true that larger animals run faster?

(b) Plot the data points. Does it appear that running speed is approximately a linear function of length?

(c) Find the equation of the regression line for R as a function of L, and explain in practical terms the meaning of its slope. (Round regression line parameters to two decimal places.) Add the plot of the regression line to the data plot in Part (b).

(d) Based on the plot in Part (c), which is faster *for its size*, the red fox or the cheetah?

9. **Antimasonic voting:** In *A Guide to Quantitative History*,[9] the authors, R. Darcy and Richard C. Rohrs, use mathematics to investigate the influence of religious zeal, as evidenced by the number C of churches in a given township, on the percent M of voting which was antimasonic in Genesee County, New York, from 1828 to 1832. The data used there is partially reproduced below.

Township	C = number of church buildings	M = percent antimasonic voting
Alabama	2	60
Attica	4	65
Stafford	5	74
Covington	6	81.7
Elbe	7	88.3

Based solely on the data above, analyze the premise that "The percent of antimasonic voting in Genesee County around 1830 had a direct dependence on the number of churches in the township." Your analysis should include a statement of why you believe there is a relationship and what that relationship might appropriately be. You are encouraged to consult the book referenced above for a deeper and more authoritative analysis.

---

[8]The table is adapted from J. T. Bonner, *Size and Cycle*, 1965, Princeton University Press, Princeton, N.J.
[9]Published by Praeger Publishers, Westport, Conn., 1995.

10. **Running ants:** A scientist collected the following data on the speed (in centimeters per second) at which ants ran at the given ambient temperature (in degrees Celsius).[10]

Temperature	25.6	27.5	30.3	30.4	32.2	33.0	33.8
Speed	2.62	3.03	3.57	3.56	4.03	4.17	4.32

(a) Find the equation of the regression line, giving the speed as a function of the temperature.

(b) Explain in practical terms the meaning of the slope of the regression line.

(c) The scientist observes the ants running at a speed of 2.5 centimeters per second. What is the ambient temperature?

11. **Energy cost of running:** Physiologists have studied the steady-state oxygen consumption (measured per unit of mass) in a running animal as a function of its velocity (i.e., its speed). They have determined that the relationship is approximately linear, at least over an appropriate range of velocities. The following table gives the velocity v (in kilometers per hour) and the oxygen consumption E (in milliliters of oxygen per gram per hour) for the rhea, a large flightless South American bird.[11] (For comparison, 10 kilometers per hour is about 6.2 miles per hour.)

Velocity v	Oxygen consumption E
2	1
5	2.1
10	4
12	4.3

(a) Find the equation of the regression line for E in terms of v.

(b) The slope of the linear function giving oxygen consumption in terms of velocity is called the *cost of transport* for the animal, since it measures the energy required to move a unit mass by one unit distance. What is the cost of transport for the rhea?

(c) Physiologists have determined the general approximate formula $C = 8.5W^{-0.40}$ for the cost of transport C of an animal weighing W grams. If the rhea weighs 22,000 grams, is its cost of transport from Part (b) higher or lower than what the general formula would predict? Based on this, is the rhea a more or less efficient runner than a typical animal its size?

(d) What would your equation from Part (a) lead you to estimate for the oxygen consumption of a rhea at rest? Would you expect that estimate to be higher or lower than the actual level of oxygen consumption of a rhea at rest?

[10]The table is taken from the data of H. Shapley, "Note on the thermokinetics of Dolichoderine ants," *Proc. Nat. Acad. Sci.* **10** (1924), 436–439.

[11]The table is based on C. R. Taylor, R. Dmi'el, M. Fedak, and K. Schmidt-Nielsen, "Energetic cost of running and heat balance in a large bird, the rhea," *Am. J. Physiol.* **221** (1971), 597–601.

3.5 SYSTEMS OF EQUATIONS

Many physical problems can be described by a system of two equations in two unknowns, and often the wanted information is found by *solving the system of equations*. As we shall see, this involves nothing more than finding the intersection of two lines, and we already know how to do that since we can find the intersection of *any* two graphs.

Graphical solutions of systems of equations

To show the method, we look at a simple example. We have $900 to spend on the repair of a gravel drive. We want to make the repairs using a mix of coarse gravel priced at $28 per ton and fine gravel priced at $32 per ton. To make a good driving surface, we need 3 times as much fine gravel as coarse gravel. How much of each will our budget allow us to buy?

As in any *story problem*, the first step is to convert the words and sentences into symbols and equations. Let's begin by writing equations using the words *tons of coarse gravel* and *tons of fine gravel*. We have a $900 budget, and we spend $28 per ton of coarse gravel and $32 per ton for fine gravel:

$$28 \times \boxed{\text{Tons of coarse gravel}} + 32 \times \boxed{\text{Tons of fine gravel}} = 900 . \qquad (3.9)$$

We also know that we must use 3 times as much fine gravel as coarse. In other words, we multiply the amount of coarse gravel by 3 to get the amount of fine gravel:

$$\boxed{\text{Tons of fine gravel}} = 3 \times \boxed{\text{Tons of coarse gravel}} . \qquad (3.10)$$

Now let's choose variable names:

$$c = \text{amount of coarse gravel}$$
$$f = \text{amount of fine gravel} .$$

We complete the process by putting these letter names into Equation (3.9) and Equation (3.10), and we arrive at a *system of equations* which we need to solve:

$$28c + 32f = 900$$
$$f = 3c .$$

To *solve the system* just means to find values of c and f that make *both equations true at the same time*. There are many ways to do this, but we can easily change this into a problem that we already know how to do with the calculator. The key step is first to solve each of the equations for one of the variables. Since the second equation is already solved for f, we do the same for the first equation, $28c + 32f = 900$:

$$28c + 32f = 900$$
$$32f = 900 - 28c$$

$$f = \frac{900 - 28c}{32}.$$

Thus we can replace the original system of equations by

$$f = \frac{900 - 28c}{32}$$
$$f = 3c.$$

We want to find where *both* of these equations are true. That is, we want to find where their graphs cross. Thus we can proceed using the crossing graphs method for solving equations that we learned in Chapter 2.

Let's recall the procedure. The first step is to enter $\boxed{3.16}$ both equations as shown in Figure 3.32. We record the appropriate variable correspondences:

$$Y_1 = \frac{900 - 28c}{32}, \text{ first expression for } f \text{ on vertical axis}$$
$$Y_2 = 3c, \text{ second expression for } f \text{ on vertical axis}$$
$$X = c, \text{ coarse gravel on horizontal axis.}$$

A common-sense estimate can help us choose our graphing range. We are buying $900 worth of gravel which costs about $30 per ton. Thus we will certainly buy no more than about 30 tons of either type. Thus, we choose a viewing window $\boxed{3.17}$ with a horizontal span from $c = 0$ to $c = 30$ and a vertical span from $f = 0$ to $f = 30$. The properly configured window is in Figure 3.33.

Figure 3.32: *Entering functions for the purchase of gravel*

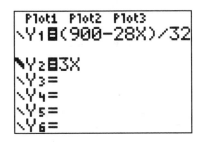

Figure 3.33: *Configuring the graphing window*

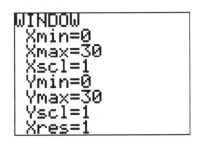

Now when we graph, we get the display 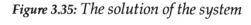 in Figure 3.34, where the thin line is the graph of $f = \dfrac{900 - 28c}{32}$ and the thick line is the graph of $f = 3c$. The solution we seek is the crossing point, which we find $\boxed{3.19}$ as we did in Chapter 2. We see from Figure 3.35 that, rounded to two decimal places, we should buy $c = 7.26$ tons of coarse gravel and $f = 21.77$ tons of fine gravel.

Figure 3.34: Graphing a system of equations for gravel

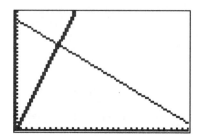

Figure 3.35: The solution of the system

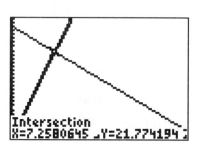

KEY IDEA 3.6: HOW TO SOLVE A SYSTEM OF TWO EQUATIONS IN TWO UNKNOWNS

Step 1: Solve both equations for one of the variables.

Step 2: The solution of the system of equations is the intersection point of the two graphs. This is found using the crossing graphs method from Chapter 2.

EXAMPLE 3.9   *A Picnic*

We have $56 to spend on pizzas and sodas for a picnic. Pizzas cost $12 each and sodas cost $0.50 each. Four times as many sodas as pizzas are needed. How many pizzas and how many sodas will our budget allow?

Solution: The cost of the picnic can be written as

$$12 \times \text{Number of pizzas} \ + \ 0.50 \times \text{Number of sodas} = \text{Cost of picnic}.$$

Since the picnic is to cost $56, we can rewrite this as

$$12 \times \text{Number of pizzas} \ + \ 0.50 \times \text{Number of sodas} = 56 \text{ dollars}.$$

If we use S for the number of sodas and P for the number of pizzas, this is

$$12P + 0.5S = 56 \ .$$

We also know that we need four times as many sodas as pizzas. In other words, to get the number of sodas we multiply the number of pizzas by 4:

$$\text{Number of sodas} = 4 \times \text{Number of pizzas}$$

In terms of the letters S and P, this is

$$S = 4P \ .$$

Thus we need to solve the system of equations

$$
\begin{aligned}
12P + 0.5S &= 56 \\
S &= 4P \ .
\end{aligned}
$$

We begin by solving the first equation for S. (Alternatively, it would be correct to solve each equation for P. We chose to solve for S since the work is already done for the second equation.) We have

$$
\begin{aligned}
0.5S &= 56 - 12P \\
\\
S &= \frac{56 - 12P}{0.5} \ .
\end{aligned}
$$

Now we need to use the crossing graphs method to solve the system of equations

$$
\begin{aligned}
S &= \frac{56 - 12P}{0.5} \\
S &= 4P \ .
\end{aligned}
$$

We enter $\boxed{3.20}$ both functions as shown in Figure 3.36 and record the appropriate variable correspondences:

$$
\begin{aligned}
\mathsf{Y_1} &= S, \text{ sodas on vertical axis} \\
\mathsf{X} &= P, \text{ pizzas on horizontal axis.}
\end{aligned}
$$

Since pizzas cost \$12 each we can't buy more than 5 of them. And that means we will buy no more than 20 sodas. Thus, to make the graph we set $\boxed{3.21}$ the horizontal span from $P = 0$ to $P = 5$ and the vertical span from $S = 0$ to $S = 20$. This configuration

Figure 3.36: Entering a system of equations for a picnic

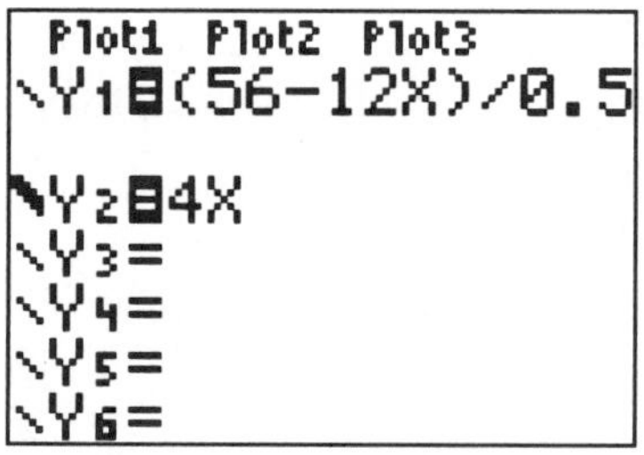

Figure 3.37: How many pizzas and sodas to buy

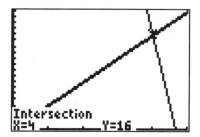

produces the graphs in Figure 3.37, where the thin line is the graph of $S = \dfrac{56 - 12P}{0.5}$, and the thick line is the graph of $S = 4P$.

We find the intersection point $\boxed{3.22}$ using the calculator, and we see from Figure 3.37 that the solution is X=4 and Y=16. Since X corresponds to P and Y corresponds to S, we conclude that we can buy 4 pizzas and 16 sodas.

An alternative algebraic solution of this system of equations is provided below.

Algebraic solutions

We can, if we wish, solve systems of equations without using the calculator. Let's see how to do that with the system of equations for the picnic in Example 3.9:

$$12P + 0.5S = 56$$
$$S = 4P .$$

The first step is to solve one equation for one of the variables. In general, we can use either equation and either variable. Sometimes, as in this case, a wise choice can save work. Since this is already done for us in the second equation, we avoid unnecessary work by choosing to solve the second equation for S:

$$S = 4P .$$

Now we put the expression $4P$ in place of S in the first equation:

$$12P + 0.5(4P) = 56 .$$

The result is that the variable S has been *eliminated*, and we are left with a linear equation involving only the single variable P. We learned in Section 2.3 how to solve such equations.

We get

$$
\begin{aligned}
12P + 0.5(4P) &= 56 \\
12P + 2P &= 56 \\
14P &= 56 \\
\\
P &= \frac{56}{14} \\
P &= 4 \, .
\end{aligned}
$$

This gives us the solution for P. We put this value for P back into the equation $S = 4P$ to get the value for S:

$$ S = 4P = 4 \times 4 = 16 \, . $$

We get the solution $P = 4$ and $S = 16$, and this agrees with our earlier solution using the calculator.

KEY IDEA 3.7: HOW TO SOLVE A SYSTEM OF TWO EQUATIONS IN TWO UNKNOWNS ALGEBRAICALLY

Step 1: Solve one of the equations for one of the variables.

Step 2: Put the expression you got in Step 1 into the other equation.

Step 3: Solve the resulting linear equation in one variable.

Step 4: To get the other variable, put the value you got from Step 3 into the expression you found in Step 1.

Exercise Set 3.5

1. **A party:** You have $36 to spend on refreshments for a party. Large bags of chips cost $2.00 and sodas cost $0.50. You need to buy five times as many sodas as bags of chips. How many bags of chips and how many sodas can you buy?

2. **Mixing feed:** A milling company wants to mix alfalfa, which contains 20% protein, and *wheat mids*,[12] which contains 15% protein, to make cattle feed.

    (a) If you make a mixture of 30 pounds of alfalfa and 40 pounds of wheat mids, how many pounds of protein are in the mixture? (<u>Hint</u>: In the 30 pounds of alfalfa there are $30 \times 0.2 = 6$ pounds of protein.)

    (b) Write a formula that gives the amount of protein in a mixture of a pounds of alfalfa and w pounds of wheat mids.

    (c) Suppose the milling company wants to make 1000 pounds of cattle feed that contains 17% protein. How many pounds of alfalfa and how many pounds of wheat mids must be used?

3. **An order for bulbs:** You have space in your garden for 55 small, flowering bulbs. Crocus bulbs cost $0.35 each and daffodil bulbs cost $0.75 each. Your budget allows you to spend $25.65 on bulbs. How many crocus bulbs and how many daffodil bulbs can you buy?

4. **American dollars and British pounds:** Assume that at the current exchange rate the British pound is worth $2.66 in American dollars. In your wallet are some one dollar bills and several British pound notes. There are 17 bills altogether, which have a total value of $30.28 in American dollars. How many American dollars and how many British pound notes do you have in your wallet?

5. **Population growth:** There are originally 255 foxes and 104 rabbits on a game reserve. The fox population grows at a rate of 33 foxes per year, and the rabbits increase at a rate of 53 rabbits per year. Under these conditions, how long does it take for the number of rabbits to catch up with the number of foxes? How many of each animal will be present at that time?

---

[12]When the wheat kernel is removed to produce flour, some parts of the whole wheat, including the bran, are often pressed into pellets known as *wheat mids* and sold as a feed ingredient.

6. **Male and female high school graduates:** The table below shows the percentage of male and female high school graduates who enrolled in college within 12 months of graduation.[13]

Year	1960	1965	1970	1975
Males	54%	57.3%	55.2%	52.6%
Females	37.9%	45.3%	48.5%	49%

(a) Find the equation of the regression line for percentage of male high school graduates entering college as a function of time.

(b) Find the equation of the regression line for percentage of female high school graduates entering college as a function of time.

(c) Assume that the regression lines you found in Part (a) and Part (b) represent trends in the data. If the trends persisted, when would you expect first to have seen the same percentage of female and male graduates entering college?

You may be interested to know that this actually occurred for the first time in 1980. The percentages fluctuated but remained very close during the 80's. In the 90's significantly more female graduates entered college than did males. In 1992, for example, the rate for males was 59.6% compared with 63.8% for females.

7. **Fahrenheit and Celsius:** If you know the temperature C in degrees Celsius, you can find the temperature F in degrees Fahrenheit from the formula

$$F = \frac{9}{5}C + 32 \,.$$

At what temperature will a Fahrenheit thermometer read exactly twice as much as a Celsius thermometer?

8. **The Concorde:** The Concorde jet leaves Kennedy Airport in New York at 1:00 p.m. local time. It arrives at Heathrow Airport in London at 10:00 p.m. London time. It sits on the ground two hours for servicing and then takes off for the return trip to New York. It arrives back at Kennedy 11:00 p.m. New York time. How long does it take the Concorde to fly from New York to London? (Suggestion: Let T be the amount of time it takes to fly from New York to London (or return), and let H be the time difference from London to New York. That is, if it is 1:00 p.m. in New York, then it is $1 + H$ p.m. in London. You can get one equation for the flight from New York to London and another for the return trip. Another approach is to relate T to the difference between the time required for a round trip and the ground time.)

[13]From the 1995 *Statistical Abstract of the United States.*

9. **A bag of coins:** A bag contains 30 coins, some dimes and some quarters. The total amount of money in the bag is $3.45. How many dimes and how many quarters are in the bag?

10. **An interesting system of equations:** What happens if you try to solve the following system of equations?

$$\begin{aligned} x + y &= 1 \\ x + y &= 2 \end{aligned}$$

Can you explain what is going on?

11. **Another interesting system of equations:** What happens if you try to solve the following system of equations?

$$\begin{aligned} x + 2y &= 3 \\ -2x - 4y &= -6 \end{aligned}$$

Can you explain what is going on?

12. **A system of three equations in three unknowns:** Consider the following system of three equations in three unknowns.

$$\begin{aligned} 2x - y + z &= 3 \\ x + y + 2z &= 9 \\ 3x + 2y - z &= 4 \end{aligned}$$

(a) Solve the first equation for z.

(b) Put the solution you got in Part (a) for z into *both* the second and third equations.

(c) Solve the system of two equations in two unknowns you found in Part (b).

(d) Write the solution of the original system of three equations in three unknowns.

13. **An application of three equations in three unknowns:** A bag of coins contains nickels, dimes, and quarters. There are a total of 21 coins in the bag, and the total amount of money in the bag is $3.35. There is one more dime than there are nickels in the bag. How many dimes, nickels, and quarters are in the bag?

3.6 CHAPTER SUMMARY

One of the simplest and most important geometric objects is the straight line, and the line is intimately tied to the idea of a *linear function*. This is a function with a constant rate of change, and the constant is known as the *slope*.

The Geometry of Lines

An important characterization of a straight line such as that determined by a roof is that it rises or falls at the same rate everywhere. This rate is known as the *slope* of the line and is commonly denoted by the letter m. Critical relationships involving the slope are summarized below.

Calculating the slope: The slope m can be calculated using

$$m = \frac{\text{Vertical change}}{\text{Horizontal change}}.$$

Fundamental property of the slope: The slope of a line tells how steeply it is increasing or decreasing.

- If m is positive, the line is rising.
- If $m = 0$, the line is horizontal.
- If m is negative, the line is falling.

Using the slope: The slope m can be used to calculate vertical change.

$$\text{Vertical change} = m \times \text{Horizontal change}.$$

Linear Functions

A linear function is one with a constant rate of change. As with lines, this is known as the slope and is denoted by m. Linear functions can be written in a special form:

$$y = mx + b,$$

where m is the slope and b is the *initial value*. The graph of a linear function is a straight line, and the slope of the linear function is the same as the slope of the line, while the initial value of the function is the same as the vertical intercept of the line.

The slope of a linear function is used much like the slope of a line.

- The slope m can be calculated from the change in y corresponding to a given change in x:

$$m = \frac{\text{Change in function}}{\text{Change in variable}} = \frac{\text{Change in } y}{\text{Change in } x}.$$

- The slope can be used to calculate the change in y resulting from a given change in x:

$$\text{Change in } y = m \times \text{Change in } x.$$

For example, the income tax liability T is, over a suitably restricted range of incomes, a linear function of the adjusted gross income I. Suppose the tax table shows a tax liability of $T = 780$ dollars when the adjusted gross income is $I = 15,000$ dollars and a tax liability of \$825 for an adjusted gross income of \$15,500.

- We can calculate the slope of the linear function $T = T(I)$ as follows.

$$m = \frac{\text{Change in tax}}{\text{Change in income}} = \frac{825 - 780}{15,500 - 15,000} = \frac{45}{500} = 0.09.$$

This means that a taxpayer in this range could expect to pay 9 cents tax on each additional dollar earned. Economists would term this the *marginal tax rate*.

- Once the slope, or marginal tax rate, is known, one can calculate the tax liability for other incomes. If the adjusted gross income is \$15,350, the additional tax over that owed on \$15,000 is given by

$$\text{Additional tax } = m \times \text{Additional income } = 0.09 \times 350 = 31.50 \text{ dollars.}$$

Thus the total tax liability is $780 + 3150 = 811.50$ dollars.

Modeling Data with Linear Functions

It is appropriate to model observed data with a linear function provided the data shows a constant rate of change. When the data for the variable is evenly spaced, this is always evidenced by a constant change in function values. For example, if a rock is dropped and its velocity is recorded each second, one might collect the following data:

$t = $ seconds	0	1	2	3	4	5
$V = $ feet per second	0	32	64	96	128	160

To see if it is appropriate to model velocity with a linear function, we look at successive differences:

Change in t	From 0 to 1	From 1 to 2	From 2 to 3	From 3 to 4	From 4 to 5
Change in V	32	32	32	32	32

Since the change in V is always the same, we conclude that V is a linear function of t. We can get a formula for this linear function by first calculating the slope:

$$m = \frac{\text{Change in } V}{\text{Change in } t} = \frac{32}{1} = 32,$$

so then

$$V = \text{Slope } \times t + \text{ Initial value } = 32t + 0 = 32t.$$

Linear Regression

Experimentally gathered data is always subject to error, and rarely will data points show exactly constant rate of change. When appropriate, we can still model the data with a linear function using the idea of *linear regression*. Linear regression gives the line with the least possible total of the squares of the vertical distances from the line to the data points. It is by far the most commonly used method of approximating linear data, and it is almost always accomplished with a computer or calculator.

The education expenditures $E = E(t)$ of the federal government in billions of dollars are given by the following table.

t = years since 1985	0	1	2	3	4
E = billions spent	27	29	30.4	30.9	32.1

The data table does not show a constant rate of change, but a plot of the data in Figure 3.38 shows that the data points nearly fall on a straight line. If we wish to model this with a linear function, we use the calculator to get the regression line $E = 1.21t + 27.46$. In Figure 3.39 we have added the regression line to the data points to show how the approximate model works.

Figure 3.38: Educational spending

Figure 3.39: Regression line model for educational spending

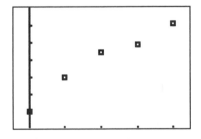

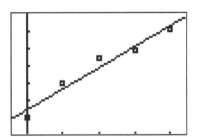

The regression line shows the linear trend, if any exists, exhibited by observed data. Extreme caution should be exercised in any use of the regression line beyond the range of observed data points.

Systems of Equations

Many physical problems are described by a system of two linear equations in two unknowns, and often the needed information is gotten by *solving the system of equations*. This can be accomplished either graphically or by hand calculation.

The graphical solution is the easier way and fits nicely with the spirit of this course. The procedure is as follows.

Step 1: Solve both equations for one of the variables.

Step 2: The solution of the system is the intersection point of the two graphs. This is found using the crossing graphs method.

For example, the purchase of c tons of coarse gravel and f tons of fine gravel might be described by the following system of equations.

$$28c + 32f = 900$$
$$f = 3c.$$

We solve the first equation for f to obtain the equivalent system

$$f = \frac{900 - 28c}{32}$$
$$f = 3c.$$

If we graph each of these functions and find the crossing point, we discover that we should buy $c = 7.26$ tons of coarse gravel and $f = 21.77$ tons of fine gravel.

CHAPTER 4 *Exponential Functions*

Exponential functions occur almost as often in nature, science, and mathematics as do linear functions. We have already encountered examples of exponential functions in looking at population growth, radioactive decay, free fall subject to air resistance, and interest on bank loans. We will look more closely at the structure of such functions and show where the formulas we presented in those applications originated.

4.1 EXPONENTIAL GROWTH AND DECAY

Recall that a linear function $y = y(x)$ with slope m is one that has a constant rate of change m. Another way of saying this is that a linear function changes by constant sums of m. That is, when x is increased by 1, we get the new value of y by adding m to the old y value. In contrast, an *exponential function* $N = N(t)$ with *base a* is one that changes by constant *multiples* of a. That is, when t is increased by 1, we get the new value of N by multiplying the current N value by a.

Exponential growth

Exponential functions occur naturally in some population growth analyses, and we look there to provide an illustration. The simplest population studies involve organisms such as bacteria which reproduce by cell division. Consider, for example, an experiment where there are initially 3000 bacteria in a petri dish, and the population doubles each hour. We let $N = N(t)$ denote the number of bacteria present after t hours. The number of bacteria present when the experiment began is $N(0) = 3000$. This is the *initial value* of the function. It is the starting value of N or the value of N when $t = 0$. During the first hour of the experiment the number of bacteria doubles, so that after one hour there are $N(1) = 2 \times 3000 = 6000$ bacteria present. During the next hour the population doubles again, so there are $N(2) = 2 \times 6000 = 12,000$ bacteria present after 2 hours. In general we can get the population N one hour in the future by multiplying the current N value by 2. This means that N is an exponential function with base 2. When the base of an exponential function is greater than 1, we will sometimes refer to the base as the *growth factor*. In this case we would say that the bacteria exhibit *exponential growth* with an hourly growth factor of 2 and an initial value of 3000.

If we calculate several values of the function N for bacteria described above and write the answer in a special form, we can see how to get a formula for N:

$$
\begin{aligned}
N(0) &= 3000 = 3000 \times 2^0 \\
N(1) &= 3000 \times 2 = 3000 \times 2^1 \\
N(2) &= 3000 \times 2 \times 2 = 3000 \times 2^2 \\
N(3) &= 3000 \times 2 \times 2 \times 2 = 3000 \times 2^3 \\
N(4) &= 3000 \times 2 \times 2 \times 2 \times 2 = 3000 \times 2^4.
\end{aligned}
$$

The pattern should be evident. To get the population N after t hours, we multiply the initial amount by the hourly growth factor 2 a total of t times, and that is the same as multiplying the initial population by 2^t:

$$ N = 3000 \times 2^t . $$

This formula makes it easy to calculate the number of bacteria present at any time t. For example, after 6 hours there are $N(6) = 3000 \times 2^6 = 192,000$ bacteria present. Or after one and a half hours there are $N(1.5) = 3000 \times 2^{1.5} = 8485$ bacteria present. (Here we have rounded our answer to the nearest integer since we don't expect to see parts of bacteria in the petri dish.)

The formula $N = 3000 \times 2^t$ which we obtained is typical of exponential functions in general. The formula for an exponential function $N = N(t)$ with base a and initial value P is

$$ N = Pa^t . $$

Exponential decay

Let's look at the bacteria experiment again, but this time suppose that an antibiotic has been introduced into the petri dish, so that rather than growing in number, each hour half of the bacteria die. Under these conditions there will be only 1500 bacteria left after one hour. After another hour, there will be 750 bacteria left. In general, we can get the number of bacteria present one hour in the future by multiplying the present population by one-half. Thus, under these conditions, N is an exponential function with base $\frac{1}{2}$. Since the population is decreasing rather than increasing, we would say that the population exhibits *exponential decay* with an hourly *decay factor* of $\frac{1}{2}$. Just as we did in the case where the population was growing, we can show a pattern that will lead us to a formula for $N = N(t)$:

$$N(0) \ = \ 3000 = 3000 \times \left(\frac{1}{2}\right)^0$$

$$N(1) \ = \ 3000 \times \frac{1}{2} = 3000 \times \left(\frac{1}{2}\right)^1$$

$$N(2) \ = \ 3000 \times \frac{1}{2} \times \frac{1}{2} = 3000 \times \left(\frac{1}{2}\right)^2$$

$$N(3) \ = \ 3000 \times \frac{1}{2} \times \frac{1}{2} \times \frac{1}{2} = 3000 \times \left(\frac{1}{2}\right)^3$$

$$N(4) \ = \ 3000 \times \frac{1}{2} \times \frac{1}{2} \times \frac{1}{2} \times \frac{1}{2} = 3000 \times \left(\frac{1}{2}\right)^4 .$$

In general, we see that $N = 3000 \times \left(\frac{1}{2}\right)^t$. This is an exponential function with initial value 3000 and base, or hourly decay factor, $\frac{1}{2}$.

In Figure 4.1 we have made the graph of the exponential function 3000×2^t. A table

---

Figure 4.1: *Exponential growth*

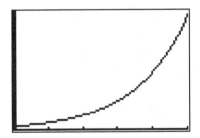

Figure 4.2: *Exponential decay*

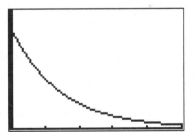

of values led us to choose a viewing window [4.1] with a horizontal span of $t = 0$ to $t = 5$ and a vertical span of $N = 0$ to $N = 100{,}000$. The graph is concave up, and this is typical of exponential growth. Growth is slow at first, but later becomes more rapid. In Figure 4.2 we have graphed $3000 \times \left(\frac{1}{2}\right)^t$ using [4.2] a horizontal span of $t = 0$ to $t = 5$ and a vertical viewing span of $N = 0$ to $N = 3500$. This graph is concave down and shows typical behavior for exponential decay. Decay is initially rapid, but the decay rate later slows.

KEY IDEA 4.1: EXPONENTIAL FUNCTIONS

A function $N = N(t)$ is exponential with base a if N changes in constant multiples of a. That is, if t is increased by 1, the new value of N is found by multiplying by a.

1. The formula for an exponential function with base a and initial value P is

$$N = Pa^t.$$

2. If $a > 1$, then N shows exponential growth with *growth factor* a. The graph of N will be similar in shape to that in Figure 4.1.

3. If $a < 1$, then N shows exponential decay with *decay factor* a. The graph of N will be similar in shape to that in Figure 4.2.

EXAMPLE 4.1 *Radioactive Decay*

Radioactive substances decay over time, and the rate of decay depends on the element. If, for example, there are G grams of heavy hydrogen H_3 in a container, then as a result of radioactive decay, one year later there will be $0.783G$ grams of heavy hydrogen left. Suppose we begin with 50 grams of heavy hydrogen.

1. Use a formula to express the number G of grams left after t years as an exponential function of t. Identify the base or yearly decay factor and the initial value.

2. How much heavy hydrogen is left after 5 years?

3. Plot the graph of $G = G(t)$ and describe in words how the amount of heavy hydrogen present changes with time.

4. How long will it take for half of the heavy hydrogen to decay?

Solution to Part 1: The initial value, 50 grams, is given. We are also told that we can get the amount G of heavy hydrogen one year in the future by multiplying the present amount by 0.783. Thus, $G = G(t)$ is an exponential function with initial value 50 grams and base, or yearly decay factor, 0.783. We conclude that the formula for G is given by

$$G = \text{Initial value} \times (\text{Yearly decay factor})^t$$
$$G = 50 \times 0.783^t.$$

Solution to Part 2: To get the amount of heavy hydrogen left after 5 years, we use the formula above to calculate $G(5)$:

$$G(5) = 50 \times 0.783^5 = 14.72 \text{ grams.}$$

Solution to Part 3: We made the graph shown in Figure 4.3 using a horizontal span of $t = 0$ to $t = 10$ and a vertical span of $G = 0$ to $G = 60$. You may wish to look at a table of values to see why we chose these settings. We note that in this graph the correspondences are

$$X \ = \ t, \text{ years on horizontal axis}$$
$$Y \ = \ G, \text{ grams left on vertical axis.}$$

The features of this graph are typical of exponential decay. The amount of heavy hydrogen decreases rapidly over the first few years, but after that the rate of decay slows.

Solution to Part 4: We want to know when there will be 25 grams of heavy hydrogen left. That is, we want to solve the equation

$$50 \times 0.783^t = 25$$

for t. We can do this using the crossing graphs method. In Figure 4.4 we have added the graph of $G = 25$ and then used $\boxed{4.3}$ the calculator to get the intersection point shown in Figure 4.4. We see that there will be 25 grams of heavy hydrogen left after about 2.83 years. This number is known as the *half-life* of heavy hydrogen because it tells how long it takes for half of it to decay.

Figure 4.3: *The decay of heavy hydrogen*

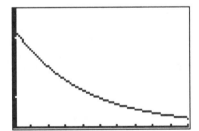

Figure 4.4: *The half-life of heavy hydrogen*

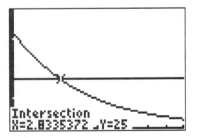

It is a fact that the half-life of a radioactive substance does not depend on the initial amount. Thus we would have gotten the same answer, 2.83 years, if we had started with 1000 grams of heavy hydrogen and asked how long it took until only 500 grams were left. See Exercise 7 at the end of this section.

Constant proportional change

The growth or decay factor is a crucial bit of information for an exponential function, but in practice its value is rarely given directly. One way in which exponential functions are commonly described involves percentage, or proportional, growth or decay. For example, the first census of the United States in 1790 showed a resident population of 3.93 million people. For certain populations, it is reasonable to expect that each year the population will grow by some fixed percentage. At least from 1790 through 1860, when the pattern was disrupted by the Civil War, this occurred in the United States, and the population grew by about 3% each year. That means that by the end of each year the population had grown to 103% of its value at the beginning of the year. Or if $U = U(t)$ represents U.S. population (in millions) t years since 1790, the value of U one year in the future can be gotten by multiplying the current value of U by 1.03. Thus U is an exponential function with initial value 3.93 and yearly growth factor 1.03. This observation allows us to write the formula for U:

$$U = \text{Initial value} \times (\text{Yearly growth factor})^t$$
$$U = 3.93 \times 1.03^t .$$

In this example of an exponential function, the population is always changing by the same percentage. This is in fact an alternative characterization of exponential functions: The percentage or proportional growth (or decay) rate is always the same. For populations, percentage rate of change is often called *per capita rate of change*. Note in this example that we get the yearly growth factor 1.03 by adding 1 to the decimal form 0.03 of the yearly percentage growth rate: $1.03 = 1 + 0.03$.

Let's examine this in the case of exponential decay. Suppose that because of disease or famine, the population in a mythical medieval land decreased at a rate of 3% per year. That means that the population at the end of each year will be 97% of its value at the beginning of the year. Thus, this is an exponential function with yearly decay factor 0.97 and yearly percentage decay rate 0.03. In this case we get the decay factor 0.97 by subtracting the percentage decay rate 0.03 from 1: $0.97 = 1 - 0.03$.

If we also know that the initial size of this medieval population was 250,000, and if we let $M = M(t)$ be the population after t years, we have collected the information we need to write a formula for M:

$$M = \text{Initial value} \times (\text{Yearly decay factor})^t$$
$$M = 250,000 \times 0.97^t .$$

The relationships between growth or decay factors and percentage rates of change we found in the two examples above are typical.

KEY IDEA 4.2: ALTERNATIVE CHARACTERIZATION OF EXPONENTIAL FUNCTIONS

A function is exponential if it shows constant percentage (or proportional) growth or decay.

1. For an exponential function with discrete (yearly, monthly, etc.) percentage growth rate r as a decimal, the growth factor is $a = 1 + r$.

2. For an exponential function with discrete percentage decay rate r as a decimal, the decay factor is $a = 1 - r$.

EXAMPLE 4.2   Compound Interest

A credit card holder begins the year owing $395.00 to a bank card. The bank card charges 1.2% interest on the outstanding balance each month. For purposes of this exercise, assume that no additional payments or charges are made and that no additional service charges are levied. Let $B = B(t)$ denote the balance of the account t months after January 1.

1. Explain why B is an exponential function of t. Identify the monthly percentage growth rate, the monthly growth factor, and the initial value. Write a formula for $B = B(t)$.

2. How much is owed after 7 months?

3. Assume that there is a limit of $450 on the card and the bank will demand a payment the first month this limit is exceeded. How long is it before a payment is required?

Solution to Part 1: For each dollar that is owed at the beginning of the month, an additional 0.012 dollar will be owed at the end of the month. Thus for each dollar owed at the beginning of the month, $1.012 will be owed at the end of the month. That means the value of B at the end of the month can be found by multiplying the balance at the beginning of the month by 1.012. Therefore B is an exponential function with monthly growth factor 1.012. The monthly percentage growth rate is the monthly interest rate, 1.2%. The initial value is $395. Thus B is given by the formula

$$B = \text{Initial value} \times (\text{Monthly growth factor})^t$$
$$B = 395 \times 1.012^t .$$

Solution to Part 2: To calculate the balance after 7 months, we put $t = 7$ into the formula we found in Part 1:

$$B(7) = 395 \times 1.012^7 = 429.40 \text{ dollars.}$$

Solution to Part 3: To find when the credit limit is reached, we need to solve

$$395 \times 1.012^t = 450$$

for t. We could solve this using the crossing graphs method, but in this case it is easier to make a table of values for $B(t)$ as we have done in Figure 4.5. The table shows that the credit limit will be exceeded in 11 months.

Figure 4.5: A table of values for a bank card

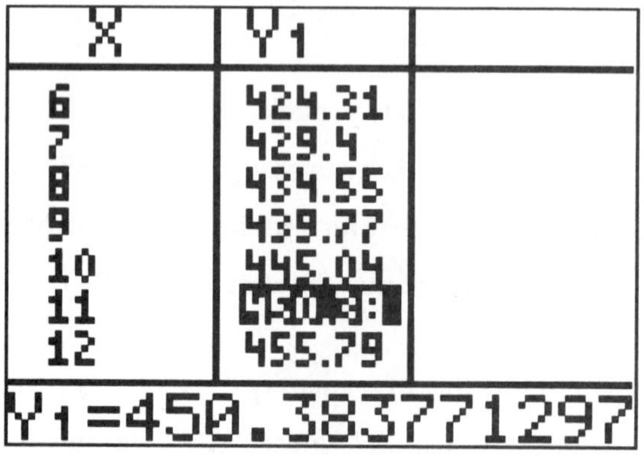

Unit conversion

Sometimes it is important to represent exponential functions in units that are not the same as those which are given. For example, we noted earlier that from 1790 to 1860, United States population grew exponentially with yearly growth factor 1.03. (That is, it grew by 3% per year.) If we are interested in modeling the population as it changed each decade, we want to know the *decade* growth factor. Since the yearly growth factor is 1.03, in 10 years we would multiply by 1.03 a total of ten times, that is, by 1.03^{10}.

Decade growth factor   =   (Yearly growth factor)^{10}

Decade growth factor   = $1.03^{10} = 1.344$, rounded to three decimal places.

Thus, if we want the population at the end of a decade, we multiply the population at the beginning of the decade by 1.344. This is the decade growth factor, and if we use d to denote decades since 1790, we have the information we need to write a formula for our exponential function:

$$U = \text{Initial value} \times (\text{Decade growth factor})^d$$

$$U = 3.93 \times 1.344^d.$$

Let's turn the problem around. Census data is normally taken each decade, and rather than being given information on a yearly basis, we might have been told that from 1790 to 1860 the United States population grew by 34.4% per decade. Then the *decade* growth factor would be 1.344. If we wanted to express the exponential function in terms of years, we would need to recover the yearly growth factor from the decade growth factor. Since the population grows by a factor of 1.344 each decade, in d decades it will grow by a factor of 1.344^d. In particular, in one year, which is one-tenth of a decade, it will grow by a factor of

$$\text{Yearly growth factor} \quad = \quad (\text{Decade growth factor})^{\frac{1}{10}}$$

$$\text{Yearly growth factor} \quad = \quad 1.344^{\frac{1}{10}} = 1.03 \text{ , rounded to two decimal places.}$$

Thus we have recovered the yearly growth factor of 1.03 and can conclude that the population grew by 3% per year.

The reasoning we have used above applies in general.

KEY IDEA 4.3: UNIT CONVERSION FOR GROWTH FACTORS

If the growth or decay factor for one period of time is a, then the growth or decay factor A for k periods of time is given by

$$A = a^k \text{ .}$$

This relationship may be illustrated with the following diagram.

GROWTH FACTOR FOR ONE PERIOD OF TIME	$\xrightarrow{\text{RAISE TO K}^{\text{TH}} \text{ POWER}}$ $\xleftarrow{\text{RAISE TO 1/K}^{\text{TH}} \text{ POWER}}$	GROWTH FACTOR FOR K PERIODS OF TIME

***EXAMPLE 4.3** Getting the Decay Factor from the Half-life*

It is standard practice to give the rate at which a radioactive substance decays in terms of its *half-life*. That is the amount of time it takes for half of the substance to decay. The half-life of carbon 14 is 5770 years.

1. If you start with 1 gram of carbon 14, how long will it take for only $\frac{1}{4}$ gram to remain?

2. What is the yearly decay factor rounded to five decimal places?

3. Find the yearly percentage decay rate and explain what it means in practical terms.

4. Assuming we start with 25 grams of carbon 14, find a formula that gives the amount of carbon 14 left after t years.

Solution to Part 1: To say that the half-life is 5770 years means that, no matter how much carbon 14 is originally in place, at the end of 5770 years, half of that amount will remain. Thus if we start with 1 gram, in 5770 years $\frac{1}{2}$ gram will remain. In 5770 more years, there will be half of that amount, one-quarter gram, remaining. It takes two half-lives, or $5770 + 5770 = 11,540$ years, for 1 gram of carbon 14 to decay to one-quarter gram of carbon 14.

Solution to Part 2: In one half-life, carbon 14 decays by a factor of $\frac{1}{2}$. Thus in h half-lives, it will decay by a factor of

$$\left(\frac{1}{2}\right)^h .$$

Since one year is $\frac{1}{5770}$ half-lives, in one year carbon 14 will decay by a factor of

$$\text{Yearly decay factor} \quad = \quad (\text{Half-life decay factor})^{\frac{1}{5770}}$$
$$\text{Yearly decay factor} \quad = \quad \left(\frac{1}{2}\right)^{\frac{1}{5770}} = 0.99988 , \text{ rounded to five decimal places.}$$

Solution to Part 3: The yearly decay factor is 0.99988. Thus the yearly percentage decay rate is $1 - 0.99988 = 0.00012$. This means that each year the amount of carbon 14 is reduced by 0.012%.

Solution to Part 4: The initial value is 25, and from Part 2 we know that the yearly decay factor is 0.99988. Thus if we let $C = C(t)$ be the amount (in grams) of carbon 14 left after t years, then

$$C \quad = \quad \text{Initial value} \times (\text{Yearly decay factor})^t$$
$$C \quad = \quad 25 \times 0.99988^t .$$

Exponential functions and daily experience

Exponential functions are common in daily news reports, and it is important that citizens spot them and understand their significance. Economic reports are often driven by exponential functions, and inflation is a prime example. When it is reported that the inflation rate is 3% per year, the meaning is that prices are increasing at a rate of 3% each year, and consequently prices are growing exponentially. Good economic news might be that "Last year's 5% inflation rate is now down to 4%." While this should indeed be encouraging, the lower rate does not change the fact that we are still dealing with an exponential function whose graph will eventually get very steep. Lowering from 5% to 4% changes the growth factor from 1.05 to 1.04 and so delays the time until the steep part of the curve is reached, but, as is shown in Figure 4.6, if the price function remains exponential in nature, then over the long term, catastrophic price increases will occur.

Figure 4.6: *5% inflation (thin graph) versus 4% inflation (thick graph)*

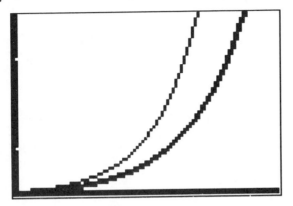

Short-term population growth is another example which is, in many settings, exponential in nature. It is the shape of the exponential curve, and not necessarily the exact exponential formula, which causes concerns about eventual overcrowding. Many environmental issues related to human and animal populations are inextricably tied to exponential functions. It may be relatively inexpensive to dispose properly of large amounts of materials at a waste cleanup site, but further cleanup may be much more expensive. That is, as we begin the cleanup process, the amount of objectionable material remaining to be dealt with may be a decreasing exponential function. Since certain toxic substances are dangerous even in very small quantities, it can be very expensive to reduce them to safe levels. Once again, it is the nature of exponential decay, not the exact formula, which contributes to the astronomical expense of environmental cleanup.

Whenever phenomena are described in terms of percentage change, they may be modeled by exponential functions, and an understanding how exponential functions behave is the key to understanding their true behavior. One of the most important features of exponential phenomena is that they may change at one rate over a period of time but change at a much different rate later on. Exponential growth may be modest for a time, but eventually the function will increase at a dramatic rate. Similarly, exponential decay may show encouragingly rapid progress initially, but such rates of decrease cannot continue indefinitely.

Exercise Set 4.1

1. **Exponential growth with given initial value and growth factor:** Write the formula for an exponential function with initial value 23 and growth factor 1.4. Plot its graph.

2. **Exponential decay with given initial value and decay factor:** Write the formula for an exponential function with initial value 200 and decay factor 0.73. Plot its graph.

3. **A population with given per capita growth rate:** A certain population has yearly per capita growth rate of 2.3%. Use a formula to express the population as an exponential function if the initial value is 3 million.

4. **Unit conversion with exponential growth:** The exponential function $N = 3500 \times 1.77^d$, where d is measured in decades, gives the number of individuals in a certain population.

   (a) What is the percentage growth rate per decade?

   (b) What is the yearly growth factor rounded to three decimal places? What is the yearly percentage growth rate?

   (c) What is the growth factor (rounded to two decimal places) for a century? What is the percentage growth rate per century?

5. **Unit conversion with exponential decay:** The exponential function $N = 500 \times 0.68^t$, where t is measured in years, shows the amount of a certain radioactive substance present.

   (a) What is the yearly percentage decay rate?

   (b) What is the monthly decay factor rounded to three decimal places? What is the monthly percentage decay rate?

   (c) What is the percentage decay rate per second? (<u>Note</u>: For this calculation, you will need to use all the decimal places that your calculator can show.)

6. **A savings account:** You initially invest $500 in a savings account which pays a yearly interest rate of 4%.

   (a) Write a formula for an exponential function giving the balance in your account as a function of the time since your initial investment.

   (b) What monthly interest rate best represents this account? Round your answer to three decimal places.

   (c) Calculate the decade growth factor.

(d) Use the formula you found in Part (a) to determine how long it will take for the account reach $740. Explain how this is consistent with your answer to Part (c).

7. **Half-life of heavy hydrogen:** We stated in Example 4.1 that the half-life of a radioactive substance does not depend on the initial amount. Using the information from Example 4.1, show that it takes the same amount of time for 100 grams of heavy hydrogen to decay to 50 grams as it does for 50 grams to decay to 25 grams. How long will it take for 100 grams of heavy hydrogen to decay to 6.25 grams? (Note: You can do this without your calculator!)

8. **How fast do exponential functions grow?** At age 25 you start to work for a company and are offered two rather fanciful retirement options.

   Retirement option 1: When you reach age 65, you will be paid a lump sum of $25,000 for each year of service.

   Retirement option 2: When you start to work, the company will deposit $10,000 into an account that pays a monthly interest rate of 1%. When you reach age 65, the account will be closed and the balance given to you.

   Which retirement option is more favorable to you?

9. **Inflation:** The yearly *inflation rate* tells the percentage by which prices increase. For example, from 1990 through 1995 the inflation rate in the United States remained stable at about 3% each year. In 1990 an individual retired on a fixed income of $36,000 per year. Assuming the inflation rate remains at 3%, how long will it take for the retirement income to deflate to half its 1990 value? (Note: To say that retirement income has deflated to half its 1990 value means that prices have doubled.)

10. **Long-term population growth:** While exponential growth can often be used to model accurately population growth for some periods of time, there are inevitably, in the long term, limiting factors which make purely exponential models inaccurate. If the U.S. population had continued to grow by 3% each year from 1790, when it was 3.93 million, until today, what would the population of the United States have been in 1990? For comparison, the population of the United States in 1990 was 248,709,873 according to census data. The population of the world was 5.292 billion people.

11. **The population of Mexico:**[1] In 1980 the population of Mexico was about 67.38 million. For the years 1980 through 1985 the population grew at a rate of about 2.6% per year.

   (a) Find a formula for an exponential function which gives the population of Mexico.

   (b) Use the function you found in Part (a) to predict when the population of Mexico will reach 90 million.

12. **Cleaning contaminated water:** A tank of water is contaminated with 60 pounds of salt. In order to bring the salt concentration down to a level consistent with EPA standards, clean water is running into the tank, and the well-mixed overflow is being collected for removal to a toxic waste site. The result is that at the end of each hour there is 22% less salt in the tank than at the beginning of the hour. Let $S = S(t)$ denote the number of pounds of salt in the tank t hours after the flushing process begins.

   (a) Explain why S is an exponential function and find its hourly decay factor.

   (b) Give a formula for S.

   (c) Make a graph of S that shows the flushing process during the first 15 hours, and describe in words how the salt-removal process progresses.

   (d) In order to meet EPA standards, there can be no more than 3 pounds of salt in the tank. How long must the process continue until EPA standards are met?

   (e) Suppose this cleanup procedure costs $8000 per hour to operate. How much does it cost to reduce the amount of salt from 60 pounds to 3 pounds? How much does it cost to reduce the amount of salt from 3 pounds to 0.1 pound?

13. **Grains of wheat on a chess board:** A children's fairy tale tells of a clever elf who extracted from a king the promise to give him one grain of wheat on a chess board square today, two grains on an adjacent square tomorrow, four grains on an adjacent square the next day, and so on, doubling the number of grains each day until all 64 squares of the chess board were used. How many grains of wheat did the hapless king contract to place on the 64th square? There are about 1.1 million grains of wheat in a bushel. Assume that a bushel of wheat sells for $4.25. What was the value of the wheat on the 64th square?

---

[1] Adapted from *Calculus* by D. Hughes-Hallett, A. M. Gleason, *et al.*, 2nd edition, 1998, John Wiley & Sons, New York.

4.2 MODELING EXPONENTIAL DATA

Just as we did with linear data, we will show how to recognize exponential data and develop the appropriate tools for constructing exponential models.

Recognizing exponential data

Underground water seepage, such as that from a toxic waste site, can contaminate water wells many miles away. Suppose that water seeping from a toxic waste site is polluted with a certain contaminant at a level of 64 milligrams per liter. Several miles away, monthly tests are made on a water well to monitor the level of this contaminant in the drinking water. In the table below we record the difference between the contaminant level of the waste site (64 milligrams per liter) and the contaminant level of the water well (in milligrams per liter).

Time in months	0	1	2	3	4	5
Contaminant level difference	64	45.44	32.26	22.91	16.26	11.55

Let t be the time in months since the leakage began and D the difference (in milligrams per liter) between the contaminant level of the waste site and that of the water well. The table shows that the difference D is narrowing, so that the water well is becoming more heavily polluted. We note further that D decreases quickly at first, but the rate of decrease slows later on. This is a feature of exponential decay, and it leads us to suspect that this data may be appropriately modeled by an exponential function. How can we test the data to see if it really is exponential in nature? If this data is decaying exponentially, then each month we should get the new difference in contaminant level by multiplying the old difference in contaminant level by the monthly decay factor:

$$\text{New } D = \text{Monthly decay factor} \times \text{Old } D .$$

Using division to rewrite this, we get

$$\frac{\text{New } D}{\text{Old } D} = \text{Monthly decay factor.} \tag{4.1}$$

Thus, a data table representing an exponential function should show common quotients if we divide each function entry by the one preceding it, assuming we have evenly spaced values for the variable. In the following table, we have calculated this quotient for each of the data entries, rounding our answers to two decimal places.

Time increment	From 0 to 1	From 1 to 2	From 2 to 3	From 3 to 4	From 4 to 5
Ratios of D	$\dfrac{45.44}{64} = 0.71$	$\dfrac{32.26}{45.44} = 0.71$	$\dfrac{22.91}{32.26} = 0.71$	$\dfrac{16.26}{22.91} = 0.71$	$\dfrac{11.55}{16.26} = 0.71$

According to Equation (4.1), this table tells us two things. First of all, the ratios are always the same, so the data is exponential. Second, the common quotient 0.71 is the monthly decay factor.

Comparing this with how we handled linear data in Chapter 3, we notice an important analogy. Linear functions are functions that change by *constant sums*, and so we detect linear data by looking for a common *difference* in function values when we use successive data points, assuming evenly spaced values for the variable. When our data set is given in variable increments of 1, this common difference is the slope of the linear model. Exponential functions are functions that change by *constant multiples*, and so we detect exponential data by looking for a common *quotient* when we use successive data points, again assuming evenly spaced values for the variable. When our data set is given in increments of 1, this common quotient is the growth (or decay) factor of the exponential model.

Constructing an exponential model

Since we know the decay factor, we need only one further bit of information to make our exponential model: the initial value. But that also appears in the table as $D(0) = 64$ milligrams per liter. Thus the formula, which will serve as our exponential model for D, is

$$D = \text{Initial value} \times (\text{Monthly decay factor})^t$$
$$D = 64 \times 0.71^t \text{ milligrams per liter.}$$

In situations such as this, where we have a model for a difference, it will be convenient to carry this a step further and get a formula for the contamination level $C = C(t)$ of the water well. Since D represents the contaminant level of the waste site, 64 milligrams per liter, minus the contaminant level C of the water well, we do this as follows:

$$D = 64 - C$$
$$C + D = 64$$
$$C = 64 - D$$
$$C = 64 - 64 \times 0.71^t \text{ milligrams per liter.}$$

Just as with linear models, we can use this formula to get information which may not be apparent from the data. Suppose, for example, that this particular contaminant is considered dangerous for drinking water when it reaches a level of 57 milligrams per liter. When will this dangerous level be reached? We want to solve the equation

$$64 - 64 \times 0.71^t = 57$$

for t.

We use the crossing graphs method to do that. We entered $\boxed{4.4}$ the contaminant function $C = 64 - 64 \times 0.71^t$ and the target level, 57 milligrams per liter, into the calculator. We record the appropriate variable correspondences:

$$Y_1 = C, \text{ contaminant level on vertical axis}$$

$$X = t, \text{ months on horizontal axis.}$$

We made the table of values in Figure 4.7 to find out how to set up the graphing window. This table shows that the critical level of contamination will occur sometime before 7 months. This led us to choose a horizontal span of $t = 0$ to $t = 8$. Since the contaminant level starts at 0 and will never be more than 64, we add a little extra room using a vertical span of $C = 0$ to $C = 75$. The resulting graphs are in Figure 4.8. We have used the calculator to get the crossing

Figure 4.7: *A table of values for contaminant levels*

Figure 4.8: *When the contaminant level reaches 57 milligrams per liter*

X	Y1	Y2
3	41.094	57
4	47.737	57
5	52.453	57
6	55.802	57
7	58.179	57
8	59.867	57
9	61.066	57

X=7

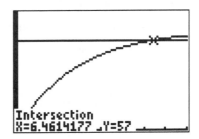

Intersection
X=6.4614177 Y=57

point $\boxed{4.5}$, and from the prompt at the bottom of Figure 4.8, we see that the contamination level of the water well will reach 57 milligrams per liter 6.46 months after the leakage began.

The graph in Figure 4.8 not only answers for us the question we asked but also provides a display that shows the classic exponential behavior for this phenomenon. As we noted earlier, the difference D between the contaminant level of the waste site and that of the water well decreases quickly at first, but the rate of decrease slows later on. As Figure 4.8 shows, this means that the contamination itself increases rapidly at first, but as the level nears that of the polluted source, the rate of increase slows.

EXAMPLE 4.4 *Making Ice*

A freezer maintains a constant temperature of 6 degrees Fahrenheit. An ice tray is filled with tap water and placed in the refrigerator to make ice. The difference between the temperature of the water and that of the freezer was sampled each minute and recorded in the table below.

Time in minutes	0	1	2	3	4	5
Temperature difference	69	66.3	63.7	61.2	58.8	56.5

1. Test to see that the data is exponential.

2. Find an exponential model for temperature difference.

3. Use your answer in Part 2 to make a model for the temperature of the cooling water as a function of time.

4. When will the temperature of the water reach 32 degrees?

Solution to Part 1: Let t be the time in minutes and D the temperature difference. To test if the data is exponential we make a table of successive quotients. (We rounded our answers to two decimal places.)

Time increment	From 0 to 1	From 1 to 2	From 2 to 3	From 3 to 4	From 4 to 5
Ratios of D	$\dfrac{66.3}{69} = 0.96$	$\dfrac{63.7}{66.3} = 0.96$	$\dfrac{61.2}{63.7} = 0.96$	$\dfrac{58.8}{61.2} = 0.96$	$\dfrac{56.5}{58.8} = 0.96$

The table of values shows a common quotient of 0.96, and we conclude that the data can be modeled by an exponential function.

Solution to Part 2: We know that the decay factor for D is the common ratio 0.96 we calculated in Part 1. We also know that when $t = 0$, then $D = 69$, and that is the initial value of D:

$$D = \text{Initial value} \times (\text{Decay factor})^t$$
$$D = 69 \times 0.96^t .$$

Solution to Part 3: Let W be the temperature of the water. The function D is the temperature W of the water minus 6, the temperature of the freezer:

$$W - 6 = D$$
$$W = 6 + D$$
$$W = 6 + 69 \times 0.96^t \text{ degrees Fahrenheit.}$$

Solution to Part 4: We want to solve the equation

$$6 + 69 \times 0.96^t = 32 \, .$$

We enter $\boxed{4.6}$ the water temperature function W and the target temperature, 32 degrees Fahrenheit. We record the variable correspondences:

$$Y_1 \;=\; W, \text{ water temperature, on vertical axis}$$

$$X \;=\; t, \text{ minutes, on horizontal axis.}$$

We made the table of values in Figure 4.9 using an increment value of 10 minutes to get an estimate of when the temperature will reach 32 degrees. The table shows that this will happen within 30 minutes. Thus we set up our graphing window using a horizontal span of $t = 0$ to $t = 30$ and a vertical span of $W = 0$ to $W = 75$. The graphs with the crossing point calculated $\boxed{4.7}$ are in Figure 4.10. We see that the temperature will reach 32 degrees in 23.9 minutes.

Figure 4.9: *A table of values for water temperature*

X	Y1	Y2
0	75	32
10	51.873	32
20	36.498	32
30	26.276	32
40	19.48	32
50	14.962	32
60	11.958	32

X=0

Figure 4.10: *When the temperature is 32 degrees*

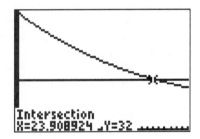

Intersection
X=23.908924 Y=32

Growth and decay factor units in exponential modeling

In the contaminated well example, the decay factor was the common quotient we calculated. This will always occur when time measurements are given in one unit increments. But when data is measured in different increments, adjustments to account for units must be made to get the right decay factor. We can show what we mean by looking at the contaminated water well example again, but this time we suppose that measurements were taken every three months. Under those conditions, the data table would have been as follows.

$t =$ months	0	3	6	9	12	15
$D =$ contaminant level difference	64	22.91	8.2	2.93	1.05	0.38

We test to see if the data is exponential by calculating successive ratios, rounding to two decimal places.

Time increment	From 0 to 3	From 3 to 6	From 6 to 9	From 9 to 12	From 12 to 15
Ratios of D	$\dfrac{22.91}{64} = 0.36$	$\dfrac{8.2}{22.91} = 0.36$	$\dfrac{2.93}{8.2} = 0.36$	$\dfrac{1.05}{2.93} = 0.36$	$\dfrac{0.38}{1.05} = 0.36$

Since the successive ratios are the same, 0.36, we conclude once again that the data is exponential. But the common ratio 0.36 is not the monthly decay factor. Rather, it is the *three month decay factor*. To get the monthly decay factor, we need to change the units. Since one month is one-third of the three month recording time, we get

$$\text{Monthly decay factor} = (\text{Three month decay factor})^{\frac{1}{3}}$$

$$\text{Monthly decay factor} = 0.36^{\frac{1}{3}} = 0.71 \text{ rounded to two decimal places.}$$

The initial value is 64, and so we arrive at the same exponential model, 64×0.71^t, as we did when we modeled the data given in one month intervals.

KEY IDEA 4.4: MODELING EXPONENTIAL DATA

1. To test if data with evenly spaced values for the variable shows exponential growth or decay, calculate successive quotients. If the quotients are all the same, it is appropriate to model the data with an exponential function. If the quotients are not all the same, some other model will be needed.

2. To model exponential data, use the common successive quotient for the growth (or decay) factor, but if the data is not measured in one unit increments, it will be necessary to make an adjustment for units. One of the data points can be used to determine the initial value.

EXAMPLE 4.5 *Finding the Time of Death*

One important topic of forensic medicine is the determination of the time of death. A method that is sometimes used involves temperature. Suppose at 6:00 p.m. a body is discovered in a basement of a building where the ambient air temperature is maintained at 72 degrees.[2] At the moment of death, the body temperature was 98.6 degrees, but after death the body cools, and eventually its temperature matches the ambient air temperature. Beginning at 6:00 p.m., the body temperature is measured and the difference $D = D(t)$ between body temperature and ambient air temperature is recorded. The measurement is repeated every 2 hours.

t = hours since 6:00 p.m.	0	2	4	6	8
D = temperature difference	12.02	8.08	5.44	3.65	2.45

---

[2]If the air temperature fluctuates, the establishment of time of death using body temperature is more difficult. It is noted in the book *Helter Skelter* by Vincent Bugliosi that body temperature was used to aid in establishing the time of death in the Charles Manson group murders of Sharon Tate and others.

1. Show that the data can be modeled by an exponential function.

2. Find an exponential model for the data that shows temperature difference as a function of hours.

3. Find a formula for a function $T = T(t)$ which gives the temperature of the body at time t.

4. What was the time of death?

Solution to Part 1: To see if the data is exponential, we calculate successive quotients, rounding to two decimal places.

Time increment	From 0 to 2	From 2 to 4	From 4 to 6	From 6 to 8
Ratios of D	$\dfrac{8.08}{12.02} = 0.67$	$\dfrac{5.44}{8.08} = 0.67$	$\dfrac{3.65}{5.44} = 0.67$	$\dfrac{2.45}{3.65} = 0.67$

Since the successive quotients are the same, we conclude that the data shows exponential decay.

Solution to Part 2: To give an exponential model, we need to know the initial value and the hourly decay factor. From the data table, the initial value is $D(0) = 12.02$. Our calculation of successive quotients in Part 1 gave us the 2 hour decay factor, 0.67. We want the 1 hour decay factor. Since 1 hour is half of 2 hours, we get

$$\text{Hourly decay factor} = (\text{2 hour decay factor})^{\frac{1}{2}}$$
$$\text{Hourly decay factor} = 0.67^{\frac{1}{2}} = 0.82 \text{, rounded to two decimal places.}$$

Thus we have

$$D = 12.02 \times 0.82^{t} .$$

Solution to Part 3: The function $D = 12.02 \times 0.82^{t}$ gives the difference between the temperature of the body and that of the air. Since we are assuming that the air has a temperature of 72 degrees, we add that to D to get the temperature of the body:

$$T = 72 + 12.02 \times 0.82^{t} .$$

Solution to Part 4: To find the time of death we need to know when the temperature was that of a living person, 98.6 degrees. Thus we need to solve the equation

$$72 + 12.02 \times 0.82^t = 98.6$$

for t. Before we begin, we note that since $t = 0$ corresponds to some time several hours after death, the value of t we are looking for is surely negative. We solve the equation using the crossing graphs method. In Figure 4.11 we have made the graph of T versus t

---

Figure 4.11: Finding the time of death

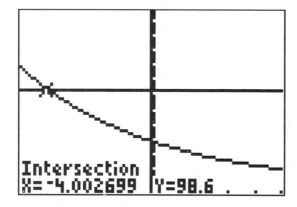

using a horizontal span of $t = -5$ to $t = 5$ hours and a vertical span of $T = 70$ to $T = 120$ degrees. You may wish to consult a table of values to see why we chose these window settings. We then added the graph of the target temperature $T = 98.6$ and found the intersection point. From the bottom of Figure 4.11 we see that death occurred just over 4 hours before the body was found. Since $t = 0$ corresponds to 6:00 p.m., we set the time of death at about 2:00 p.m.

Exercise Set 4.2

1. **Making an exponential model:** Show that the following data is exponential and find a formula for an exponential model.

t	0	1	2	3	4	5
$f(t)$	3.8	3.95	4.11	4.27	4.45	4.62

2. **An exponential model with unit adjustment:** Show that the following data is exponential and find a formula for an exponential model. (<u>Note</u>: It will be necessary to make a unit adjustment. For this problem round your answers to three decimal places.)

t	0	4	8	12	16	20
$g(t)$	38.3	28.65	21.43	16.04	11.99	8.97

3. **Data that is not exponential:** Show that the following data is not exponential.

t	0	1	2	3	4	5
$h(t)$	4.9	26.6	91.7	200.2	352.1	547.4

4. **Linear and exponential data:** One of the two tables below shows linear data and the other shows exponential data. Identify which is which, and find models for both.

Table A

t	0	1	2	3	4	5
$f(t)$	6.7	7.77	9.02	10.46	12.13	14.07

Table B

t	0	1	2	3	4	5
$g(t)$	5.8	7.53	9.26	10.99	12.72	14.45

5. **An investment:** You have invested money in a savings account which pays a fixed monthly interest on the account balance. The following table shows the account balance over the first five months.

Time in months	0	1	2	3	4	5
Savings balance	$1750	$1771	$1792.25	$1813.76	$1835.52	$1857.55

(a) How much money was originally invested?

(b) Show that the data is exponential and find an exponential model for the account balance.

(c) What is the monthly interest rate?

(d) What is the yearly interest rate?

(e) Suppose you made this investment on the occasion of the birth of your daughter. Your plan is to leave the money in the account until she starts college at age 18. How large a college fund will she have?

(f) How long does it take your money to double in value? How much longer does it take it to double in value again?

6. **A bald eagle murder mystery:** At 3:00 p.m. a park ranger discovered a dead bald eagle which had been impaled by an arrow. Only two archers were found in the region. The first archer is able to establish that between 11:00 a.m. and 1:00 p.m. he was in a nearby diner having lunch. The second archer can show that he was in camp with friends between 9:00 a.m. and 11:00 a.m. The air temperature in the park has remained at a constant 62 degrees. Beginning at 3:00 p.m. the difference $D = D(t)$ between the temperature of the dead eagle and that of the air was measured and recorded in the table below. (Here t is the time in hours since 3:00 p.m.) This table, together with the fact that the body temperature of a living bald eagle is 105 degrees, exonerates one of the archers, but the other may remain suspect. Which archer's innocence is established?

t = hours since 3:00 p.m.	0	1	2	3	4	5
D = temperature difference	26.83	24.42	22.22	20.22	18.40	16.74

7. **A skydiver:** When a skydiver jumps from an airplane, his downward velocity increases until the force of gravity matches air resistance. The velocity at which this occurs is known as the *terminal velocity*. It is the upper limit on the velocity a free falling skydiver will attain (in a stable, spread position), and for an average size man its value is about 176 feet per second (or 120 miles per hour). A skydiver jumped from an airplane, and the difference $D = D(t)$ between the terminal velocity and his downward velocity in feet per second was measured in five second intervals and recorded in the following table.

t = seconds into free fall	0	5	10	15	20	25
D = velocity difference	176	73.61	30.78	12.87	5.38	2.25

(a) Show that the data is exponential and find an exponential model for D. (Round all your answers to two decimal places.)

(b) What is the percentage decay rate per second for the velocity difference of the skydiver? Explain in practical terms what this number means.

(c) Let $V = V(t)$ be the skydiver's velocity t seconds into free fall. Find a formula for V.

(d) How long would it take the skydiver to reach 99% of terminal velocity?

8. **The half-life of U^{239}:** Uranium 239 is an unstable isotope of uranium which decays rapidly. In order to determine the rate of decay, one gram of U^{239} was placed in a container, and the amount remaining was measured at one minute intervals and recorded in the table below.

Time in minutes	0	1	2	3	4	5
Grams remaining	1	0.971	0.943	0.916	0.889	0.863

(a) Show that this is exponential data and find an exponential model. (For this problem, round all your answers to three decimal places.)

(b) What is the percentage decay rate each minute? What does this number mean in practical terms?

(c) What is the half-life of U^{239}?

9. **An inappropriate linear model for radioactive decay:** *This is a continuation of Exercise 8.* Physicists have established that radioactive substances display constant percentage decay, and thus radioactive decay is appropriately modeled exponentially. This exercise is designed to show how using data without an understanding of the phenomenon which generated it can lead to inaccurate conclusions.

(a) Plot the data points from Exercise 8. Do they appear almost to fall on a straight line?

(b) Find the equation of the regression line and add its graph to the one you made in Part (a).

(c) If you used the regression line as a model for decay of U^{239}, how long would it take for the initial one gram to decay to half that amount? Compare this with your answer to Part (c) of Exercise 8.

(d) The linear model represented by the regression line makes an absurd prediction concerning the amount of uranium 239 remaining after 1 hour. What is this prediction?

10. **Wages:** A worker is reviewing his pay increases over the past several years. The table below shows the hourly wage W (in dollars) he earned as a function of time t, measured in years since the beginning of 1990.

Time t	1	2	3	4
Wage W	15.30	15.60	15.90	16.25

(a) By calculating ratios, show that the data in this table is exponential. (Round the quotients to two decimal places.)

(b) What is the yearly growth factor for the data?

(c) The worker can't remember what hourly wage he earned at the beginning of 1990. Assuming that W is indeed an exponential function, determine what that hourly wage was.

(d) Find a formula giving an exponential model for W as a function of t.

(e) What percentage raise did the worker receive each year?

(f) Assuming an increase in prices of 34% over the decade of the 1990's, use your model to determine whether the worker's wage increases will have kept pace with inflation.

4.3 MODELING NEARLY EXPONENTIAL DATA

As in the case of linear data, rarely can experimentally gathered data be modeled exactly by an exponential function, but such data can in many cases be closely approximated by an exponential model. A key step in analyzing exponential data is to establish its relationship with linear data. This is done using the *logarithm* function.

From linear to exponential data

To establish the link between linear and exponential data, we look at an example. You should verify that the data in Table 4.1 is linear by showing that successive differences give a common value of 0.42. This is the slope for the linear data.

Table 4.1: *A table of linear data*

x	0	1	2	3	4	5
$y = y(x)$	1	1.42	1.84	2.26	2.68	3.1

In Figure 4.12 we have entered ⎣4.8⎦ this data and then plotted ⎣4.9⎦ it in Figure 4.13. As expected, the points fall on a straight line.

Figure 4.12: *Linear data entered in the calculator*

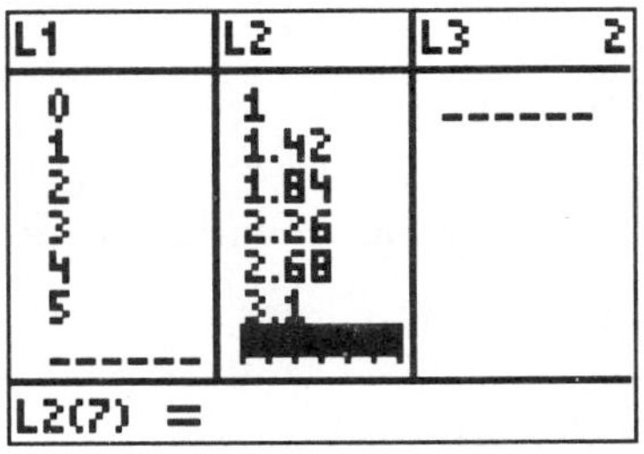

Figure 4.13: *A plot of linear data*

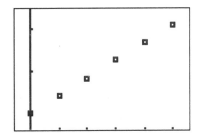

Let's see what happens if we replace each y value in the table by e^y. This process is often referred to as *exponentiating* the data. The resulting table (with calculations rounded to four decimal places) is in Table 4.2. (We have chosen to round to four decimal places on this occasion so that our hand-generated work will match what happens on the calculator later.)

Table 4.2: *Exponentiating linear data*

x	0	1	2	3	4	5
e^y	2.7183	4.1371	6.2965	9.5831	14.5851	22.1980

In Figure 4.14 we have added 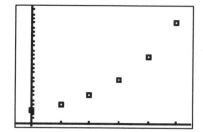 the data from Table 4.2 to the calculator data entry screen and then plotted 4.11 it in Figure 4.15. The plot in Figure 4.15 has a familiar and not

Figure 4.14: *Entering exponentiated data*

Figure 4.15: *Plot of exponentiated data*

surprising shape; it looks like exponential data. You should verify that the data in Table 4.2 is indeed exponential by showing that successive quotients give a common value of 1.5219. This is the growth factor for the exponential data.

We should also take note of the relationship between the slope 0.42 of the linear data and the growth factor 1.5219 of the exponentiated data. Since we used e^x to generate the data in Table 4.2, the following calculation is not surprising:

$$e^{\text{slope of linear data}} = e^{0.42} = 1.5219 = \text{ growth factor of exponential data.}$$

This same relationship holds for the initial value 1 for the linear data and the initial value 2.7183 for the exponential data, and you should verify the following calculation:

$$e^{\text{linear initial value}} = e^1 = 2.7183 = \text{ exponential initial value.}$$

The relationships we saw here are true in general, and the following schematic is presented to summarize these relationships and make them easy to remember.

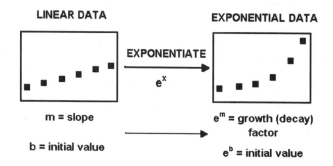

LINEAR DATA

EXPONENTIAL DATA

EXPONENTIATE

e^x

m = slope

b = initial value

e^m = growth (decay) factor

e^b = initial value

From exponential to linear data: the logarithm

The key tool needed to reverse the link, that is, to convert exponential data to linear data, is a special function known as the *natural logarithm*. This is denoted by $\ln x$, and it is closely related to the exponential function e^x. In formal mathematical terms, $\ln x$ and e^x are *inverses*. If we apply the exponential function to a number x, that number is put as an exponent for e, giving e^x. The logarithm does just the opposite; the logarithm applied to e^x removes the e and just leaves the exponent x. That is, $\ln e^x = x$. Thus, for example, $\ln e^2 = 2$. Informally, the logarithm undoes exponentiation. More formally, $\ln x$ is the power of e that gives x. The values of such expressions as $\ln 7$ are in general very difficult to calculate by hand, and it is appropriate to get the evaluation from the calculator ⌐4.12⌐ as we have done in Figure 4.16. This new function, the natural logarithm, is no more complicated than is the exponential function e^x, and a little experience with it will make it seem more familiar.

Figure 4.16: *Calculating* $\ln 7$

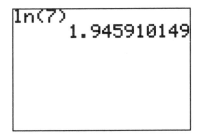

Figure 4.17: *Exponential data entered in the calculator*

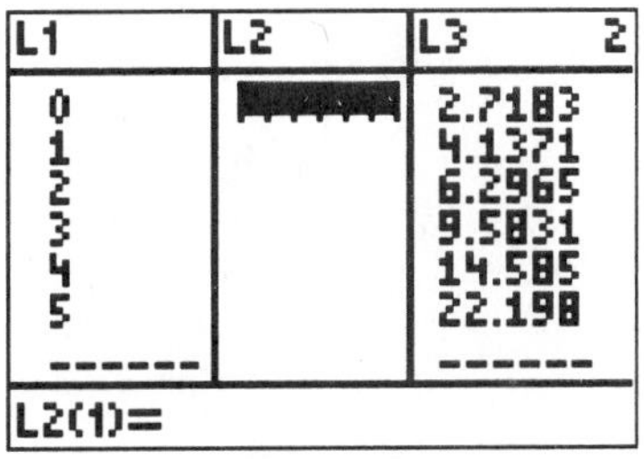

Let's see now how the logarithm works to straighten out exponential data. This time we start with the exponential data in Table 4.2. In Figure 4.17 we have entered the exponential data in the third column. We know already that if we plot this data, we get the picture in Figure 4.15. We want to see what happens if we take the logarithm of each entry in the third column, as we have done ⌐4.13⌐ in Figure 4.18. In this figure, each entry in the second column is the logarithm of the corresponding entry in the third column; for example, 1 is the logarithm of 2.7183. The key thing to note here is that as a result, we recover in the second column exactly the linear data from Table 4.1. This shows clearly how the logarithm reverses the effects of exponentiation. That is, the logarithm of exponential data is linear. Thus if we plot it, we will get the picture of linear data in Figure 4.13. When we go in this direction, from exponential to linear data using the logarithm, we get expected relationships among initial values, slope, and growth (or decay) factor:

$$\ln(\text{growth factor}) = \ln 1.5219 \;=\; 0.42 = \text{slope for linear data}$$

$$\ln(\text{exponential initial value}) = \ln 2.7183 \;=\; 1 = \text{linear initial value}.$$

Figure 4.18: *Using* ln x *to recover linear data from exponential data*

```
L1        L2        L3        2
0         ▮         2.7183
1         1.42      4.1371
2         1.84      6.2965
3         2.26      9.5831
4         2.68      14.585
5         3.1       22.198
-------   -------   -------
L2(1)=1
```

As before, we provide the following schematic, which is intended to summarize our findings and give an easy way of remembering them.

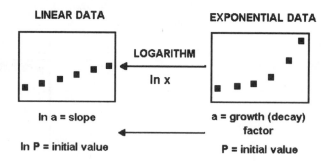

KEY IDEA 4.5: THE CONNECTION BETWEEN EXPONENTIAL AND LINEAR DATA

From linear to exponential: Exponentiating linear data produces exponential data.

$$e^{\text{slope}} = \text{growth (decay) factor.}$$
$$e^{\text{linear initial value}} = \text{exponential initial value.}$$

From exponential to linear: The logarithm of exponential data is linear.

$$\ln(\text{growth (decay) factor}) = \text{slope.}$$
$$\ln(\text{exponential initial value}) = \text{linear initial value.}$$

Exponential regression

We want to see how the link we have established between linear and exponential data can be used to make important analyses. The following data taken from the *1996 Information Please Almanac* shows U.S. population from 1800 to 1860, just prior to the Civil War.

Date	1800	1810	1820	1830	1840	1850	1860
Population in millions	5.31	7.24	9.64	12.87	17.07	23.19	31.44

Let t be the time in years since 1800 and N the population in millions. The table for N as a function of t is then

t	0	10	20	30	40	50	60
N	5.31	7.24	9.64	12.87	17.07	23.19	31.44

Since the data table shows population growth, it is not unreasonable to suspect that it might be exponential in nature. If you calculate successive quotients of N, you will find that it is not exactly exponential, but as with linear regression, it may still be appropriate to make an exponential model which approximates the data. That is the first question we want to answer: "Is it reasonable to approximate the data with an exponential model?"

First we enter $\boxed{4.14}$ the data in the calculator as we have done in Figure 4.19. Next we plot $\boxed{4.15}$ it to get a look at the overall trend. Figure 4.20 shows the classic shape of exponential growth, and the idea that the data may be exponential is reinforced.

Figure 4.19: *Entering pre-Civil War population data*

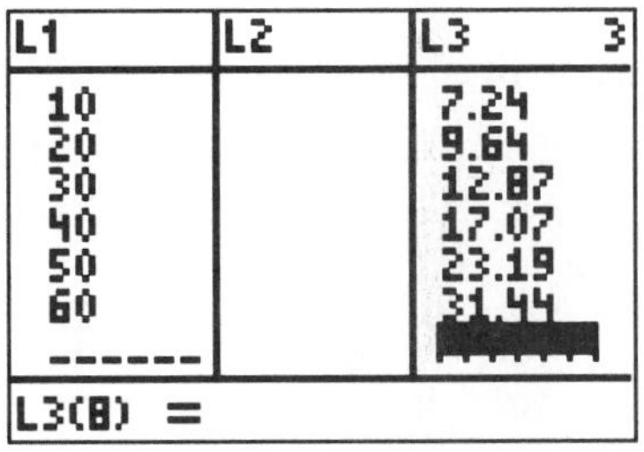

Figure 4.20: *A plot of pre-Civil War data*

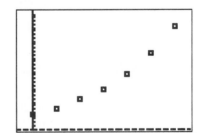

However, a caution is in order here. It is very difficult in general to be sure that data is exponential by looking at its plot because there are other types of data that will give a similar appearance. The key to determining if it is appropriate to model data by an exponential function is the logarithm/exponential link we have established between exponential and linear data. If this data is indeed exponential, then if we take the logarithm, the result should be linear data, and the plot of this data should show the points (approximately) falling on a straight line. [3] This is how we want to proceed. In Figure 4.21 we have entered the logarithm $\boxed{4.16}$ of the data in the second column, and in Figure 4.22 we have plotted $\boxed{4.17}$ it.

---

[3] Traditionally this has been accomplished by plotting the data points on a semi-logarithmic scale or on *semi-logarithmic graphing paper*.

Figure 4.21: *Automatic entry of the log-arithm of the data*

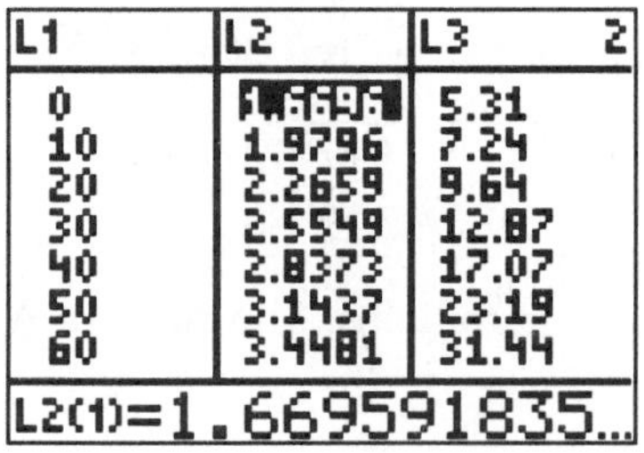

Figure 4.22: *The logarithm of the data appears to be linear*

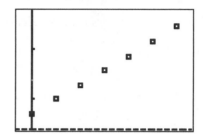

Figure 4.22 gives striking evidence that the logarithm of population data is nearly linear, and hence that the original population data is nearly exponential.

Now that we have convincing evidence that the original data set is exponential, we want to find an exponential function to model it. We find that there is another advantage to having "straightened out" the data with the logarithm; this allows us to apply what we know about linear data and make $\boxed{4.18}$ a regression line. The regression line parameters for $\ln N$ are shown in Figure 4.23. Let's write down what all the letters there mean. We have rounded to four decimal places:[4]

$$
\begin{aligned}
\mathsf{Y_1} &= \ln N, \text{ the logarithm of population} \\
\mathsf{X} &= t, \text{ years since 1800} \\
a &= 0.0294 = \text{ slope of the regression line} \\
b &= 1.6747 = \text{ initial value of regression line.}
\end{aligned}
$$

Thus we can write the formula for the regression line as

$$\ln N = 0.0294t + 1.6747 .$$

Figure 4.23: *Regression line data for* $\ln N$

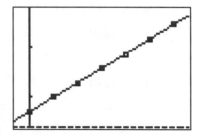

Figure 4.24: *Regression line added to logarithm of data*

---

[4]In exploiting the logarithm/exponential link, it is often necessary to use more digits than our standard convention of two.

In Figure 4.24 we have added $\boxed{4.19}$ the regression line to the plot of the logarithm of the data, and this reinforces our perception that these points lie nearly in a straight line.

There is one final step to make an exponential model for N. We used the logarithm to linearize the data and get a regression line for $\ln N$. To go back to N and the original data, we should exponentiate. In performing this step it is crucial to remember how slope, initial value, and growth factors are affected by this process.

$$\text{Slope of } \ln N = 0.0294 \quad \longrightarrow \quad \text{Growth factor of } N = e^{0.0294} = 1.0298$$

$$\text{Initial value of } \ln N = 1.6747 \quad \longrightarrow \quad \text{Initial value of } N = e^{1.6747} = 5.3372.$$

Thus we can approximate N with the exponential function

$$N = 5.3372 \times 1.0298^t .$$

In Figure 4.25 we have added the graph $\boxed{4.20}$ of this function to the original data plot of N in Figure 4.20. The level of agreement between the exponential curve and the data points is striking.[5]

Figure 4.25: *Exponential model for U.S. population data*

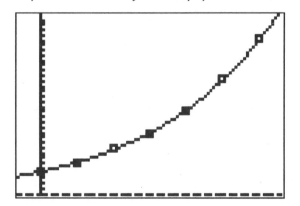

The process we have followed here is known as *exponential regression*, and we use the resulting formula in the same ways that we used the regression line formula for linear data. For example, the growth factor 1.0298 for N tells us immediately that from 1800 to 1860, U.S. population grew at a rate of 2.98% per year, or by about 34.1% per decade. We might also use the formula to estimate the population in 1870. To do this, we would put 70 in for t:

$$\text{Exponential regression estimate for 1870 } = 5.3372 \times 1.0298^{70} = 41.69 \text{ million.}$$

---

[5]While exponential models are often used to study population growth, such close agreement between the model and real data is unusual.

There is some uncertainty about the actual U.S. population in 1870. The 1870 census reported the number as 38.56 million. This figure was later revised to 39.82 million because it was thought that southern population had been undercounted. Whether we use the original or the revised 1870 census estimate, it is clear that U.S. population grew a good deal less than we would have expected from the exponential trend established from 1800 to 1860. Such a discrepancy leads us to seek a historical explanation, and the most obvious culprit is the Civil War. In fact the death and disruption of the Civil War may have had long-lasting effects on U.S. population growth. From 1790 to 1860, population grew at a steady rate of around 34.1% per decade. From 1860 to 1870 population grew by only 27%, and this rate steadily declined to its historical low of 7.2% from 1930 to 1940, the decade of the Great Depression.

KEY IDEA 4.6: EXPONENTIAL REGRESSION

When there is reason to believe that data for a function $N = N(t)$ might be approximately modeled by an exponential function, exponential regression can be performed as follows.

Step 1: Take the logarithm of the data for N and plot it. If N represents approximately exponential data, these points will lie nearly in a straight line.

Step 2: Get the regression line for $\ln N$.

Step 3: Get the exponential regression formula for N from the regression line for $\ln N$ by exponentiating. The key relationships are

$$e^{(\text{slope of regression line})} = \text{growth (decay) factor.}$$
$$e^{(\text{initial value of regression line})} = \text{initial value for } N.$$

EXAMPLE 4.6 *Colonial Population Growth*

The following data table shows colonial population from 1610 to 1670, with t the number of years since 1610 and $C = C(t)$ the population in thousands.[6] The source for the data, *The 1996 Information Please Almanac*, cautions that records from this period are spotty and so the numbers should be considered as estimates.

t = years since 1610	0	10	20	30	40	50	60
C = population in thousands	0.35	2.3	4.6	26.6	50.4	75.1	111.9

1. Plot the data to get an overall view of its nature.

2. Plot the logarithm of the data. Does the plot indicate that the original data might be modeled by an exponential function?

---

[6] Historical records from these periods ignored Native Americans.

3. Use exponential regression to construct a model for C. (Round regression line parameters to three decimal places.) Add the graph of the exponential model to the plot of population data.

4. Compare the population growth rate as given by exponential regression during colonial times with that in the 60 years preceding the Civil War.

Solution to Part 1: We enter $\boxed{4.21}$ the data in the calculator as shown in Figure 4.26 and then plot $\boxed{4.22}$ it as in Figure 4.27. The points in Figure 4.27 show some features of

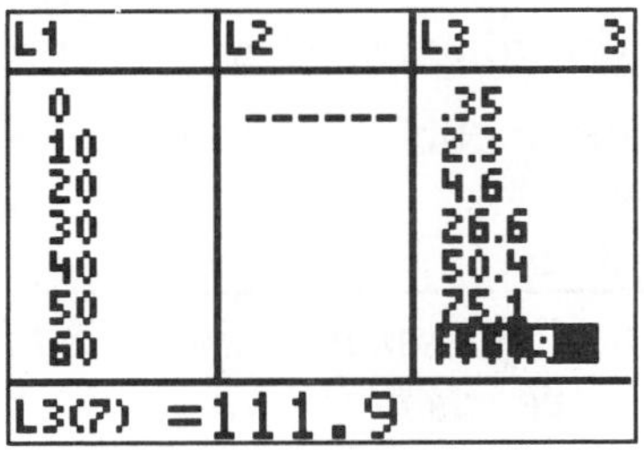

Figure 4.26: *Entering colonial data*

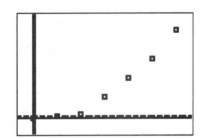

Figure 4.27: *Plot of colonial data*

exponential growth; it is slow at first and faster later on, but the picture certainly leaves room for doubt.

Solution to Part 2: We get the logarithm $\boxed{4.23}$ of the data as shown in Figure 4.28. When we plot the logarithm of the data as we did in Figure 4.29, we see that the points may generally line up, but they show some erratic behavior. The widely dispersed distribution of the points should cause us to be somewhat skeptical of a linear model for this data and hence of an exponential model for the original data.

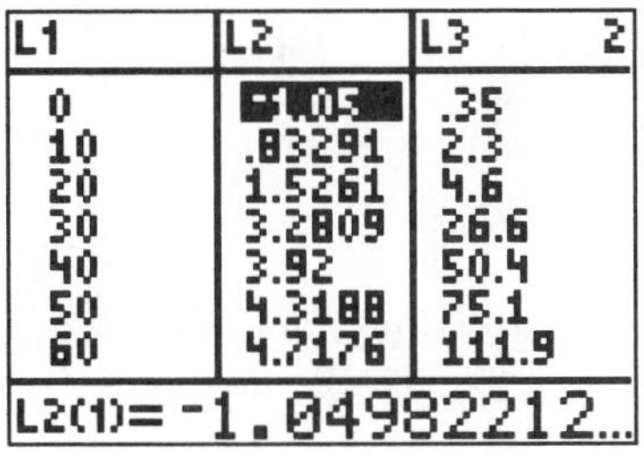

Figure 4.28: *Getting the logarithm of the data*

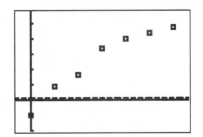

Figure 4.29: *Plotting* $\ln C$: *The distribution is erratic*

Solution to Part 3: We get the regression line parameters for $\ln C$ as shown in Figure 4.30.
Let's record the appropriate correspondences:

$$Y_1 \;=\; \ln C, \text{ logarithm of population}$$

$$X \;=\; t, \text{ years since 1610}$$

$$a \;=\; 0.095, \text{ slope of regression line}$$

$$b \;=\; -0.351, \text{ initial value of regression line.}$$

Thus the regression line for $\ln C$ is given by

$$\ln C = 0.095t - 0.351 \; .$$

In Figure 4.31 we have added $\boxed{4.24}$ the graph of the regression line for $\ln C$ to the plot of
the logarithm of the data. The points are scattered about the line, but agreement remains
dubious.

Figure 4.30: *Regression line parameters for* $\ln C$

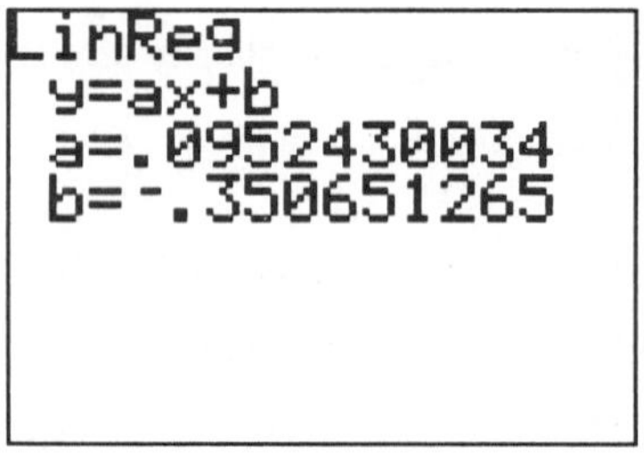

Figure 4.31: *Regression line added to plot of* $\ln C$ *versus* t

```
LinReg
 y=ax+b
 a=.0952430034
 b=-.350651265
```

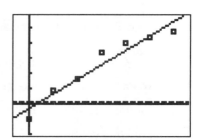

We exponentiate to get the model for C:

$$\text{Growth factor} \;=\; e^{0.095} = 1.1$$

$$\text{Initial value} \;=\; e^{-0.351} = 0.704 \; .$$

Using these values, we get the exponential model for C:

$$C = 0.704 \times 1.1^{t} \; .$$

Finally, in Figure 4.32, we add $\boxed{4.25}$ the graph of the exponential model to the plot of
data points for C.

Solution to Part 4: According to the exponential model we have constructed, colonial population grew at a rate of about 10% per year as compared with 2.98% per year from 1800
to 1860. But since the colonial model is so inaccurate, the estimate of 10% is unreliable.

Figure 4.32: *Exponential model added to plot of C versus t*

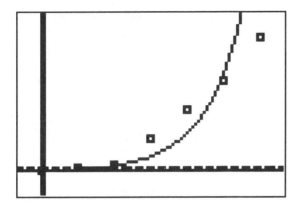

A note about the growth rate

We should note that in some applications, when exponential regression is used, the slope of the regression line for the logarithm of the data is used as the proportional growth (or decay) rate. For example, in our study of U.S. population growth from 1800 to 1860, we found that N grew during this period at a rate of about 2.98% per year, a per capita growth rate of 0.0298 per year as a decimal. The regression line $\ln N = 0.0294t + 1.16747$ has a slope of 0.0294 per year, and so 0.0294 is an alternative measure of the per capita growth rate. Of course, in this example the difference between the two measures of the per capita growth rate (0.0298 and 0.0294) is small.

In the next section we will re-examine the procedure of exponentiating the slope to get the growth (or decay) factor and see how to interpret the slope as a growth rate.

Exercise Set 4.3

Special rounding instructions: For this exercise set, round all regression line parameters to three decimal places.

1. **Using the definition of natural logarithm:** Recall that $\ln x$ is the power of e that gives x.

    (a) Without using your calculator, find the value of $\ln e^4$.

    (b) Use your calculator to show that $e^6 = 403.43$, rounded to two decimal places. Use this calculation to give the value of $\ln 403.43$. Verify your answer by calculating the logarithm with the calculator.

2. **The common logarithm:** The *common logarithm* or *logarithm to the base 10* works just like the natural logarithm[7] except that it uses 10 in place of e. Scientists and mathematicians normally use $\log x$ to denote the common logarithm. Formally, $\log x$ is the power of 10 that gives x.

    (a) What is $\log 10$? (<u>Hint</u>: What power of 10 gives 10?)

    (b) What is $\log 1000$?

3. **Cable TV:** The following table shows the percentage of American households with cable TV for the years from 1976 through 1984.

Date	1976	1977	1978	1979	1980	1981	1982	1983	1984
Percent with cable TV	15.1	16.6	17.9	19.4	22.6	28.3	35	40.5	43.7

    (a) Plot the natural logarithm of the data points. Does this plot make it look reasonable to approximate the original data with an exponential function?

    (b) Find the regression line for the natural logarithm of the data and add its graph to the plot of the logarithm.

    (c) Use exponential regression to construct an exponential model for the original cable TV data.

    (d) Plot the original data points and the exponential model.

    (e) What was the yearly percentage growth rate from 1976 through 1984 for the percentage of homes with cable TV?

---

[7]Historically the common logarithm was used before the natural logarithm. It can be used to do all the things we did in this chapter if e is replaced by 10 at each step. Many modern scientists and mathematicians prefer the natural logarithm, but use of the common logarithm has by no means disappeared.

(f) In 1984 an executive had a plan which could make money for the company, provided that at least 65% of American homes could be expected to have cable TV by
1987. Based solely on an exponential model for the data in the table, would it be
reasonable for the executive to implement the plan?

4. **Auto parts production workers:** The following table, taken from the *1994 U.S. Industrial
Outlook*, shows the average hourly wages for American auto parts production workers
from 1987 through 1994.

Date	1987	1988	1989	1990	1991	1992	1993	1994
Hourly wage	$13.79	$14.72	$14.99	$15.35	$15.70	$16.15	$16.50	$16.85

(a) Plot the natural logarithm of the data. Does it appear that it is reasonable to model
auto parts worker wages using an exponential function?

(b) Find the equation of the regression line for the natural logarithm of the data.

(c) Make an exponential model for auto parts worker wages.

(d) What was the yearly percentage growth rate in average hourly wages for auto parts
producers during this period?

(e) From 1987 through 1994, inflation was about 3.8% per year. If hourly wages beginning at $13.79 in 1987 had kept pace with inflation, what would be the average
hourly wage in 1994?

(f) What percentage raise should a worker receiving a wage of $16.85 in 1994 get in
order to bring wages in line with inflation?

5. **National health care spending:** The following table shows national health care costs measured in billions of dollars.

Date	1950	1960	1970	1980	1990
Costs in billions	12.7	26.9	75	248	600

(a) Plot the natural logarithm of the data. Does it appear that the data on health care spending can be appropriately modeled by an exponential function?

(b) Find the equation of the regression line for the logarithm of the data and add its graph to the plot in Part (a).

(c) By what percent per year were national health care costs increasing during the period from 1950 through 1990?

(d) Find an exponential function that approximates the original data for health care costs.

(e) How much money does the model you found in Part (d) predict will be spent on health care in the year 2000?

6. **A bad data point:** A scientist sampled data for a natural phenomenon which she has good reason to believe is appropriately modeled by an exponential function $N = N(t)$. A laboratory assistant reported that there may have been an error in recording one of the data points but is not certain which one. Which is the suspect data point? Explain your reasoning.

t	0	1	2	3	4	5	6
N	21.3	37.5	66.0	102.3	204.4	359.7	663.1

7. **Grazing rabbits:** The amount A of vegetation (measured in pounds) eaten in a day by a grazing animal is a function of the amount V of food available (measured in pounds per acre).[8] Even if vegetation is abundant, there is a limit, called the *satiation level*, to the amount the animal will eat. The following table shows for rabbits the difference D between the satiation level and the amount A of food eaten for a variety of values of V.

$V = $ vegetation level	27	36	89	134	245
$D = $ satiation level $- A$	0.16	0.12	0.07	0.05	0.01

(a) Draw a plot of $\ln D$ against V. Does it appear that D is approximately an exponential function of V?

(b) Find the equation of the regression line for $\ln D$ against V and add its graph to the plot in Part (a).

---

[8]This exercise is based on the work of J. Short, "The functional response of kangaroos, sheep and rabbits in an arid grazing system," *Journal of Applied Ecology* **22** (1985), 435–447.

(c) Find an exponential function that approximates D.

(d) The satiation level of a rabbit is 0.18 pound per day. Use this together with your work in Part (c) to find a formula for A.

(e) Find the vegetation level V for which the amount of food eaten by the rabbit will be 90% of its satiation level.

8. **Growth in length:** In the fishery sciences it is important to determine the length of a fish as a function of its age. One common approach, the von Bertalanffy model, uses a decreasing exponential function of age to describe the growth in length yet to be attained; in other words, the difference between the maximum length and the current length is supposed to decay exponentially with age. The following table shows the length L (in inches) at age t (in years) of the North Sea sole.[9]

$t = $ age	1	2	3	4	5	6	7	8
$L = $ length	3.7	7.5	10.0	11.5	12.7	13.5	14.0	14.4

The maximum length attained by the sole is 14.8 inches.

(a) Make a table showing, for each age, the difference D between the maximum length and the actual length L of the sole.

(b) Find the equation of the regression line for $\ln D$ against t.

(c) Find the exponential function that approximates D.

(d) Find a formula expressing the length L of a sole as a function of its age t.

(e) Draw a graph of L against t.

(f) If a sole is 11 inches long, how old is it?

(g) How long should a 9-year-old sole be?

---

[9]The table is from the work of A. Bückmann, as described by R. J. H. Beverton and S. J. Holt, *On the Dynamics of Exploited Fish Populations*, Fishery Investigations, Series 2, Volume 19, 1957, Ministry of Agriculture, Fisheries and Food, London.

9. **Nearly linear or exponential data:** One of the two tables below shows data which is better approximated with a linear function, and the other shows data which is better approximated with an exponential function. Make plots to identify which is which, and then use the appropriate regression to find models for both.

Table A

t	1	2	3	4	5
$f(t)$	3.62	23.01	44.26	62.17	83.25

Table B

t	1	2	3	4	5
$g(t)$	3.62	5.63	8.83	13.62	21.22

10. **Atmospheric pressure:** The following table gives a measurement of atmospheric pressure (in grams per square centimeter) at the given altitude (in kilometers).[10]

Altitude	5	10	15	20	25
Atmospheric pressure	569	313	172	95	52

(For comparison, one kilometer is about 0.6 mile, and one gram per square centimeter is about 2 pounds per square foot.)

(a) Plot the natural logarithm of the data, and find the equation of the regression line for the natural logarithm of the data.

(b) Make an exponential model for the data on atmospheric pressure.

(c) What is the atmospheric pressure at an altitude of 30 kilometers?

(d) Find the atmospheric pressure on Earth's surface. This is termed *standard atmospheric pressure.*

(e) At what altitude is the atmospheric pressure equal to 25% of standard atmospheric pressure?

---

[10]This exercise is based on *Space Mathematics* by B. Kastner, published by NASA, 1985.

4.4 ANOTHER LOOK AT EXPONENTIAL REGRESSION

In the preceding section we saw how to determine whether data is approximately exponential by exploiting the fact that the logarithm converts exponential data to linear data. To convert the regression line for the logarithm of the data to an approximate exponential model, we need to exponentiate the linear initial value and the slope. In this section we explain this conversion in terms of an important property of the exponential and the logarithm. With this property in hand, we proceed to interpret the slope in the context of compound interest and see a useful alternative form for exponential functions.

How a change in $\ln x$ affects x

Remember that $\ln x$ is the power of e that gives x. For example, $\ln(e^2)$ is the power of e that gives e^2. Thus $\ln(e^2) = 2$. It is easy to get natural logarithms for powers of e, and we have recorded several in Table 4.3.

Table 4.3: A brief table of natural logarithms

x	e	e^2	e^3	e^4	e^5
$\ln x$	1	2	3	4	5

There is an interesting pattern shown in Table 4.3 that we should note. Each time x is multiplied by e, the value of the natural logarithm increases by 1. Turning this around, we can say that when we add 1 to $\ln x$, then x is multiplied by e. In general, if we add k to $\ln x$, then the effect on x is to multiply it k times by e. That is, x gets multiplied by e^k. This observation is important for understanding how we do exponential regression.

KEY IDEA 4.7: HOW A CHANGE IN $\ln x$ AFFECTS x

If k is added to $\ln x$, then the effect on x is to multiply it by e^k.

Now we re-examine the procedure for converting a linear formula for $\ln N$ to an exponential formula for N. To find an exponential formula we need to know the initial value and the growth (or decay) factor. If the linear initial value for $\ln N$ is b, then $\ln N(0) = b$. This says that b is the power of e that gives $N(0)$, so $N(0) = e^b$. That explains why we exponentiate the linear initial value to get the exponential initial value.

Now we turn to the procedure for finding the growth (or decay) factor for N, which is the amount by which N is multiplied when we add 1 to the variable. Now if we add 1 to the variable for the linear function $\ln N$, then the effect on this function is to add a quantity equal to the slope. But, by our observation above, this in turn has the effect of *multiplying N by e^{slope}*. That explains why the growth (or decay) factor is the exponential of the slope.

To illustrate this point of view we look at an example.

EXAMPLE 4.7   *An Account Balance*

You invest $50 in an account whose balance is given by the following table. Here t is the time in years since you open the account, and B is the balance in dollars.

t = years since account opened	0	1	2	3	4
B = balance in dollars	50.00	55.26	61.07	67.49	74.59

1. Find the equation of the regression line for the natural logarithm of the data.

2. What yearly growth factor and yearly percentage growth rate best represent this account?

3. Find a formula for an exponential function that approximates the data for B in the table.

Solution to Part 1: We enter $\boxed{4.26}$ the data in the calculator and then get the logarithm $\boxed{4.27}$ of the data as shown in Figure 4.33. We then find the regression line parameters for $\ln B$ as shown in Figure 4.34. We record the appropriate correspondences:

$$Y_1 \;=\; \ln B, \text{ logarithm of balance}$$

$$X \;=\; t, \text{ years since opening account}$$

$$a \;=\; 0.1000, \text{ slope of regression line}$$

$$b \;=\; 3.9120, \text{ initial value of regression line.}$$

Figure 4.33: *Logarithm of balance data*

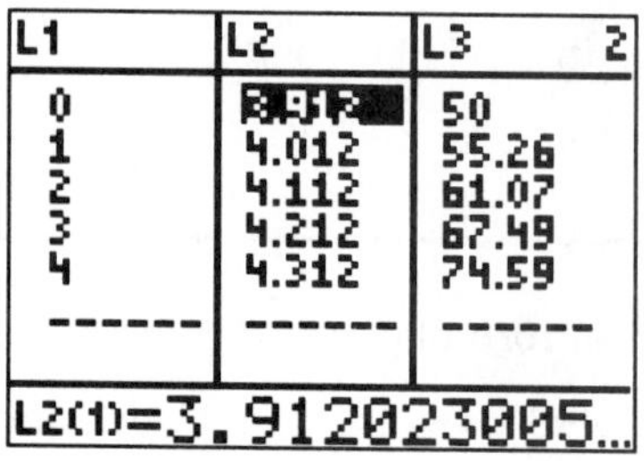

Figure 4.34: *Regression line parameters for* $\ln B$

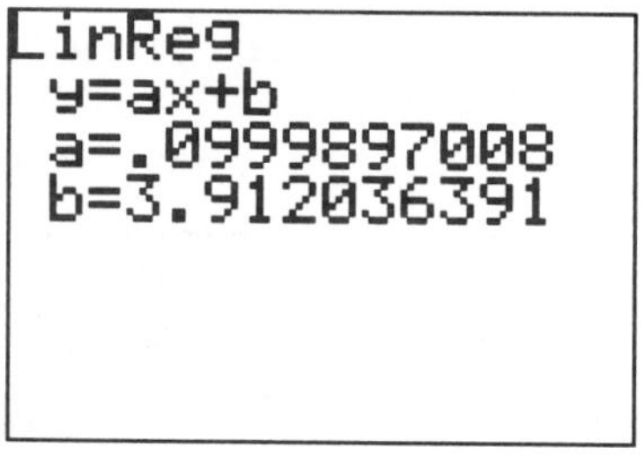

Thus the regression line for $\ln B$ is given by

$$\ln B = 0.1000t + 3.9120 \,.$$

Solution to Part 2: The slope of the regression line is 0.1000, and thus each for each year that passes, an amount of 0.1000 is added to $\ln B$. By our earlier observation, this causes B to be multiplied by $e^{0.1000}$, or about 1.1052, each year. Since this is the number we multiply by B to get the next year's balance, 1.1052 is the yearly growth factor for B. Thus the yearly percentage growth rate is 10.52%.

Solution to Part 3: In Part 2 we found the yearly growth factor for B to be 1.1052, so we just need to find the initial value for B. Since the initial value of the regression line for $\ln B$ is 3.9120, in our model for B itself we take the initial value to be $e^{3.9120} = 50.00$. The resulting formula for our approximate exponential model is

$$B = 50 \times 1.1052^t.$$

We stress that the procedure for converting the regression line for the logarithm of the data to an approximate exponential model is the same here as it was in the preceding section. We have just explained the procedure by taking a closer look at the link between linear and exponential functions.

Continuous compounding and the APR

We have seen that we exponentiate the slope of the regression line for the logarithm of data to get the growth (or decay) factor for the original data. At the end of the preceding section we indicated that the slope itself can be used as a growth (or decay) rate. To explain this further we study this slope in the context of compound interest.

Recall that financial institutions report the annual interest rate they offer as the APR, sometimes called the *nominal* interest rate. To find out how the account will actually grow, we need to know the effective annual rate, or EAR. (Financial institutions often refer to this as the *annual percentage yield*, or APY.) In Example 4.7 above, the EAR is just the yearly percentage growth rate of 10.52%. We want to find in financial terms the connection between this yearly percentage growth rate and the slope 0.1000 (or just 0.1) of the regression line in Example 4.7.

Suppose we open an account with a financial institution which advertises an APR (as a decimal) equal to this slope, 0.1. To find the yearly growth factor, and so get the EAR, we need to know how often interest is compounded. If the interest is compounded only once a year, then the yearly percentage growth rate is just the decimal APR of 0.1, so the yearly growth factor is $1 + 0.1$, or 1.1. What is the yearly growth factor if interest is compounded twice a year (i.e., semiannually)? Then the $\frac{1}{2}$-year percentage growth rate is $\frac{0.1}{2}$ (as a decimal), so the $\frac{1}{2}$-year growth factor is $1 + \frac{0.1}{2}$. To find the *yearly* growth factor we need to perform a unit conversion: One year is two half-year periods, so the yearly growth factor is $\left(1 + \frac{0.1}{2}\right)^2$, or

1.1025. In a similar way, if interest is compounded four times a year (i.e., quarterly), then the $\frac{1}{4}$-year percentage growth rate is $\frac{0.1}{4}$ (as a decimal), so the $\frac{1}{4}$-year growth factor is $1 + \frac{0.1}{4}$, and thus the yearly growth factor is $\left(1 + \frac{0.1}{4}\right)^4$, about 1.1038. The pattern is evident: If the decimal APR is 0.1 and interest is compounded n times a year, then the yearly growth factor for our account is

$$\left(1 + \frac{0.1}{n}\right)^n.$$

Now we ask what happens if interest is compounded continuously, in other words, at every instant. To get an idea, we see what happens in the above formula as the number n of compounding periods gets larger and larger. We use the calculator to make a table of values for the function $G = G(n)$ defined by $G = \left(1 + \frac{0.1}{n}\right)^n$, which gives the yearly growth factor as a function of the number n of compounding periods in a year. In Figure 4.35 we show the first part of the table $\boxed{4.28}$ for G, and in Figure 4.36 we show a later selection from the table $\boxed{4.29}$ which includes the case of daily compounding ($n = 365$). We note that the yearly growth factor for daily compounding $G(365)$ is about 1.1052 and that in fact all the values for G past this point in the table equal this value (to four decimal places).[11] We conclude that the yearly growth factor for a decimal APR of 0.1 with continuous compounding is 1.1052. Then the yearly percentage growth rate is 10.52%.

Figure 4.35: *Yearly growth factor for less frequent compounding*

X	Y₁	
1	1.1	
2	1.1025	
3	1.1034	
4	1.1038	
5	1.1041	
6	1.1043	
7	1.1044	
X=1		

Figure 4.36: *Yearly growth factor for more frequent compounding*

X	Y₁	
215	1.1051	
265	1.1052	
315	1.1052	
365	1.1052	
415	1.1052	
465	1.1052	
515	1.1052	
X=215		

We have seen that a financial institution which offers a decimal APR of 0.1, or 10%, with continuous compounding actually promises an EAR of 10.52% (the yearly percentage growth rate), since then the yearly growth factor is 1.1052. If we look again at the calculations in Part 2 of Example 4.7, we see the financial interpretation of the slope: In that example we exponentiated the slope, 0.1, to find the yearly growth factor, 1.1052. This means that the slope of the regression line for $\ln B$ is the APR for continuous compounding, since the corresponding yearly growth factor equals the exponential of the slope.

---

[11] One noteworthy conclusion based on this table is that there is a limit to the increase of the yearly growth factor (and hence the EAR) as the number of compounding periods increases. One might have expected the EAR to increase without limit for more and more frequent compounding.

The observations we have made hold for all exponential functions describing growth by compounding of interest: For a given APR (as a decimal) with continuous compounding, the yearly growth factor is $a = e^{\text{APR}}$. Further, since this equation says that the APR is the power of e that gives a, we can use the definition of the natural logarithm to find an equation giving the APR in terms of the growth factor. We get that $\text{APR} = \ln a$.

KEY IDEA 4.8: CONTINUOUS COMPOUNDING AND APR

If B is an exponential function of time (in years) describing an account balance growing by continuous compounding of interest, then the slope of the linear function $\ln B$ is the decimal APR, and the yearly growth factor can be obtained by exponentiating:

$$\text{Growth factor} = e^{\text{APR}} .$$

This can be rewritten as

$$\text{APR} = \ln(\text{Growth factor}) .$$

Regardless of the frequency of compounding, the EAR is the yearly percentage growth rate. For continuous compounding, then, as a decimal,

$$\text{EAR} = e^{\text{APR}} - 1 .$$

Exponential growth rate and an alternative form

We have seen the advantage of analyzing nearly exponential data by using the logarithm as a link to linear functions. Once we have an approximate linear model for the logarithm of the data, we still have to perform two steps: exponentiate the initial value of the linear model to get the initial value of the exponential model, and then exponentiate the slope of the linear model to get the growth (or decay) factor. Now we will see how to bypass the second of these steps and go directly to a useful alternative form of the exponential model.

Let's return to the account we studied in Example 4.7, where the initial balance is $50. We found the regression line for the logarithm of the balance data to be

$$\ln B = 0.1t + 3.912 .$$

This says that to find $\ln B$ at any t, we add to its initial value of 3.912 an amount of $0.1t$. But then, by our earlier observation, to find B itself at any t, we *multiply* its initial value of 50 by $e^{0.1t}$. In symbols,

$$B = 50 \times e^{0.1t} .$$

We now have two forms for this exponential function B giving the account balance: The original form is

$$B = 50 \times 1.1052^t ,$$

and the alternative form is

$$B = 50e^{0.1t}.$$

The connection between the two forms is made by remembering that $e^{0.1}$ is about 1.1052, and

$$1.1052^t = \left(e^{0.1}\right)^t = e^{0.1t}.$$

We can write a general exponential function $N = N(t)$ either as $N = Pa^t$ or in the alternative form $N = Pe^{rt}$. The connection between the two forms is based on the relation $a = e^r$. Since this last equation says that r is the power of e that gives a, we can rewrite that equation as $r = \ln a$.

This alternative form for exponential functions has appeared repeatedly in earlier chapters with a variety of applications. As we have just seen, the alternative form arises naturally for exponential functions describing continuous compounding of interest, where r is the APR (as a decimal). The alternative form is used in other contexts as well, where this interpretation of r in terms of the APR for continuous compounding is not as natural. For example, in the context of population ecology it is standard to write the formula for a population growing exponentially in the form Pe^{rt}. Ecologists often call the number r the *exponential growth rate* (or *exponential rate of increase*); it is a measure of the per capita growth rate of a population assumed to be growing continuously. We will adopt this terminology for general exponential functions as well: If any exponential function is written in the alternative form Pe^{rt}, we call r the exponential growth rate. We can think of the exponential growth rate in general as being analogous to the special case of the APR for continuous compounding of interest. Note, though, that the general exponential growth rate is written in decimal form, not as a percentage.

KEY IDEA 4.9: ALTERNATIVE FORM FOR EXPONENTIAL FUNCTIONS

If N is an exponential function of t with formula $N = Pa^t$, then an alternative form for N is

$$N = Pe^{rt}.$$

The connection is given by $a = e^r$, or $r = \ln a$. The number r is the exponential growth rate and equals the slope of the linear function $\ln N$.

EXAMPLE 4.8 *Population Growth*

An animal population is introduced to a reserve and grows according to the following table. Here time is measured in years since the population was introduced to the reserve, and the population is measured in hundreds of animals.

Time in years	0	1	2	3	4
Population in hundreds	43	67	100	135	195

1. Find the equation of the regression line for the natural logarithm of the data.

2. What exponential growth rate per year best describes the population?

3. Find an approximate exponential model in the alternative form.

4. What is the yearly growth factor for the model you found in Part 3 above? Write the exponential model in standard form.

5. For another animal population growing exponentially, the yearly growth factor is given as 1.26. What is the exponential growth rate per year for this population?

Solution to Part 1: Let t be the time in years since the population was introduced and $N = N(t)$ the population size in hundreds. We enter the data in the calculator, get the logarithm of the data, and then find the regression line parameters for $\ln N$. We record the result using the appropriate correspondences:

$$\begin{aligned}
\mathsf{Y_1} &= \ \ln N, \text{ logarithm of population} \\
\mathsf{X} &= \ t, \text{ years since introduction} \\
\mathsf{a} &= \ 0.372, \text{ slope of regression line} \\
\mathsf{b} &= \ 3.805, \text{ initial value of regression line.}
\end{aligned}$$

Thus the regression line for $\ln N$ is given by

$$\ln N = 0.372t + 3.805 \ .$$

Solution to Part 2: The exponential growth rate is the slope of the regression line, so $r = 0.372$ per year. This is measured on an annual basis since t is in years.

Solution to Part 3: We need to find the initial value for N. Since the initial value of the regression line for $\ln N$ is 3.805, in our model for N itself we take the initial value to be $e^{3.805} = 44.93$ hundred animals. The resulting approximate exponential model in the alternative form is

$$N = 44.93e^{0.372t} \ .$$

Solution to Part 4: The yearly growth factor a is given by $a = e^r = e^{0.372}$, or about 1.45, so the standard form for N is

$$N = 44.93 \times 1.45^t.$$

Solution to Part 5: The yearly growth factor for this new population is $a = 1.26$, so the exponential growth rate is $r = \ln a = \ln 1.26$, or about 0.23 per year.

One advantage to using the alternative form is that unit conversion is simplified: For the function N in Example 4.8 above, if we want to measure time in months instead of years, then we just divide the exponential growth rate per year, 0.372 , by 12 to get the exponential growth rate per month. We get $\dfrac{0.372}{12} = 0.031$ for the exponential growth rate per month. Thus if m is time in months, we have the formula $N = 44.93e^{0.031m}$. If we had started with the standard form $N = 44.93 \times 1.45^t$, then we would have needed to find the monthly growth factor for N by computing $1.45^{\frac{1}{12}}$. You should check that the result is 1.031 and that this monthly growth factor is the exponential $e^{0.031}$ of the corresponding exponential growth rate per month.

Exercise Set 4.4

1. **A given APR:** You open an account with a financial institution that promises an APR of 4.5%. If your account balance grows by continuous compounding of interest, what is the yearly growth factor for the balance?

2. **An account:** By studying the records for your savings account, you find that the balance grows by 4.5% each year. If interest is compounded continuously, what is the APR for this account?

3. **An investment:** You open an account by investing $250 with a financial institution that advertises an APR of 5.25%, with continuous compounding. What account balance would you expect one year after making your initial investment?

4. **A better investment:** You open an account by investing $250 with a financial institution that advertises an APR of 5.75%, with continuous compounding.

   (a) Find an exponential formula for the balance in your account as a function of time. In your answer give both the standard form and the alternative form for an exponential function.

   (b) What account balance would you expect 5 years after your initial investment? Answer this question using both of the forms you found in Part (a). Which do you think gives a more accurate answer? Why?

5. **Two formulas:** Your savings account grows by continuous compounding of interest. The balance B (in dollars) is a function of the time t in years since you opened the account.

   (a) If the formula for the balance is $B = 150e^{0.035t}$, find the APR and the EAR for your account.

   (b) Suppose now that the formula for the balance is $B = 150 \times 1.035^t$. Find the EAR and the APR for your account.

6. **A given EAR:** A financial institution promises an EAR of 5.5%, but it gives no information about how often interest is compounded.

   (a) If you open an account with this institution by making an initial deposit, what is the yearly growth factor for the exponential function describing your account balance?

   (b) If the institution were to compound interest only once a year, what APR would it advertise?

   (c) If the institution compounds interest continuously, what APR should it advertise?

7. **Hourly wages:** The hourly wage W (in dollars) earned by a certain worker is given as a function of time t (in years since the beginning of 1995). The formula is $W = 7.75 \times e^{0.044t}$.

   (a) What was the worker's hourly wage at the beginning of 1995?

   (b) What is the exponential growth rate per year for W?

   (c) What is the yearly growth factor for W?

   (d) Write the formula for W in the standard form for exponential functions.

   (e) What percentage raise does the worker receive each year?

8. **Spread of a disease:** A certain disease is spreading rapidly in a country. The cumulative number of people infected (in thousands) is given as a function of time t (in years) by the formula $N = 2.5 \times 1.15^t$.

   (a) What is the yearly growth factor for N?

   (b) What is the exponential growth rate per year for N?

   (c) Write the formula for N in the alternative form for exponential functions.

9. **Magazine circulation:** The following table gives the circulation (in hundreds) of a magazine as a function of the time in years since the magazine was started.

Time (years)	1	3	4	6	7
Circulation (hundreds)	29.5	175	436	2655	6534

   (a) Plot the natural logarithm of the data, find the equation of the regression line for the natural logarithm of the data, and add its graph to the plot.

   (b) Write a formula for an approximate model of the circulation, using the alternative form for exponential functions.

   (c) What is the exponential growth rate per year for the circulation?

(d) What is the exponential growth rate per month for the circulation?

(e) What is the monthly growth factor for the circulation?

10. **Radioactive decay:** The amount remaining (in grams) of a radioactive substance was measured and recorded in the table below. Here time is measured in days since the experiment began.

Time in days	0	2	3	5	6
Grams remaining	2.00	1.81	1.73	1.57	1.49

(a) Find the equation of the regression line for the natural logarithm of the data. (Round the regression line parameters to three decimal places.)

(b) Write a formula for an approximate model of the amount remaining, using the alternative form for exponential functions.

(c) What is the exponential growth rate per day for the amount remaining?

(d) What is the half-life of this radioactive substance?

4.5 CHAPTER SUMMARY

Exponential functions are almost as pervasive as are linear functions. They are commonly used to describe population growth, radioactive decay, free fall subject to air resistance, bank loans, inflation, and many other familiar events. Their defining property is similar to that of linear functions. Linear functions are those with a constant rate of change while exponential functions have a constant percentage or proportional rate of change. Any phenomenon which can be described in terms of yearly (monthly, daily, etc.) percentage growth (or decay) is properly modeled with an exponential function.

Exponential Growth and Decay

A linear function with slope m changes by constant sums of m. That is, if the variable is increased by 1 unit, the function is increased by adding m units. By contrast, an exponential function with base a changes by constant multiples of a. That is, when the variable is increased by 1, the function value is multiplied by a. Exponential functions are of the form

$$N = Pa^t,$$

where a is the *base* or *growth/decay factor*, and P is the initial value. When a is larger than 1, the exponential function grows rapidly, eventually becoming exceptionally large. When the base a is less than 1, the exponential function decays rapidly toward zero.

Many times exponential functions are described in terms of constant percentage growth or decay. If r is a decimal showing a percentage increase, then the growth factor is $1 + r$; if r shows a percentage decrease, then the decay factor is $1 - r$.

The growth factor for an exponential function is tied to a time period. It is sometimes important to know the growth factor for a different time period. We do this with unit conversion.

Unit conversion for exponential growth factors: If the growth factor for one period of time is a, then the growth factor A for k periods of time is given by $A = a^k$.

As an example, U.S. population in the early 1800's grew at a rate of about 2.9% per year. Thus population growth is an exponential function with yearly growth factor 1.029. The appropriate exponential function is of the form $N = P \times 1.029^t$, where t is measured in years. If we want to express population growth in terms of decades, we need a new growth factor:

$$1.029^{10} = 1.331.$$

Thus we can say that U.S. population in the early 1800's grew at a rate of 33.1% per decade, and $N = P \times 1.331^d$, where d is measured in decades.

Modeling Exponential Data

Data which is evenly spaced for the variable is linear if the function values show constant successive differences. It is exponential if it shows constant *successive quotients*. When the increment for the data is 1 unit, the common successive quotient is the growth (or decay) factor for the exponential function.

A simple example is provided by a hypothetical well which is being contaminated by seepage of water containing a pollutant at a level of 64 milligrams per liter. Each month the difference D between 64 and the contaminant level of the well is recorded.

t = months	0	1	2	3	4	5
D = contaminant level difference	64	45.44	32.26	22.91	16.26	11.55

The first quotient is calculated as $\dfrac{45.44}{64} = 0.71$. It can be verified that each successive quotient in the table gives this same value. We conclude that the data is exponential with decay factor 0.71. Since the initial value is 64, we obtain $D = 64 \times 0.71^t$.

Modeling Nearly Exponential Data

As with linear data, sampling error and other factors make it rare that experimentally gathered data can be modeled exactly by an exponential function. But as with linear functions, it may be appropriate to use the exponential model which most closely approximates the model. The key to making this happen is the *logarithm function*. The natural logarithm function $\ln x$ is the *inverse* of e^x. In other words, it is the function which "undoes" exponentiation: $\ln e^x = x$. This means that if the logarithm is applied to the function values of exponential data, the result will be linear, and linear regression can be applied. The required procedure is as follows.

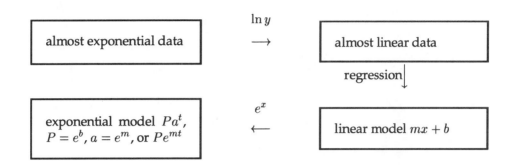

CHAPTER 5 *A Survey of Other Common Functions*

While linear and exponential functions probably are the mathematical functions which are most commonly found in applications, other important functions occur as well. We will look closely at *power functions* in the first two sections. The third section is devoted to *periodic functions*, while the last section provides a survey of *quadratic functions*, other *polynomials*, and *rational functions*.

5.1 POWER FUNCTIONS

Recall that an exponential function has the form $f(x) = Pa^x$, where the base a is fixed while the exponent x varies. For a power function, these are reversed: the base varies while the exponent remains constant, so a power function has the form $f(x) = cx^k$. The number k is called the *power*, the most significant part of a power function, and the coefficient c is equal to $f(1)$. For most applications, we are interested only in positive values of the variable x and of c, but we allow k to be any number.

We can use the graphing calculator to illustrate how power functions work and the role of k. You will be afforded the opportunity to make a graphical exploration of the role of c in the exercises at the end of this section. We look first at what happens when the power k is positive. In Figure 5.1 we show the graphs of x, x^2, x^3, and x^4. For the viewing window, we have used both a horizontal and a vertical span of 0 to 3. The first thing we note is that when the power is positive, the graph of the power function is increasing. We see also that when $k = 1$ the graph is a straight line. This tells us that a power function with power 1 is a linear function. Finally we observe that, for values of x larger than 1 (the common crossing point in Figure 5.1), larger powers cause the power function to grow faster.

In Figure 5.2, we look at what happens when the power k is negative. Here we show the graphs of x^{-1}, x^{-2}, x^{-3}, and x^{-4} using both a horizontal and a vertical span of 0 to 3. We see that for negative powers, power functions decrease toward 0. Furthermore, negative powers which are larger in size cause the graph to approach the horizontal axis more rapidly than do negative powers which are smaller in size.

Figure 5.1: *Power functions with positive powers are increasing functions*

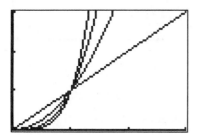

Figure 5.2: *Power functions with negative powers decrease toward 0*

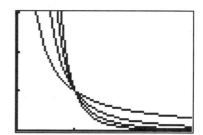

KEY IDEA 5.1: POWER FUNCTIONS

For a power function $f(x) = cx^k$ with c and x positive:

1. If k is positive, then f is increasing. Larger positive values of k cause f to increase faster.

2. If k is negative, then f decreases toward zero. Negative values of k which are larger in size cause f to decrease faster.

EXAMPLE 5.1 *Distance Fallen as a Function of Time*

When a rock is dropped from a tall structure, it will fall $D = 16t^2$ feet in t seconds.

1. Make a graph that shows the distance the rock falls versus time if the building is 70 feet tall.

2. How long does it take the rock to strike the ground?

Solution to Part 1: The first step is to enter the function ⌐5.1⌐ in the calculator function list and to record the variable associations:

$$Y_1 \;=\; D, \text{ distance on vertical axis}$$
$$X \;=\; t, \text{ time on horizontal axis.}$$

Since the building is 70 feet tall, we allow a bit of extra room and set the vertical span from 0 to 100. Since our everyday experience tells us that it will only take a few seconds for the rock to reach the ground, we set the horizontal span from 0 to 5 ⌐5.2⌐ . The graph appears in Figure 5.3.

Solution to Part 2: We want to know the value of t when the rock strikes the ground, that is, when $D = 70$ feet. Thus we need to solve the equation

$$16t^2 = 70.$$

We proceed with the crossing graphs method. Enter the target distance, 70, on the function list and plot to see the picture in Figure 5.4. We use the calculator to find the crossing point $\boxed{5.3}$ at $t = 2.09$ seconds as shown in Figure 5.4.

Figure 5.3: *Graph of distance fallen versus time*

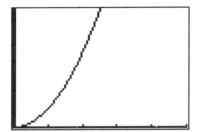

Figure 5.4: *When the rock strikes the ground*

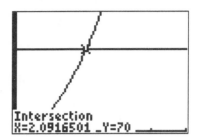

Homogeneity property of power functions

Many times the qualitative nature of a function is as important as the exact formula that describes it. It is, for example, crucial to physicists to understand that distance fallen is proportional to the square of the time (as opposed to an exponential, linear, or some other type of function). This is an important qualitative observation about how gravity acts on objects near the surface of the Earth. To make clear what we mean, we look at an important mathematical property of power functions known as *homogeneity*.

Suppose that in a power function $f(x) = cx^k$, the value of x is increased by a factor of t. What happens to the value of f? To help answer this question, let's first look at some specific examples.

How tripling the side of a square affects area: The area A of a square of side s is equal to the square of s. Thus $A = s^2$. Suppose a square initially has sides of length 4 feet. If the length of the sides of the square is tripled, how is the area affected? To answer this, we calculate the original area and compare it to the area after the sides of the square are tripled:

$$\text{Original area } = 4^2 = 16 \text{ square feet.}$$

To get the new area, we use $s = 12$:

$$\text{New area } = 12^2 = 144 \text{ square feet.}$$

Thus, if the side of the square is tripled, the area increases from 16 square feet to 144 square feet, by a factor of 9. The key thing to note here is that 9 is 3^2. Increasing the side by a factor of 3 increased the volume by a factor of 3^2, that is, 3 raised to the same power as the function.

How doubling the radius of a sphere affects the volume: A slightly more complicated calculation will show the same phenomenon. The volume V inside a sphere, such as a tennis ball or basketball, depends on its radius r and is proportional to the cube of the radius. Specifically, from elementary geometry we know that $V = \dfrac{4\pi}{3}r^3$. Suppose a balloon initially has a radius of 5 inches. If air is pumped into the balloon until the radius doubles, what is the effect on the volume? We proceed as before, calculating the original volume and comparing it with the volume after the radius is doubled. Initially the radius is $r = 5$. Accuracy is important in this calculation, and so in this case we report all the digits given by the calculator:

$$\text{Original volume } = \frac{4\pi}{3}5^3 = 523.5987756.$$

To get the new volume, we use $r = 10$:

$$\text{New volume } = \frac{4\pi}{3}10^3 = 4188.790205.$$

To understand how the volume has changed, we divide:

$$\frac{\text{New volume}}{\text{Old volume}} = \frac{4188.790205}{523.5987756} = 8.$$

Thus, increasing the radius by a factor of 2 results in an increase in volume by a factor of 8. Once again, the key thing to observe is that 8 happens to be 2^3. Summarizing, we see that if the radius is increased by a factor of 2, then the volume is increased by a factor of 2^3, that is, 2 raised to the power of the function.

The phenomenon we observed in these two examples is in fact characteristic of power functions, and we can show this using some elementary properties of exponents. Let's return to our original question. Suppose $f = cx^k$. If x is increased by a factor of t, what is the effect on f? We have

$$\text{Old value } = cx^k.$$

To get the new value, we replace x by tx:

$$\text{New value } = c(tx)^k = ct^kx^k = t^k(cx^k) = t^k \times \text{Old value}.$$

Thus, just as we observed in our examples, increasing x by a factor of t increases f by a factor of t^k.

KEY IDEA 5.2: HOMOGENEITY PROPERTY OF POWER FUNCTIONS

For a power function $f = cx^k$, if x is increased by a factor of t, then f is increased by a factor of t^k.

EXAMPLE 5.2   *Number of Species Versus Available Area*

Ecologists have studied how the number S of species of a given group existing in a closed environment (often an island) varies with the area A that is available.[1]   They make use of the approximate *species-area relation*

$$S = cA^k$$

to estimate, among similar habitats, the number of species as a function of available area. For birds on islands in the Bismarck Archipelago near New Guinea, the value of k is estimated to be about $k = 0.18$.

1. If one island in the Bismarck Archipelago were twice as large as another, how many more species of birds would it have? Interpret your answer in terms of percentages.

2. If there are 50 species of birds on an island in the Bismarck Archipelago of area 100 square kilometers, find the value of c, and then make a graph of the number of species as a function of available area for islands in the 50 to 200 square kilometer range. (Here we measure A in square kilometers.)

3. It is thought that a power relation $S = cA^k$ also applies to the Amazon rain forest, whose size is being reduced by, among other things, burning in preparation for farming. Ecologists use the estimate $k = 0.30$ for the power. If the rain forest were reduced in area by 20%, by what percentage would the number of surviving species be expected to decrease?

4. Experience has shown that, in a certain chain of islands, a 10% reduction in area leads to a 4% reduction in the number of species. Find the value of k in the species-area relation. What would be the result of a 25% reduction in the usable area of one of these islands?

Solution to Part 1: The number of species S is a power function of the area A, with power $k = 0.18$. If the variable A is doubled, then, by the homogeneity property of power functions, the function S increases by a factor of $2^{0.18} = 1.13$. Thus the number of species of birds increases by a factor of 1.13.

In terms of percentages, doubling A is a 100% increase, and changing S by a factor of 1.13 represents a 13% increase. Thus increasing the area by 100% has the effect of increasing the number of species by 13%.

---

[1]See the article "Island Biogeography and the Design of Natural Reserves" by J. Diamond and R. M. May, in R. M. May (ed.), *Theoretical Ecology*, 2nd edition, 1981, Sinauer Associates, Sunderland, MA. See also the references therein.

Solution to Part 2: We know that $S = cA^{0.18}$ and that an area of 100 square kilometers supports 50 species. That is, $S = 50$ when $A = 100$:

$$50 = c \times 100^{0.18}$$
$$50 = c \times 2.29.$$

This is a linear equation, and we can solve for c by hand calculation:

$$c = \frac{50}{2.29} = 21.83.$$

We conclude that, in the Bismarck Archipelago, the number of species is related to area by $S = 21.83A^{0.18}$. We first enter $\boxed{5.4}$ this on our calculator function list and record variable correspondences:

$$Y_1 = S, \text{ number of species on vertical axis}$$
$$X = A, \text{ area on horizontal axis.}$$

We are asked to make the graph for islands from 50 to 200 square kilometers in area. Thus we use a horizontal span of 50 to 200. For the vertical span, we look at the table of values in Figure 5.5. Allowing a little extra room, we use a vertical span of 40 to 60. The completed graph is in Figure 5.6.

Figure 5.5: *A table of values for number of species*

X	Y₁
50	44.144
75	47.486
100	50.01
125	52.059
150	53.796
175	55.31
200	56.655

X=50

Figure 5.6: *Number of species versus area in the Bismarck Archipelago*

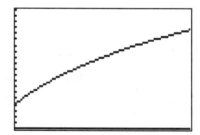

Solution to Part 3: To say that there is a 20% decrease in area means that the new area is 80% of its original value. Thus A has changed by a factor of 0.8. Using the homogeneity property of power functions, we conclude that S, the number of species, changes by a factor of $0.8^{0.30} = 0.94$. Thus, 94% of the original species can be expected to survive, and so the number of species has been reduced by 6%.

Solution to Part 4: We know that a 10% reduction in the area of an island will result in a 4% reduction in the number of species. In other words, if A is changed by a factor of 0.9, then S is changed by a factor of 0.96. The homogeneity property of power functions tells us that $0.96 = 0.9^k$. We can use the crossing graphs method to solve this equation as is shown in Figure 5.7, where we have used a horizontal span of 0 to 1 and a vertical span of 0.8 to 1.2. We read from the prompt at the bottom of the figure that $k = 0.39$, rounded to two decimal places.

Figure 5.7: *Finding the power in the species-area relation*

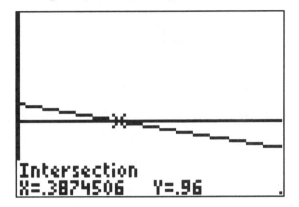

If the area is reduced by 25%, we are changing A by a factor of 0.75. Thus S changes by a factor of $0.75^{0.39} = 0.89$. That is an 11% reduction in the number of species.

ᴐmparing exponential and power functions

Over limited ranges, the graphs of exponential functions and power functions may appear to be similar. We see this in Figure 5.8, where we have graphed the exponential function 2^x and the power function x^2 using a horizontal span of 1 to 4 and a vertical span of 0 to 20. As the figure shows, the two graphs are nearly identical on this span. This sometimes makes it difficult to determine if observed data should be modeled with an exponential function or with a power function. We will return to the problem of how one might make an appropriate choice of model in the next section. For now, we want to show the consequences of an inappropriate choice. Figure 5.8 shows the similarity of the functions over the displayed range, but Figure 5.9 shows a dramatic difference if we view the graphs with a horizontal span of 1 to 7 and a vertical span of 0 to 70. The exponential function 2^x is the graph in Figure 5.9 which rises rapidly above the graph of the power function.

If you view these graphs on an even larger horizontal span, the differences will be even more dramatic. This behavior is typical of the comparison of any exponential function with base larger than one with any power function, no matter how large the power. Power functions may look similar to exponential functions over a limited range, and they may even grow

Figure 5.8: A limited span where a power function and an exponential function appear similar

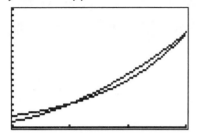

Figure 5.9: The characteristic dominance of exponential functions over power functions in the long term

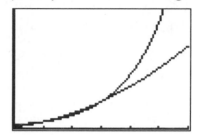

more rapidly for brief periods. But eventually, it always happens that exponential functions grow many times faster than power functions. This makes the consequences of choosing the wrong model quite serious. For example, if we were to choose an exponential model for federal spending when a power model was in fact appropriate, it would lead us to predict federal spending at levels many orders of magnitude too large. Or, if we chose a power model when in fact an exponential model was appropriate, we would be led to predict future federal spending at levels which are far too low.

The fact that exponential functions eventually grow many times faster than do power functions is one of the most important qualitative distinctions between these two types of functions.

KEY IDEA 5.3: EXPONENTIAL FUNCTIONS GROW FASTER THAN POWER FUNCTIONS

Over a sufficiently large horizontal span, an exponential function (with base larger than one) will increase much more rapidly than a power function.

EXAMPLE 5.3 *Exponential Versus Power Models*

In Chapter 4 we modeled early U.S. population using the exponential model $N = 5.34 \times 1.03^t$, where t is the number of years since 1800, and N is population in millions. Recall that an exponential model was chosen not only because it fit the data, but also because populations are expected (at least for brief periods) to grow at a constant percentage rate. Suppose we had used the power model $P = 0.04t^{1.5} + 5.34$, where t is years since 1800, and P is U.S. population in millions.[2]

1. Graph the exponential model and the power model over the years from 1800 to 1830. Do the models yield similar predictions over this time period?

2. Graph the exponential model and the power model over the years from 1800 to 1870. Do the models yield similar predictions over this time period? The actual population in 1860 was 31.44 million. Which model gives the more accurate estimate for 1860?

Solution to Part 1: First we enter $\boxed{5.5}$ the two formulas in the calculator function list and record variable correspondences:

$$Y_1 = N, \text{ exponential model population on vertical axis}$$

$$Y_2 = P, \text{ power model population on vertical axis}$$

$$X = t, \text{ years since 1800 on horizontal axis.}$$

We want the horizontal span to go from 0 to 30. We consult the table $\boxed{5.6}$ of values in Figure 5.10 to get a vertical span.

Figure 5.10: *A table of values for two population models from 1800 to 1830*

X	Y₁	Y₂
0	5.34	5.34
5	6.1905	5.7872
10	7.1765	6.6049
15	8.3195	7.6638
20	9.6446	8.9177
25	11.181	10.34
30	12.962	11.913

X=0

Figure 5.11: *Comparing population models from 1800 to 1830*

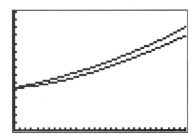

Allowing a bit of extra room, we set the vertical span from 0 to 15. The graphs appear in Figure 5.11, and we note that they lie very close together. This shows that the two

---

[2]Strictly speaking, this is not a power model since the extra constant term 5.34 is added. This addition is often made to satisfy an initial condition.

models give very similar predictions from 1800 to 1830, a fact which is borne out by the table of values in Figure 5.10.

Solution to Part 2: We want to change the horizontal span so that it goes from 0 to 70. The table of values in Figure 5.12 led us to choose a vertical span of 0 to 45, and we used these settings to make the graphs in Figure 5.13. The exponential model is the graph which is on top in Figure 5.13. The graph shows that the exponential model predicts a larger population in the mid-1800's than does the power model. If you look on a larger horizontal span you will observe greater separation in the curves, as is characteristic of the more rapid growth of exponential functions. Consulting the table of values in Figure 5.12, we see that the exponential model predicts an 1860 population of 31.46 million while the power model predicts an 1860 population of only 23.93 million. Clearly the exponential model is much closer to the true population of 31.44 million.

Figure 5.12: *A table of values for two population models from 1800 to 1870*

X	Y1	Y2
10	7.1765	6.6049
20	9.6446	8.9177
30	12.962	11.913
40	17.419	15.459
50	23.41	19.482
60	31.461	23.93
70	42.281	28.766

X=70

Figure 5.13: *Comparing population models from 1800 to 1870*

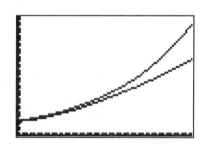

Exercise Set 5.1

1. **The role of the coefficient in power functions with positive power:** Consider the power functions $f(x) = cx^2$. On the same screen, make graphs of f versus x for $c = 1$, $c = 2$, $c = 3$, and $c = 4$. We suggest a horizontal span of 0 to 5. A table of values will be helpful in choosing a vertical span. Based on the plots you make, discuss the effect of the coefficient c on a power function when the power is positive.

2. **The role of the coefficient in power functions with a negative power:** Consider the power functions $f(x) = cx^{-2}$. On the same screen, make graphs of f versus x for $c = 1$, $c = 2$, $c = 3$, and $c = 4$. We suggest a horizontal span of 0 to 5. A table of values will be helpful in choosing a vertical span. Based on the plots you make, discuss the effect of the coefficient c on a power function when the power is negative.

3. **Speed and stride length:** The speed at which certain animals run is a power function of their stride length, and the power is $k = 1.7$. If one animal has a stride length three times as long as another, how much faster does it run?

4. **Weight and length:** A biologist has discovered that the weight of a certain fish is a power function of its length. He also knows that when the length of the fish is doubled, then its weight increases by a factor of 8. What is the power k?

5. **Length of skid marks versus speed:** If a car skids to a stop, the length L in feet of the skid marks is related to the speed S in miles per hour of the car by the power function $L = \dfrac{1}{30h}S^2$. Here the constant h is the *friction coefficient*, which depends on the road surface.[3] For dry concrete pavement, the value of h is about 0.85.

   (a) If a driver going 55 miles per hour on dry concrete jams on the brakes and skids to a stop, how long will be skid marks be?

   (b) A policeman investigating an accident on dry concrete pavement finds skid marks 230 feet long. The speed limit in the area is 60 miles per hour. Is the driver in danger of getting a speeding ticket?

   (c) This part of the problem applies to any road surface, and so the value of h is not known. Suppose you are driving at 60 miles per hour but because of approaching darkness wish to slow to a speed that will cut your emergency stopping distance in half. What should your new speed be? (<u>Hint</u>: You should use the homogeneity

---

[3]See *Accident Reconstruction* by J. C. Collins, 1979, Charles C. Thomas Publisher, Springfield, Il. Collins notes that the friction coefficient is actually reduced at higher speeds.

property of power functions here. By what factor should you change your speed to insure that L changes by a factor of 0.5?)

6. **Dropping rocks on other planets:** It is a consequence of Newton's law of gravitation that near the surface of any planet, the distance D fallen by a rock in time t is given by $D = ct^2$. That is, distance fallen is proportional to the square of the time, no matter what planet one may be on. But the value of c depends on the mass of the planet. For Earth, if time is measured in seconds and distance in feet, the value of c is 16.

   (a) Suppose a rock is falling near the surface of a planet. What is the comparison in distance fallen from 2 seconds to 6 seconds into the drop? (Hint: This question may be rephrased as follows: "If time increases by a factor of 3, by what factor will distance increase?")

   (b) For objects falling near the surface of Mars, if time is measured in seconds and distance in feet, the value of c is 6.4. If a rock is dropped from 70 feet above the surface of Mars, how long will it take for the rock to strike the ground?

   (c) On Venus, a rock dropped from 70 feet above the surface takes 2.2 seconds to strike the ground. What is the value of c for Venus?

7. **Newton's law of gravity:** According to Newton's law of gravity, the gravitational attraction between two massive objects such as planets is proportional to d^{-2}, where d is the distance between the centers of the objects. Specifically, the gravitational force F between such objects is given by $F = cd^{-2}$, where d is the distance between their centers. The value of the constant depends on the masses of the two objects and on the *universal gravitational constant*.

   (a) Suppose the force of gravity is causing two large asteroids to move toward each other. What is the effect on the gravitational force if the distance between their centers is halved? What is the effect on the gravitational force if the distance between their centers is reduced to one quarter of its original distance?

   (b) Suppose that for a certain pair of asteroids whose centers are 300 kilometers apart, the gravitational force is 2,000,000 newtons. (One newton is about one quarter of a pound.) What is the value of c? Find the gravitational force if the distance between the centers of these asteroids is 800 kilometers.

   (c) Using the value of c you found in Part (b), make a graph of gravitational force versus distance between the centers of the asteroids for distances from 0 to 1000 kilometers. What happens to the gravitational force when the asteroids are close together? What happens to the gravitational force when the planets are far apart?

8. **Geostationary orbits:** For communications satellites to work properly, they should appear from the surface of the Earth to remain stationary. That is, they should orbit the Earth exactly once each day. For *any* satellite, the *period P* (the length of time required to complete an orbit) is determined by its mean distance A from the center of the Earth. For a satellite of negligible mass, P and A are related by a power function $A = cP^{\frac{2}{3}}$.

(a) The moon is 239,000 miles from the center of the Earth and has a period of about 28 days. How high above the center of the Earth should a geostationary satellite be? (<u>Hint</u>: You want the distance A for a satellite with period $\frac{1}{28}$th that of the moon. The homogeneity property of power functions is applicable.)

(b) The radius of the Earth is about 3963 miles. How high above the surface of the Earth should a geostationary satellite be?[4]

9. **Giant ants and spiders:** Many science fiction movies feature animals such as ants, spiders, or apes growing to monstrous sizes and threatening defenseless Earthlings. (Of course they are in the end defeated by the hero and heroine.) Biologists use power functions as a rough guide to relate body weight and cross-sectional area of limbs to length or height. Generally, weight is thought to be proportional to the cube of length, while cross-sectional area of limbs is proportional to the square of length. Suppose an ant, having been exposed to "radiation," is enlarged to 500 times its normal length. (Such an event can only occur in Hollywood fantasy. Radiation is simply incapable of causing such a reaction.)

(a) By how much will its weight be increased?

(b) By how much will the cross-sectional area of its legs be increased?

(c) Pressure on a limb is weight divided by cross-sectional area. By how much has the pressure on a leg of the giant ant increased? What do you think is likely to happen to the unfortunate ant? [5] <u>Note</u>: The factor by which pressure increases is given by

$$\frac{\text{Factor of increase in weight}}{\text{Factor of increase in area}}.$$

---

[4]The actual height is 22,300 miles above the surface of the Earth. You will get a slightly different answer because we have neglected the mass of the moon.

[5]Similar arguments about creatures of extraordinary size go back at least to Galileo. See his famous book *Two New Sciences*, published in 1638.

10. **Kepler's third law:** By 1619 Johannes Kepler had completed the first accurate mathematical model describing the motion of planets around the sun. His model consisted of three laws which for the first time in history allowed accurate predictions of future locations of planets. Kepler's third law related the period (the length of time required for a planet to complete a single trip around the sun) to the mean distance D from the planet to the sun. In particular, he stated that the period P is proportional to $D^{1.5}$.

(a) Neptune is about 30 times as far from the sun as is the Earth. How long does it take Neptune to complete an orbit around the sun? (<u>Hint</u>: The period for Earth is 1 year. If the distance is increased by a factor of 30, by what factor will the period be increased?)

(b) The period of Mercury is about 88 days. The Earth is about 93 million miles from the sun. How far is Mercury from the sun? (<u>Hint</u>: The period of Mercury is different from that of Earth by a factor of $\dfrac{88}{365}$.)

5.2 MODELING DATA WITH POWER FUNCTIONS

In this section, we will learn how to construct power function models much as we did exponential models in the previous chapter.

The connection between power data and linear data

Linear regression makes it relatively easy to construct linear models, and the procedure is so reliable that in constructing other types of models, scientists and mathematicians often seek links with linear data so that regression may be used. For example, to construct an exponential model for observed phenomena as we did in the previous chapter, the needed link was the logarithm. We began with exponential data, used the logarithm of function values to convert it to linear data, applied regression to get a linear model for the transformed data, and finally converted the linear model to an exponential model for the original data. In order to make power function models, a similar link between power data and linear data is needed. As we shall see, we can get it by using the logarithm in a slightly different way.

To see the connection, we look at the familiar power function $f(x) = 3x^2$. In Figure 5.14 we have entered $x = 1, 2, \dots, 7$ in the third column and $f(x) = 3 \times 1^2, 3 \times 2^2, \dots, 3 \times 7^2$ in the fourth column ⬚5.7 . Thus in columns 3 and 4 of Figure 5.14 we see data from the power function $f(x) = 3x^2$. In Figure 5.15 we have entered $\boxed{5.8}$ the natural logarithm of x in the first column and the natural logarithm of $f(x)$ in the second $\boxed{5.9}$ column.

Figure 5.14: *Data for x and $3x^2$*   **Figure 5.15:** *Data for $\ln x$ and $\ln(3x^2)$*

L2	L3	L4 4
------	1	3
	2	12
	3	27
	4	48
	5	75
	6	108
	7	147
L4(1)=3		

L1	L2	L3 2
0	1.0986	1
.69315	2.4849	2
1.0986	3.2958	3
1.3863	3.8712	4
1.6094	4.3175	5
1.7918	4.6821	6
1.9459	4.9904	7
L2(1)=1.098612288...		

If we plot $\boxed{5.10}$ the data for $\ln(3x^2)$ versus the data for $\ln x$, we see in Figure 5.16 that the points fall on a straight line, which indicates that a linear relation holds. This is the link we had hoped to find. If f is a power function of x, then $\ln f$ is a linear function of $\ln x$.

Getting a power model from data

In practice, what is often wanted is to begin with observed data, check to see if it is appropriately modeled with a power function, and then actually build the model. Looking more closely at what we have already done, we can see how to do that. You may wish to refer to the *Keystroke Guide* for the exact keystrokes used to produce the following work.

We want to take the data we have and see how to recover the power function $f = 3x^2$ which generated it. We have arranged things so that data for $\ln x$ is already in column 1 and data for $\ln f$ is in column 2. Thus we can get the regression line parameters $\boxed{5.11}$ shown in Figure 5.17 in the usual way. We see that $\ln f$ is related to $\ln x$ by the linear function $\ln f = 2\ln x + 1.1$.

Figure 5.16: *$\ln f$ versus $\ln x$*   **Figure 5.17:** *Regression line parameters*

```
LinReg
y=ax+b
a=2
b=1.098612289
```

Observe that the slope 2 for the regression line turns out to be the same as the power for $f(x) = 3x^2$. Also, 1.1 is the vertical intercept of the line, and $3 = e^{1.1}$ is the c value for the power function. The connection we see here is in fact characteristic of the link between linear and power functions.

KEY IDEA 5.4: POWER FUNCTIONS ON A LOGARITHMIC SCALE

If $\ln f(x)$ is a linear function of $\ln x$ with slope k, then f is a power function with power k.

The number c in the formula $f(x) = cx^k$ is $c = e^b$, where b is the vertical intercept of the line.

EXAMPLE 5.4 *The Volume inside a Sphere*

The following table of values gives the volume V in cubic inches inside a sphere of radius r inches.

Radius r	1	2	3	4	5
Volume V	4.19	33.51	113.10	268.08	523.60

1. Plot the graph of $\ln V$ versus $\ln r$ to determine if it is reasonable to think that the volume is related to the radius by a power function.[6]

2. Use regression to find a formula for the volume of a sphere as a function of the radius.

3. To check your work, plot the graph of the original data points together with the function you found in Part 2.

Solution to Part 1: The first step is to enter $\boxed{5.12}$ the data into the calculator. Figure 5.18 shows the radius in column 3 and the corresponding volume in column 4, and Figure 5.19 shows $\boxed{5.13}$ the logarithm of the radius in column 1 and the logarithm of the volume in column 2.

Figure 5.18: *Radius and volume of spheres*

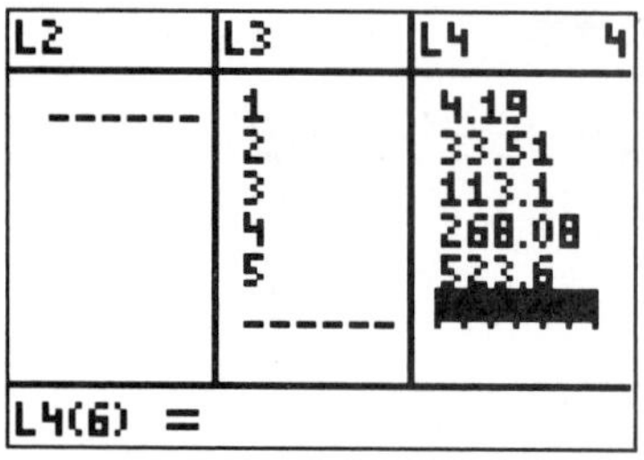

Figure 5.19: *Logarithm of radius and volume*

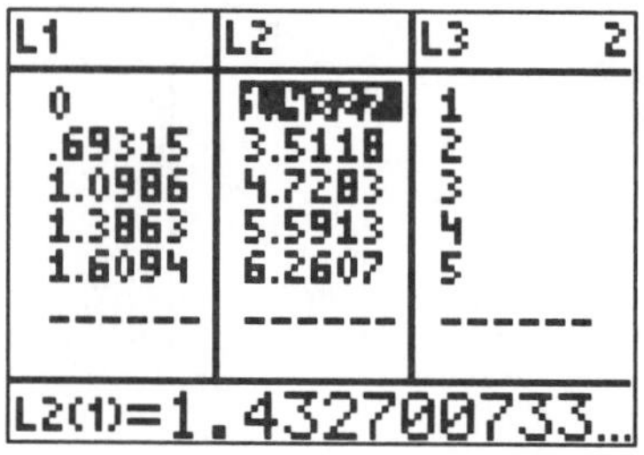

---

[6]As we noted in the preceding section, we know from elementary geometry that such a relationship is indeed valid and that $V = \dfrac{4\pi}{3}r^3$.

In Figure 5.20 we have plotted ⌐5.14⌐ the logarithm of volume versus the logarithm of radius. The points clearly fall on a straight line, and this is evidence that the volume V is indeed a power function of the radius r.

Solution to Part 2: We now proceed to calculate the regression line parameters ⌐5.15⌐ as shown in Figure 5.21. Rounding to two decimal places, we see that the linear function relating the logarithm of volume to the logarithm of the radius is $\ln V = 3\ln r + 1.43$. The power we need is the slope 3 of this line, and the value of c is $e^{1.43} = 4.18$. We conclude that $V = 4.18r^3$.

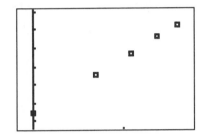

Figure 5.20: *Logarithm of volume versus logarithm of radius*

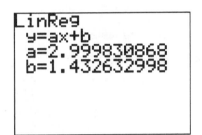

Figure 5.21: *Regression line parameters*

Solution to Part 3: To check our work, we plotted ⌐5.16⌐ the original data for volume versus radius in Figure 5.22 and added the graph of $4.18r^3$ in Figure 5.23. This shows excellent agreement between the given data and the power model we constructed.

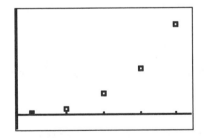

Figure 5.22: *Plot of data for volume versus radius*

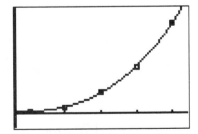

Figure 5.23: *Adding the power function model*

EXAMPLE 5.5   *Kepler's third law*

Johannes Kepler was the first to give an accurate description of the motion of the planets about the sun. He presented his model in the form of three laws. His third law states, "It is absolutely certain and exact that the ratio which exists between the periodic times of any two planets is precisely the ratio of the $\frac{3}{2}$th power of the mean distances [of the planet to the sun]."[7] Kepler formulated his third law by examining carefully the recorded measurements of the astronomer Tycho Brahe, but he never gave any derivation of it from other principles. That is to say, Kepler believed there was a power relation between the period P of a planet (the time required to complete a revolution about the sun) and its mean distance D from the sun, and he found the correct power by looking at data. We will use more accurate data and much more powerful calculation techniques than were available to Kepler to arrive at similar conclusions. The following table gives distances measured in millions of miles and periods measured in years for the nine planets.

Planet	Distance D	Period P
Mercury	36.0	0.24
Venus	67.1	0.62
Earth	92.9	1
Mars	141.7	1.88
Jupiter	483.4	11.87
Saturn	886.1	29.48
Uranus	1782.7	84.07
Neptune	2793.1	164.90
Pluto	3666.1	249

1. Plot the logarithms of the data points and determine if it is reasonable to model this data using a power function.

2. Use a formula to express P as a power function of D.

3. If one planet were twice as far from the sun as another, how would their periods compare?

Solution to Part 1: The first step is to enter [5.17] the data into the graphing calculator as we have done in Figure 5.24. Distance appears in the third column, and period appears in the fourth. In Figure 5.25 the logarithm of distance [5.18] is in the first column and the logarithm of period is in the second column. When we plot the logarithms as we have done in Figure 5.26, we see that the points line up nicely, supporting the idea that distance and period are related by a power function.

---

[7]From *The Harmonies of the World*, translated by Charles Glenn Wallis in the *Great Books*. Cited by Victor J. Katz in *A History of Mathematics*, 1993, HarperCollins, New York.

Figure 5.24: *Data for orbital period versus distance from the sun*

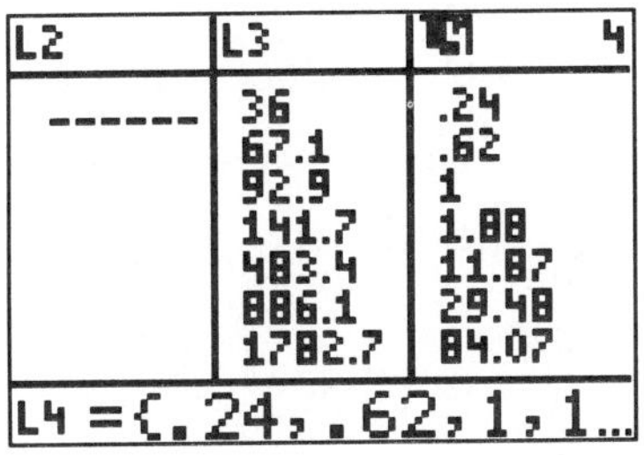

Figure 5.25: *Logarithms of planetary data*

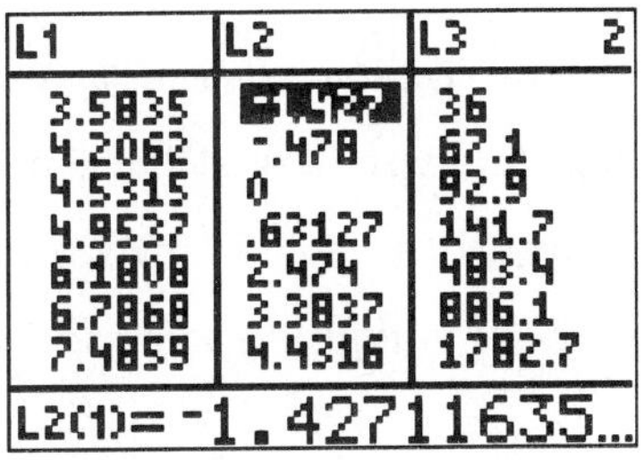

L2	L3	L4	4
------	36	.24	
	67.1	.62	
	92.9	1	
	141.7	1.88	
	483.4	11.87	
	886.1	29.48	
	1782.7	84.07	

L4 ={.24,.62,1,1...

L1	L2	L3	2
3.5835		36	
4.2062	-.478	67.1	
4.5315	0	92.9	
4.9537	.63127	141.7	
6.1808	2.474	483.4	
6.7868	3.3837	886.1	
7.4859	4.4316	1782.7	

L2(1)= -1.42711635...

Solution to Part 2: In Figure 5.27 we have calculated the regression line for $\ln P$ versus $\ln D$. We see that $\ln P = 1.5 \ln D - 6.8$. Thus, the power we need is $1.5 = \frac{3}{2}$, precisely the power proposed by Kepler. The c value for the power function is $e^{-6.8} = 0.0011$. We conclude that if we measure periods in years and distances in miles, then $P = 0.0011 D^{1.5}$.

Figure 5.26: *Plot of logarithms of data*

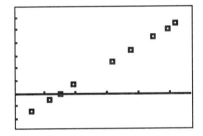

Figure 5.27: *Regression line parameters*

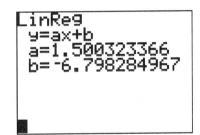

LinReg
y=ax+b
a=1.500323366
b= -6.798284967

Solution to Part 3: Here we use the homogeneity property of power functions. If the distance is increased by a factor of 2, then period will be increased by a factor of $2^{1.5} = 2.83$. Thus, if one planet is twice as far away from the sun as another, its period will be 2.83 times as long.

Almost power data

As with linear and exponential models, many times observed data cannot be modeled exactly by a power function, but linear regression allows us to make power models that approximately fit the data.

EXAMPLE 5.6 *Generation Time as a Power Function of Length*

The *generation time* for an organism is the time it takes to reach reproductive maturity. Biologists have observed that generation time depends on size, and in particular on the length of an organism. Table 5.1 gives the length L measured in feet and generation time T measured in years for various organisms.[8]

Table 5.1: Length and generation time

Organism	Length L	Generation time T
House fly	0.023	0.055 (20 days)
Cotton deermouse	0.295	0.192 (70 days)
Tiger salamander	0.673	1
Beaver	2.23	2.8
Grizzly bear	5.91	4
African elephant	11.5	12.3
Yellow birch	72.2	40
Giant sequoia	262	60

1. By plotting $\ln T$ against $\ln L$, determine if it is reasonable to model T as a power function of L.

2. Find a formula for the regression line of $\ln T$ against $\ln L$.

3. Find a formula which models T as a power function of L and plot the function along with the given data.

4. If one organism is 5 times as long as another, what would be the expected comparison of generation times? (Suggestion: Use the homogeneity property of power functions.)

Solution to Part 1: In the third column we enter 5.19 the L values, and in the fourth column we enter 5.20 the T values. The result is shown in Figure 5.28. Next we put the values of $\ln L$ in the first column 5.21 and the values of $\ln T$ in the second column 5.22 , as shown in Figure 5.29.

---

[8]This data is adapted from J. T. Bonner, *Size and Cycle*, 1965, Princeton University Press, Princeton. We have presented only a sample of the 46 organisms for which Bonner gives data.

Figure 5.28: *Data for L and T*

L2	L3	L4	4
------	.023	.055	
	.295	.192	
	.673	1	
	2.23	2.8	
	5.91	4	
	11.5	12.3	
	72.2	40	

L4 = {.055,.192,1...

Figure 5.29: *Data for ln L and ln T*

L1	L2	L3	2
-3.772	-2.9	.023	
-1.221	-1.65	.295	
-.396	0	.673	
.802	1.0296	2.23	
1.7766	1.3863	5.91	
2.4423	2.5096	11.5	
4.2794	3.6889	72.2	

L2 = {-2.90042209...

Now we plot $\boxed{5.23}$ the data for $\ln T$ against $\ln L$. The result is shown in Figure 5.30. The data does not lie exactly in a straight line but is approximately linear, so it is reasonable to model T as a power function[9] of L.

Solution to Part 2: The regression line parameters $\boxed{5.24}$ for the data in Figure 5.30 are shown in Figure 5.31. Rounding, we get

$$\ln T = 0.8 \ln L + 0.066 .$$

Figure 5.30: *Plotting the data for* $\ln L$ *against* $\ln T$

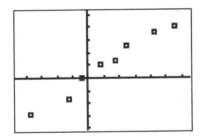

Figure 5.31: *Finding the regression line*

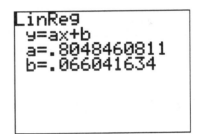

LinReg
y=ax+b
a=.8048460811
b=.066041634

Solution to Part 3: Since the slope of the regression line for $\ln T$ as a function of $\ln L$ is 0.8, that is also the power we use for our power model. Thus $T = cL^{0.8}$, and we need to find c. We get this from the vertical intercept 0.066 found in Part 2: $c = e^{0.066}$, which is about 1.1, so the power model is $T = 1.1 L^{0.8}$. In Figure 5.32 we have plotted the original data, and we added our power model in Figure 5.33.

We note that the generation time for the giant sequoia, the last data point in Figure 5.33, seems to lie well below the power function model. In Figure 5.34 we have traced the graph and set the cursor to X=262. We see that the power model shows a generation time of 94 years in contrast to the actual time of 60 years. In some settings, this might

---

[9]It is important to note that in a setting such as this, exact modeling is unreasonable to expect, and it is striking that any kind of consistent relationship can be found among such a diverse collection of species.

Figure 5.32: *Plot of data for generation time versus length*

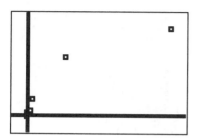

Figure 5.33: *Adding the power model*

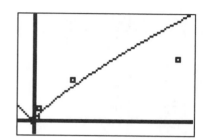

cause us to question the validity of the model. Such questions are always appropriate, but in this case we are happy with a model which can be a starting point for further analysis rather than an exact relationship as we had for Kepler's third law. We would be more concerned if the value given by the model were off by a factor of 10. In fact, one reason for the difference is the extreme variation in both length and generation time which is to some degree characteristic of data to which power function models are applied.[10] The 262 foot sequoia is over 10,000 times longer than the house fly, and its 60 year generation time is over 1000 times longer than that of the house fly. This is the cause of the bunching of data in Figure 5.33. In Figure 5.35 the plot is shown with a horizontal span of 0 to 6 and a vertical span of 0 to 5. It shows more of the smaller data points but leaves out the elephant, birch, and sequoia.

Figure 5.34: *A closer look at the sequoia*

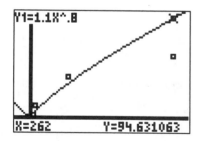

Figure 5.35: *Changing the window to a smaller range*

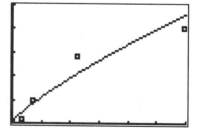

Solution to Part 4: The homogeneity property of a power function such as $T = 1.1L^{0.8}$ tells us that if length L is increased by a factor of 5, then generation time T will be increased by a factor of $5^{0.8} = 3.63$. Thus, we would expect the longer organism to have a generation time 3.63 times as long as the shorter organism.

---

[10]In contrast, data which should be modeled exponentially may show relatively small variation in the horizontal direction but much larger variation in the vertical direction.

Graphing on a logarithmic scale: common versus natural logarithms

Throughout this text we have used the natural logarithm ln x. But many scientists prefer the common logarithm log x which is associated with 10 rather than e, and this may be the logarithm which you encounter in applications. Traditionally, scientists have tested data to see if a power model is appropriate by plotting points on *log log* graphing paper, which has both the horizontal and the vertical axis marked in powers of 10. Figure 5.36 shows such a plot which might have been used in Example 5.6.

Figure 5.36: Generation time against length, on a logarithmic scale

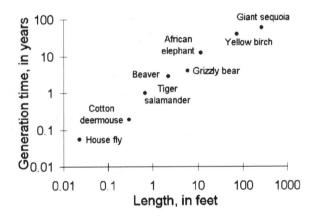

Graphing with such paper is really the same as the procedure we have used here except that the common logarithm is used instead of the natural logarithm.[11] The only substantive difference is that when calculating the c value for a power function, we use 10^b rather than e^b, where b is the vertical intercept of the regression line for the logarithm. To show this, we work through the Kepler example once more using the common logarithm rather than the natural logarithm. As we shall see, the final answers are the same.

---

[11] If the graphing paper were marked in powers of e rather than powers of 10, the procedure would be exactly the same.

Alternate solution of Example 5.5 using the common logarithm: We enter the data for D and P in the third and fourth columns just as we did in Example 5.5, but this time we put the common logarithm of D in the first column and the common logarithm of P in the second column $\boxed{5.25}$. We plot the common logarithm of the data as shown in Figure 5.37 and note that this is similar to Figure 5.26.

Next we get the regression line parameters shown in Figure 5.38 exactly as we did in Example 5.5. We see that $\log P = 1.5 \log D - 2.95$. The power we use is the slope 1.5 of this line, the same as with the natural logarithm. But to get the c value, we use 10 rather than e. Since $10^{-2.95} = 0.0011$, we arrive at the same answer $P = 0.0011 D^{1.5}$ as we did in Example 5.5.

Figure 5.37: *Plot of common logarithms of planetary data*

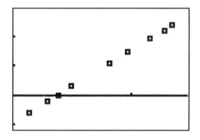

Figure 5.38: *Regression line parameters for common logarithms*

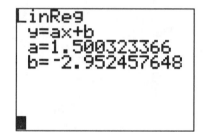

Exercise Set 5.2

1. **Hydroplaning:** On wet roads, under certain conditions the front tires of a car will *hydroplane*, or run along the surface of the water. The critical speed V at which hydroplaning occurs is a function of p, the tire inflation pressure.[12] The following table shows hypothetical data for p (in pounds per square inch) and V (in miles per hour).

Tire inflation pressure p	20	25	30	35
Critical speed V for hydroplaning	46.3	51.8	56.7	61.2

   (a) Find a formula for the regression line of $\ln V$ against $\ln p$.

   (b) Find a formula which models V as a power function of p.

   (c) In the rain a car (with tires inflated to 35 pounds per square inch) is traveling behind a bus (with tires inflated to 60 pounds per square inch), and both are moving at 65 miles per hour. If they both hit their brakes, what might happen?

2. **Growth rate versus weight:** Ecologists have studied how a population's intrinsic exponential growth rate r is related to the body weight W for herbivorous mammals.[13] In Table 5.2, W is the adult weight measured in pounds, and r is per year.

Table 5.2: Weight and exponential growth rate

Animal	Weight W	r
Short-tailed vole	0.07	4.56
Norway rat	0.7	3.91
Roe deer	55	0.23
White-tailed deer	165	0.55
American elk	595	0.27
African elephant	8160	0.06

   (a) Make a plot of $\ln r$ against $\ln W$. Is it reasonable to model r as a power function of W?

   (b) Find a formula which models r as a power function of W, and draw a graph of this function.

---

[12]See *Accident Reconstruction* by J. C. Collins, 1979, Charles C. Thomas Publisher, Springfield, Il. The critical speed for hydroplaning also increases with the amount of tire tread available.

[13]G. Caughley and C. J. Krebs, "Are big mammals simply little mammals writ large?" *Oecologia* **59** (1983), 7–17.

3. **Speed in flight versus length:** Table 5.3 gives the length L (in inches) of a flying animal and its maximum speed F (in feet per second) when it flies.[14] (For comparison, 10 feet per second is about 6.8 miles per hour.)

Table 5.3: Length and flying speed

Animal	Length L	Flying speed F
Fruit fly	0.08	6.2
Horse fly	0.51	21.7
Ruby-throated hummingbird	3.2	36.7
Willow warbler	4.3	39.4
Flying fish	13	51.2
Bewick's swan	47	61.7
White pelican	62	74.8

(a) Based on this table, is it generally true that larger animals fly faster?

(b) Find a formula which models F as a power function of L.

(c) Make the graph of the function in Part (b).

(d) Is the graph you found in Part (c) concave up or concave down? Explain in practical terms what your answer means.

(e) If one bird is 10 times longer than another, how much faster would you expect it to fly? (Use the homogeneity property of power functions.)

4. **Speed swimming versus length:** *This is a continuation of Exercise 3.* Table 5.4 gives the length L (in inches) of a swimming animal and its maximum speed S (in feet per second) when it swims.

(a) Find a formula for the regression line of $\ln S$ against $\ln L$. (Round the slope to one decimal place.)

(b) Find a formula which models S as a power function of L. Based on the power you found, what special type of power function is this?

(c) Add the graph of S against L to the graph of F you drew in Exercise 3.

(d) Is flying a significant improvement over swimming if an animal is one foot long (the approximate length of a flying fish)?

[14]The tables in this exercise and the next are adapted from J. T. Bonner, *op. cit.*

Table 5.4: Length and swimming speed

Animal	Length L	Swimming speed S
Bacillus	9.8×10^{-5}	4.9×10^{-5}
Paramecium	0.0087	0.0033
Water mite	0.051	0.013
Flatfish larva	0.37	0.38
Goldfish	2.8	2.5
Dace	5.9	5.7
Adélie penguin	30	12.5
Dolphin	87	33.8

(e) Would flying be a significant improvement over swimming for an animal 20 feet long?

(f) A blue whale is about 85 feet long, and its maximum speed swimming is about 34 feet per second. Based on these facts, do you think the trend you found in Part (b) continues indefinitely as the length increases?

5. **Metabolism:** Physiologists who study warm-blooded animals are interested in the *basal metabolic rate*, which is one measure of the energy needed for survival. It can be measured from the volume and composition of expired air. Table 5.5 gives the weight W (in pounds) of an animal and its basal metabolic rate B (in kilocalories per day).[15]

Table 5.5: Weight and basal metabolic rate

Animal	Weight W	Basal metabolic rate B
Rat	0.38	20.2
Pigeon	0.66	30.8
Hen	4.3	106
Dog	25.6	443
Sheep	101	1220
Cow	855	6421

(a) Find a formula which models B as a power function of W.

(b) Define the *metabolic weight* of a warm-blooded animal to be $W^{0.75}$ if W is its weight in pounds. Direct comparison of food intake among animals of different sizes is not

[15]The table is adapted from M. Kleiber, "Body size and metabolism," *Hilgardia* 6 (1932), 315–353. See also his book *The Fire of Life*, revised edition, 1975, Robert E. Krieger Publishing Company, Huntington, New York.

easy. Clearly large animals will consume more food than small ones, and to compensate for this we should take account of the energy needed for survival, that is, the basal metabolic rate B. To compare maximum daily food intake among animals of different sizes, we divide this food intake by the metabolic weight.

    i. Explain why dividing by the metabolic weight is more meaningful than dividing by the weight in comparing maximum daily food intake.

    ii. Find the metabolic weight of a 2.76-pound animal and that of a 126.8-pound animal.

    iii. An ecologist found the maximum daily food intake of a 2.76-pound rabbit to be about 0.18 pound and that of a 126.8-pound merino sheep to be about 2.8 pounds.[16]   Divide each intake by the corresponding metabolic weight. How do the daily consumption levels compare on this basis?

6. **Metabolism and surface area:** *This is a continuation of Exercise 5.* Table 5.6 gives the weight W (in pounds), the basal metabolic rate B (in kilocalories per day), and the surface area A (in square inches) of a variety of marsupials.[17]

Table 5.6: Weight, basal metabolic rate, and surface area

Animal	Weight W	Basal metabolic rate B	Surface area A
Fat-tailed marsupial mouse	0.03	2.16	12.4
Brown marsupial mouse	0.08	4.18	20.5
Long-nosed bandicoot	1.5	37.0	122
Brush-tailed possum	4.4	71.8	260
Tammar wallaby	10.6	159	465
Red kangaroo	71.6	643	1796

(a) Find a formula which models B as a power function of W for this group of marsupials.

(b) How does the formula you found in Part (a) compare with the one you found in Part (a) of the preceding exercise? What does this tell you about the metabolic rates of marsupials?

(c) Make a plot of $\ln A$ against $\ln W$.

[16]J. Short, "Factors affecting food intake of rangelands herbivores." In G. Caughley, N. Shepherd, and J. Short, eds., *Kangaroos*, 1987, Cambridge University Press, Cambridge, England.

[17]The data in this table is adapted from T. J. Dawson and A. J. Hulbert, "Standard metabolism, body temperature, and surface areas of Australian marsupials," *Am. J. Physiol.* **218** (1970), 1233–1238.

(d) Find a formula for the regression line of $\ln A$ against $\ln W$, and add this line to the plot you found in Part (c).

(e) The surface area of a 0.26-pound sugar glider (a marsupial similar to a flying squirrel) is about 95 square inches. Use your plot in Part (d) to compare this with the trend for surface area versus weight in the table. Can you explain the deviation?

(f) Find a formula which models A as a power function of W.

(g) Often the power $k = \frac{2}{3}$ is used to model A as a power function of W. Compare this with the power you found in Part (f).

(h) In early studies of metabolic rates, scientists often assumed that the basal metabolic rate of an animal was proportional to its surface area. Is this assumption supported by the data in the table?

7. **Proportions of trees:** Table 5.7 gives the diameter d and height h (both in feet) of some "champion" trees (largest American specimens) of a variety of shapes.[18]

Table 5.7: Diameter and height of champion trees

Tree	Diameter d	Height h
Plains cottonwood	2.9	80
Hackberry	5.7	113
Weeping willow	6.2	95
Ponderosa pine	8.6	162
Douglas-fir	14.4	221

(a) Make a plot of $\ln h$ against $\ln d$.

(b) Find a formula for the regression line of $\ln h$ against $\ln d$, and add this line to the plot you found in Part (a).

(c) Which is taller *for its diameter*: the plains cottonwood or the weeping willow?

(d) Find a formula which models h as a power function of d.

(e) It has been determined that the critical height at which a column made from green wood of diameter d (in feet) would buckle under its own weight is $140d^{2/3}$ feet.

   i. How does your answer to Part (d) compare with this formula?

   ii. Are any of the trees in the table taller than their critical buckling height?

---

[18]This exercise is based on the work of T. McMahon, "Size and shape in biology," *Science* **179** (1973), 1201–1204. He considers data for 576 trees, primarily from the American Forestry Association lists of champions.

8. **Weight versus length:** The following table shows the relationship between the length L (in centimeters) and the weight W (in grams) of the North Sea plaice (a type of flatfish).[19]

L	28.5	30.5	32.5	34.5	36.5	38.5	40.5	42.5	44.5	46.5	48.5
W	213	259	308	363	419	500	574	674	808	909	1124

(a) Find a formula which models W as a power function of L. (Round the power to one decimal place.)

(b) If one plaice were twice as long as another, how much heavier than the other should it be?

9. **Self-thinning:** When seeds of a plant are sown at high density in a plot, the seedlings must compete with each other. As time passes, individual plants grow in size, but the density of the plants that survive decreases.[20] This is the process of *self-thinning*. In one experiment, horseweed seeds were sown on October 21, and the plot was sampled on successive dates. The results are summarized in Table 5.8, which gives for each date the density p (in number per square meter) of surviving plants and the average dry weight w (in grams) per plant.

Table 5.8: Density and weight of surviving plants

Date	Density p	Weight w
November 7	140,400	1.6×10^{-4}
December 16	36,250	7.7×10^{-4}
January 30	22,500	0.0012
April 2	9100	0.0049
May 13	4510	0.018
June 25	2060	0.085

(a) Explain how the table illustrates the phenomenon of self-thinning.

(b) Find a formula which models w as a power function of p.

(c) If the density decreases by a factor of $\frac{1}{2}$, what happens to the weight?

(d) The *total plant yield* y per unit area is defined to be the product of the average weight per plant and the density of the plants: $y = w \times p$. As time goes on, the average

[19]The table is from Lowestoft market samples in 1946, as described by R. J. H. Beverton and S. J. Holt, *On the Dynamics of Exploited Fish Populations*, Fishery Investigations, Series 2, Volume 19, 1957, Ministry of Agriculture, Fisheries and Food, London.

[20]This exercise is based on the work of K. Yoda, T. Kira, H. Ogawa, and K. Hozumi, "Self-thinning in overcrowded pure stands under cultivated and natural conditions," *J. Biol. Osaka City Univ.* **14** (1963), 107–129. See also Chapter 6 of *Population Biology of Plants* by John L. Harper, 1977, Academic Press, London.

weight per plant increases while the density decreases, so it's unclear whether the total yield will increase or decrease. Use the power function you found in Part (b) to determine whether the total yield increases or decreases *with time*. Check your answer using the table.

10. **Species-area relation:** Ecologists have studied the relationship between the number S of species of a given taxonomic group within a given habitat (often an island) and the area A of the habitat.[21] They have discovered a consistent relationship: Over similar habitats, S is approximately a power function of A, and for islands the powers fall within the range 0.2 to 0.4. Table 5.9 gives, for some islands in the West Indies, the area (in square miles) and the number of species of amphibians and reptiles.

Table 5.9: *Area and number of species of amphibians and reptiles*

Island	Area A	Number S of species
Cuba	44,000	76
Hispaniola	29,000	84
Jamaica	4200	39
Puerto Rico	3500	40
Montserrat	40	9
Saba	5	5

(a) Find a formula which models S as a power function of A.

(b) Is the graph of S against A concave up or concave down? Explain in practical terms what your answer means.

(c) The species-area relation for the West Indies islands can be expressed as a rule of thumb: If one island is 10 times larger than another, then it will have _____ times as many species. Use the homogeneity property of the power function you found in Part (a) to fill in the blank in this rule of thumb.

(d) In general, if the species-area relation for a group of islands is given by a power function, the relation can be expressed as a rule of thumb: If one island is 10 times larger than another, then it will have _____ times as many species. How would you fill in the blank? (Hint: The answer depends only on the power.)

---

[21]See the discussion in Example 5.2 of Section 5.1. The data in this exercise is adapted from Philip J. Darlington, Jr., *Zoogeography*, 1980 reprint, Robert E. Krieger Publishing Company, Huntington, New York.

11. **Cost of transport:** Physiologists have discovered that steady-state oxygen consumption (measured per unit of mass) in a running animal increases linearly with increasing velocity. The slope of this line is called the *cost of transport* of the animal, since it measures the energy required to move a unit mass by one unit distance. Table 5.10 gives the weight W (in grams) and the cost of transport C (in milliliters of oxygen per gram per kilometer) of seven animals.[22]

Table 5.10: *Weight and cost of transport*

Animal	Weight W	Cost of transport C
White mouse	21	2.83
Kangaroo rat	41	2.01
Kangaroo rat	100	1.13
Ground squirrel	236	0.66
White rat	384	1.09
Dog	2600	0.34
Dog	18,000	0.17

(a) Based on the table, does the cost of transport generally increase or decrease with increasing weight? Are there any exceptions to this trend?

(b) Make a plot of $\ln C$ against $\ln W$.

(c) Find a formula for the regression line of $\ln C$ against $\ln W$, and add this line to the plot you found in Part (b).

(d) The cost of transport for a 20,790-gram emperor penguin is about 0.43 milliliter of oxygen per gram per kilometer. Use your plot in Part (c) to compare this with the trend for cost of transport versus weight in the table. Does this confirm the stereotype of penguins as awkward waddlers?

(e) Find a formula which models C as a power function of W.

12. **Exponential growth rate and generation time:** In this exercise, we will examine the relationship between a population's intrinsic exponential growth rate r and the generation time T, that is, the time it takes an organism to reach reproductive maturity. The following table gives values of the generation time and the exponential growth rate for a selection of lower organisms.[23] The basic unit of time is a day.

---

[22]The data in this table is taken from C. R. Taylor, K. Schmidt-Nielsen, and J. L. Raab, "Scaling of energetic cost of running to body size in mammals," *Am. J. Physiol.* **219** (1970), 1104–1107. For an extensive collection of data with references, see M. A. Fedak and H. J. Seeherman, "Reappraisal of energetics of locomotion shows identical cost in bipeds and quadrupeds including ostrich and horse," *Nature* **282** (1979), 713–716.

[23]The table is adapted from E. R. Pianka, *Evolutionary Ecology*, 4th edition, 1988, Harper & Row, New York.

Organism	Generation time T	r
Bacterium *E. coli*	0.014	60
Protozoan *Paramecium aurelia*	0.5	1.24
Spider beetle *Eurostus hilleri*	110	0.01
Golden spider beetle	154	0.006
Spider beetle *Ptinus sexpunctatus*	215	0.006

(a) By plotting $\ln r$ against $\ln T$, determine if it is reasonable to model r as a power function of T.

(b) Find a formula for the regression line of $\ln r$ against $\ln T$. (Round the slope to one decimal place.)

(c) Find a formula which models r as a power function of T, and make a graph of this function. (Use a horizontal span from 0 to 10.)

(d) In Example 5.6 we found a power relation between generation time T (in years) and length L. Use this relation and the results of Part (c) to find a formula for r as a function of length L for this group of lower organisms. (Convert the power relation from Example 5.6 to measure time in days, and remember that $T^{-1} = \dfrac{1}{T}$.)

13. **Using the common logarithm:** Solve Exercise 8 using the common logarithm.

5.3 PERIODIC FUNCTIONS AND RIGHT TRIANGLE TRIGONOMETRY

Many natural phenomena such as tidal activity and radio waves display a repetitive or *periodic* behavior, and we will use a special function, the sine function, to model them.

Period and amplitude

Tides at coastal areas vary in height with season and location because of weather, funneling effects of the ocean floor, and other local factors. But in the open ocean, tides range from a high of about 3 feet above sea level to a low of 3 feet below sea level, and they repeat roughly once each day. This is a prime example of a periodic activity, and the *period* is the length of time it takes the process to repeat: in this case, the period is one day. The height, 3 feet, is often called the *amplitude*. If we take the horizontal axis to be sea level and make a graph of tidal height (in feet) versus time (in days), we get the graph in Figure 5.39, and the repetitive nature of tidal activity is apparent there. In Figure 5.39 we have marked the period, and in Figure 5.40 we have marked the amplitude.

Figure 5.39: The period of open ocean tides is 1 day

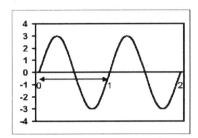

Figure 5.40: The amplitude of open ocean tides is 3 feet

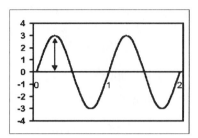

EXAMPLE 5.7 *Hours of Daylight*

The Earth's seasons are caused by a 23.4 degree tilt of the Earth's axis to the orbital plane, and this tilt is also responsible for the variance in the number of hours of daylight. At the equator there are 12 hours of daylight at the vernal equinox in March and at the autumnal equinox in September. There are about 3 additional hours of daylight at the summer solstice in June and about 3 fewer hours of daylight at the winter solstice in December.

1. Marking the horizontal axis as time in months and the vertical axis as additional hours of daylight, and beginning at the vernal equinox in March, make a graph of additional hours of daylight versus time over two years.

2. Identify the amplitude and period of the graph you made in Part 1.

Solution to Part 1: Since there are no extra hours of daylight at the vernal equinox, we start our graph at the origin. By the summer solstice in June there are 3 additional hours of daylight, and so we make the graph increase to a height of 3 at that time. From this point the graph should decrease to 0 at the autumnal equinox, decrease to minus 3 at the winter solstice, and finally return to the horizontal axis at the vernal equinox in March. The picture repeats over the second year and is shown in Figure 5.41.

Solution to Part 2: Since the graph repeats each year, the period is 12 months. The graph goes from +3 to −3, and so the amplitude is 3 hours.

Figure 5.41: *Additional hours of daylight at the equator*

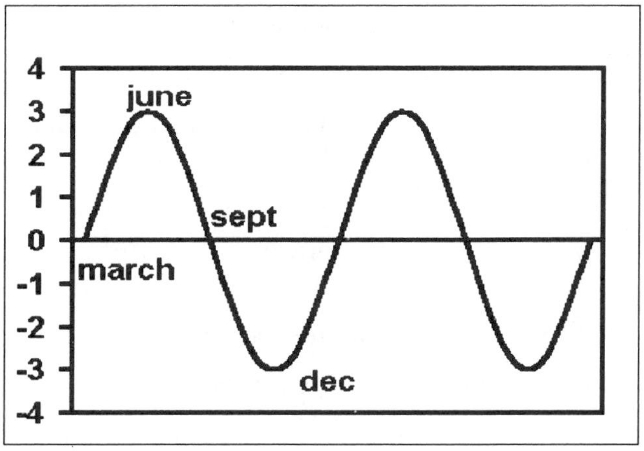

Using the sine function to model periodic phenomena

The sine function, usually written as $\sin x$, is used to model many periodic phenomena. In fact, sine functions are so common in this type of modeling that graphs such as the one shown in Figure 5.41 are often referred to as *sine waves*. In Figure 5.42 we have used the calculator [5.26] to make a graph[24] of $\sin x$ using a horizontal span of 0 to 720 and a vertical span of -2 to 2.

We note that the graph in Figure 5.42 has the shape we want, but in order to make use of it, we need to determine its period and amplitude and learn how to adjust those. To find the period of the sine function, we have moved the cursor to the end of the first cycle, and we read from the bottom of the screen that the sine function has a period of 360. You can do this by asking the calculator to locate the second crossing point. To get the amplitude, we need to find the height of one of the peaks. In Figure 5.43 we have moved the cursor to the top of the first peak, and we read the height 1 from the prompt at the bottom of the screen. Thus the sine function has a period of 360 and an amplitude of 1.

We can get sine waves of varying amplitudes and periods using $A\sin(Bx)$ for various values of A and B. To see how different values of A and B affect this function, we graphed $\sin x$, $2\sin x$, and $3\sin x$ in Figure 5.44. We note that the amplitude of $\sin x$ is 1, the amplitude of $2\sin x$ is 2, and the amplitude of $3\sin x$ is 3. In general, A is the amplitude of $A\sin x$.

---

[24]How the sine function is handled by your calculator depends on how your calculator is measuring angles, using *degrees* or *radians*. We will use degrees, and you should be sure your calculator is set [5.27] to work in the same way.

Figure 5.42: *The period of the sine function is 360*

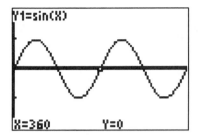

Figure 5.43: *The amplitude of the sine function is 1*

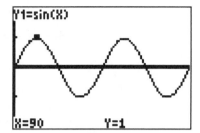

To understand how to adjust the period, we first note that $\sin(360x)$ has period 1 because $360x$ is 360 when x is 1. In Figure 5.45 we have graphed $\sin(360x)$ with a thick curve and $\sin(\frac{360}{2}x)$ with a thin curve using a horizontal span from 0 to 2. This plot verifies our earlier observation that $\sin(360x)$ has period 1, and we see that $\sin(\frac{360}{2}x)$ has period 2. In general, the function $\sin(\frac{360}{P}x)$ has period P.

Figure 5.44: *The amplitude of $A\sin x$ is A*

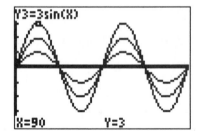

Figure 5.45: *The period of $\sin(\frac{360}{P}x)$ is P*

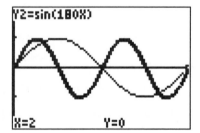

KEY IDEA 5.5: ADJUSTING THE SINE WAVE

To make a sine wave with given period and amplitude use

$$\text{amplitude} \times \sin\left(\frac{360}{\text{period}}\,x\right).$$

EXAMPLE 5.8 *Modeling Additional Daylight with a Sine Wave*

Make a sine wave which models additional hours of daylight as described in Example 5.7. According to this model, how many additional hours of daylight are there in August at the time 5 months after the vernal equinox? Assuming there are 30 days in each month, how many days after the vernal equinox can one expect there to be exactly 2 additional hours of daylight?

Solution: Since the amplitude is 3 hours, and the period is 12 months, we use

$$\text{amplitude} \times \sin \left(\frac{360}{\text{period}}\, t \right) = 3 \sin \left(\frac{360}{12}t \right) = 3\sin(30t).$$

Here t is the time in months since the vernal equinox. The graph, with a horizontal span of 0 to 24 and a vertical span of -4 to 4, is shown in Figure 5.46. To find additional hours of daylight at the specified time in August, we set the cursor at $X= 5$. We see from Figure 5.46 that we can expect 1.5 additional hours of daylight at that time.

To find out when there are 2 additional hours of daylight, we add the horizontal line $y = 2$ to the picture as we have done in Figure 5.47. The crossing graphs method shows that this happens 1.39 months after the vernal equinox. That is about 42 days after the vernal equinox.

Figure 5.46: *A sine model of additional hours of daylight*

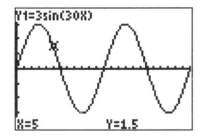

Figure 5.47: *When there are 2 additional hours of daylight*

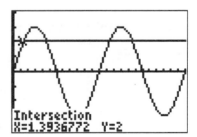

Surveying and right triangles

A *right triangle* is a triangle with a right angle, that is, an angle of 90 degrees. Some important applications of mathematics, such as surveying, involve right triangles, and the sine function, as well as two similar periodic functions, the *cosine* and *tangent* functions, are needed to make calculations with right triangles. The field of *trigonometry* was developed for the study of how the sides and angles of right triangles are related.

The standard labels associated with right triangles are shown in Figure 5.48. To indicate that an angle is a right angle, it is common to mark the angle as we have done in that figure. Using one of the *acute* (less than 90 degree) angles as a reference point, we have common names given to the sides of a right triangle: opposite, adjacent, and hypotenuse. In Figure 5.48 we have labeled one of the acute angles θ. The *opposite* side is the side opposite the angle. The *adjacent* side is the side that touches both the reference angle and the right angle. The *hypotenuse* is the side which does not touch the right angle.

Figure 5.48: *Right triangles and periodic functions*

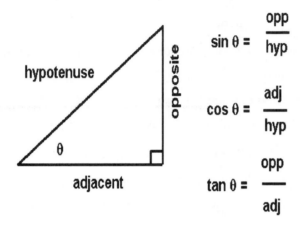

The sine of the angle θ can be calculated from the opposite side of the triangle and the hypotenuse:

$$\sin \theta = \frac{\text{opposite}}{\text{hypotenuse}}.$$

New periodic functions, in particular the cosine and tangent, can be made using other combinations of sides:

$$\cos \theta = \frac{\text{adjacent}}{\text{hypotenuse}}$$
$$\tan \theta = \frac{\text{opposite}}{\text{adjacent}}.$$

In Figure 5.49 we have used the calculator to make a graph of the cosine [5.28] function, and in Figure 5.50 we have used the calculator to make a graph of the tangent [5.29] function. (Both have a horizontal span of 0 to 720 and a vertical span of −2 to 2.) We see that the cosine function looks much like the sine function except that the wave is shifted. The tangent function in Figure 5.50 is a periodic function of an entirely different shape.

Figure 5.49: *The cosine function*

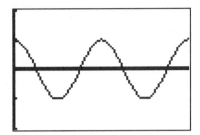

Figure 5.50: *The tangent function*

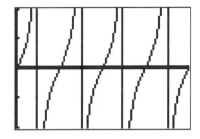

EXAMPLE 5.9   *The Height of a Building*

A surveyor stands 300 horizontal feet from a building and aims a transit at the top of the build-ing as shown in Figure 5.51. The transit measures an angle of 21 degrees from the horizontal.

Figure 5.51: *Measuring with a transit*

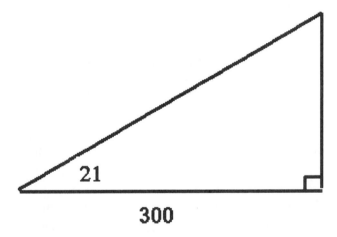

1. How tall is the building?

2. How long would a cable that reached from the surveyor directly to the top of the build-ing need to be?

Solution to Part 1: Our interest in this case is in the height of the building, and that is the side opposite the reference angle. We know the length of the adjacent side, so we should use the tangent function, since it deals with those two sides:

$$\tan 21 = \frac{\text{opposite}}{\text{adjacent}} = \frac{\text{opposite}}{300}.$$

We use x to denote the length of the opposite side. The equation $\tan 21 = \dfrac{x}{300}$ can be solved for x with the calculator using the crossing graphs method, but it is also a linear equation and is easily solved by hand calculation. We multiply each side by 300 and then use the calculator to find $\tan 21$:

$$\tan 21 = \frac{x}{300}$$
$$300 \tan 21 = x$$
$$115.16 = x.$$

We conclude that the height of the building is 155.16 feet.

Solution to Part 2: The distance from the surveyor directly to the top of the building is the hypotenuse of the triangle in Figure 5.51. We will show two ways[25] to proceed.

Solution using the cosine function: We know the adjacent side is 300 feet, and we want to know the hypotenuse. The cosine function involves these two quantities:

$$\cos 21 = \frac{\text{adjacent}}{\text{hypotenuse}} = \frac{300}{\text{hypotenuse}}.$$

We use x to denote the length of the hypotenuse. As before, we can solve the equation $\cos 21 = \frac{300}{x}$ using the crossing graphs method or by hand calculation:

$$\cos 21 = \frac{300}{x}$$
$$x \cos 21 = 300 \quad \text{(Multiply both sides by } x.)$$
$$x = \frac{300}{\cos 21} \quad \text{(Divide both sides by } \cos 21.)$$
$$x = 321.34.$$

We conclude that a cable reaching from the surveyor to the top of the building would need to be 321.34 feet long.

Solution using the sine function: Once again, we want the height, but this time we use the length 115.16 feet of the opposite side we calculated in Part 1:

$$\sin 21 = \frac{\text{opposite}}{\text{hypotenuse}} = \frac{115.16}{\text{hypotenuse}}.$$

Now we solve for x, the length of the hypotenuse:

$$\sin 21 = \frac{115.16}{x}$$
$$x \sin 21 = 115.16$$
$$x = \frac{115.16}{\sin 21}$$
$$x = 321.35.$$

The small difference between the two answers is due to rounding in Part 1.

---

[25]Since we already know two sides of the triangle, the third side could also be calculated from the Pythagorean theorem, which says that the square of the hypotenuse is the sum of the squares of the other two sides.

Exercise Set 5.3

1. **Period and amplitude:** Plot the graph of $4\sin(5x)$ and determine its period and amplitude.

2. **A sine wave with given period and amplitude:** Make the graph of a sine wave with period 10 and amplitude 4.

3. **Tides:** Suppose tides vary from 5 feet above to 5 feet below sea level and that successive low tides occur each 20 hours. Make a graph of the height in feet of the water (in relation to sea level) as a function of time t in hours, starting when the water is at sea level and rising. Determine the water level at $t = 7$ hours.

4. **A spring:** When an object suspended from a spring vibrates up and down, the displacement of the object from its rest position can be graphed against time. For the first part of this problem, we will make the unrealistic assumption that, once set into motion, the spring continues to move up and down at the same rate without ever slowing down.

   (a) Suppose a weight is suspended from a spring, and at its rest position an initial upward velocity is imparted, setting the weight in motion. The weight rises to its highest point at 8 inches above rest position, then falls to its lowest point at 8 inches below rest position, then rises back through rest position, and then the cycle repeats. Assume the characteristics of the spring and the weight are such that the cycle takes 1.6 seconds. Make a graph of the displacement of the weight above rest position as a function of time t in seconds, starting when the weight is set in motion. When is the first time the weight drops to 4 inches below rest position?

   (b) In the physical situation described in Part (a), we know that the spring will not continue to vibrate at the same rate forever. Rather the motion will slow and eventually stop. This is usually described by adding an exponential *damping* term to the function. Thus if $f(t)$ is your function from Part (a), then the function $0.8^t f(t)$ will give a more accurate picture of what really happens. Make a graph of this function, and under these assumptions determine the displacement of the weight above its rest position 5 seconds after the spring is set in motion.

5. **Vibrating strings:** If you plucked a guitar string and then were able to freeze the wire in time, it would have the shape of a sine wave. In describing such phenomena, physicists usually refer to the period of the wave as the *wavelength*. The amplitude of such a wave is the initial displacement you give the wire when you pluck it. Suppose you pluck a guitar string, giving it an initial displacement of 5 centimeters. If the resulting wavelength is 20 centimeters, make a graph of the shape of the frozen guitar string. What is the displacement of the guitar string 50 centimeters from the base of the string?

6. **The 3, 4, 5 right triangle:** Elementary geometry can be used to show that any triangle with sides 3, 4, and 5 is a right triangle. This is known as the 3, 4, 5 right triangle. The ancient Egyptians knew this fact and used it to make square corners for their building projects.

   (a) Find the sine of each of the two acute angles in a 3, 4, 5 right triangle.

   (b) Use your work in Part (a) to find the acute angles in a 3, 4, 5 right triangle.

7. **A building:** You are facing a building which is 150 feet high, and you must elevate your transit 20 degrees to view the top of the building. What is the distance in horizontal feet between you and the building?

8. **The width of a river:** You stand on the north bank of a river and look due south at a tree on the opposite bank. Your helper on the opposite bank measures 35 yards due east to a second tree. You must rotate your transit through an angle of 12 degrees to point toward the second tree. How wide is the river?

9. **A cannon:** If you elevate a certain cannon so that it makes an angle of t degrees with the ground, then the cannonball will strike the ground $\dfrac{m^2 \sin(2t)}{g}$ feet downrange, where $g = 32$ feet per second per second is acceleration due to gravity, and m is the muzzle velocity in feet per second. If the muzzle velocity is 300 feet per second, what angle would you use to make the cannonball land 1000 feet downrange?

10. **Dallas to Fort Smith:** Dallas is 190 miles due south of Oklahoma City, and Fort Smith is 140 miles due east of Oklahoma City. An airplane flies on a direct trip from Dallas to Fort Smith.

   (a) What is the tangent of the angle that the flight path makes with Interstate 35, which runs due north from Dallas to Oklahoma City?

   (b) What is the angle that the flight path makes with Interstate 35?

   (c) How far does the airplane fly?

5.4   POLYNOMIAL AND RATIONAL FUNCTIONS

There are many other types of functions which occur in mathematics and its applications. In this section, we will look at polynomials and rational functions and at the traditional use of the quadratic formula in solving second-degree polynomial equations.

A *polynomial* function is a function whose formula can be written as a sum of power functions where each of the powers is a nonnegative whole number. For example, $x^2 + 3x + 5$ is a polynomial because it is a sum of the power functions x^2, $3x$, and $5 = 5x^0$. Another example of a polynomial is $(2x - 1)^2$ since it can be written as $4x^2 - 4x + 1$. On the other hand, $\sqrt{x}$ and $\dfrac{1}{x}$ are not polynomials since the powers of x are not nonnegative whole numbers. Also 2^x, e^x, and $\ln x$ cannot be written as sums of power functions and so are not polynomials either.

Polynomial functions are useful in many applications of mathematics when a single power function is not sufficient. For example, when a rock is dropped, the distance D in feet it falls in t seconds is given by the power function $D = 16t^2$. But if the rock is thrown downward with an initial velocity of 5 feet per second, then the distance it falls is no longer governed by a power function. Instead, the polynomial function $D = 16t^2 + 5t$ is needed to describe the distance traveled by the rock.

Another use for polynomial functions is to provide polynomial models for data. As we'll see, the graphs of polynomial functions usually have important geometric properties needed in models. For example, polynomials are the simplest functions having a minimum, or having a point of inflection and a maximum. The simplicity of polynomials makes them attractive models for data having a minimum or a maximum or other geometric features.

Quadratics and parabolas

The most commonly occurring polynomials, other than linear functions, are the *quadratics*. These are polynomials where the highest power occurring is a 2, that is, the highest power term is a square. The polynomial $D = 16t^2 + 5t$ mentioned above is an example of a quadratic. The graph of a quadratic is known as a *parabola* and has a distinctive shape which occurs often in nature. In Figure 5.52 we show the parabola which is the graph of $x^2 - 3x + 7$, and in Figure 5.53 we show the parabola which is the graph of $-x^2 + 3x + 7$. A positive sign on x^2 makes the parabola open upward, as in Figure 5.52, and a negative sign on x^2 makes the parabola open downward, as in Figure 5.53.

Figure 5.52: *The graph of $x^2 - 3x + 7$, an upward-opening parabola*

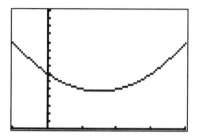

Figure 5.53: *The graph of $-x^2 + 3x + 7$, a downward-opening parabola*

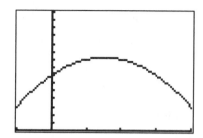

EXAMPLE 5.10   *The flight of a cannonball*

A cannonball fired from a cannon will follow the path of the parabola[26] which is the graph of $y = y(x)$, where $y = -16(1 + s^2)\left(\dfrac{x}{v_0}\right)^2 + sx$. Here s is the slope of inclination of the cannon barrel, v_0 is the initial velocity in feet per second, and the variable x is the distance downrange in feet. Suppose a cannon is elevated with a slope of 0.5 (corresponding to an angle of inclination of about 27 degrees) and the cannonball given an initial velocity of 250 feet per second. (See Figure 5.54.)

Figure 5.54: *Cannonball's inclination and initial velocity*

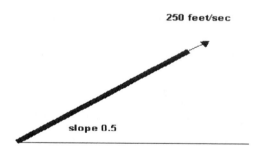

1. How far will the cannonball travel?

2. What is the maximum height the cannonball will reach?

3. Suppose the cannon is firing at a fort whose 20-foot-high wall is 1540 feet downrange. Show that the cannonball will not clear the fort wall. There are two possible adjustments that might be made so that the cannonball will clear the wall: increase the initial velocity

---

[26]There is nothing special about a cannonball here. If air resistance is neglected, then any object, such as a rifle bullet, golf ball, baseball, or basketball, ejected near the surface of the Earth (or any other planet for that matter), will follow a parabolic path.

(add powder) or change the slope of inclination.

(a) How would you adjust the initial velocity so that the wall will be cleared?

(b) How would you adjust the slope of inclination so that the wall will be cleared?

Solution to Part 1: Since the slope of inclination is 0.5 and the initial velocity is 250 feet per second, the function we wish to graph is

$$-16\left(1 + 0.5^2\right)\left(\frac{x}{250}\right)^2 + 0.5x = -20\left(\frac{x}{250}\right)^2 + 0.5x.$$

After consulting a table of values, we chose a horizontal span of 0 to 1800 and a vertical span of -50 to 220 to make the graph in Figure 5.55. We want to know how far downrange the cannonball will land, so we use the calculator to get the crossing point $x = 1562.5$ feet shown in Figure 5.55.

Solution to Part 2: To get the maximum height of the cannonball, we use the calculator to get the peak of the parabola as shown in Figure 5.56. We see that the cannonball will reach a maximum height of 195.31 feet at 781.25 feet downrange.

Figure 5.55: *Where the cannonball will land*	**Figure 5.56**: *The maximum height of the cannonball*

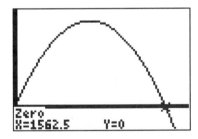

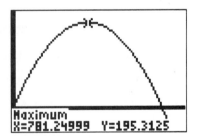

Solution to Part 3: In Figure 5.57 we have added the horizontal line $y = 20$ to the picture and calculated the intersection point. We see that the cannonball will be 20 feet high at 1521.42 feet downrange. Since the 20-foot-high wall is several feet farther downrange, the cannonball will not clear it.

For Part (a), we want to choose an initial velocity v_0 so that the height $-16\left(1 + 0.5^2\right)\left(\frac{x}{v_0}\right)^2 + 0.5x$ will be 20 when the distance x is 1540. That is, we want to solve the equation

$$-20\left(\frac{1540}{v_0}\right)^2 + 0.5 \times 1540 = 20$$

for v_0. We can do this using the crossing graphs method as shown in Figure 5.58. Since the cannonball doesn't need much increase in initial velocity to clear the wall, we use a

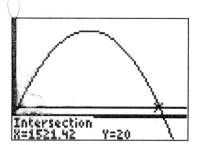

Figure 5.57: *The cannonball won't clear the 20-foot-wall*

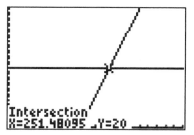

Figure 5.58: *Adjusting initial velocity*

horizontal span of 240 to 260 and a vertical span of 0 to 40. We see that an initial velocity of 251.48 feet per second is needed to clear the wall.

For Part (b), we want to leave the initial velocity at 250 feet per second and change the slope of inclination, so we need to solve the equation

$$-16(1 + s^2)\left(\frac{1540}{250}\right)^2 + s \times 1540 = 20$$

for s. Once again, we don't need to change the slope much, so we use a horizontal span of 0.4 to 0.6 and a vertical span of 0 to 40. When we calculate the intersection, we find that a slope of 0.51 is required.

Linear rates of change and second-order differences

We know that a linear function is one with a constant rate of change and that an exponential function is one with a constant proportional rate of change. Quadratic functions can also be characterized in terms of rates of change; they are functions with a linear rate of change. To see this, consider a falling rock. Its velocity t seconds after it is released is given by the linear function $-32t$ feet per second. (Here the minus sign is used since we measure distance with "up" as positive, and so falling is negative). The rock's distance will be $-16t^2$ feet at the end of t seconds.[27] Since velocity is the rate of change in distance, we see that in this case, the distance is a quadratic, and its rate of change, the velocity, is linear. Carrying this one step further, since velocity is linear, its rate of change is constant. That is, for a quadratic function *the second-order rate of change is constant*. In fact, a table of data (with evenly spaced values for the variable) is quadratic whenever the *second-order differences* in function values are constant. The following more complex example will illustrate this further.

---

[27]Since the velocity changes linearly from 0 to $-32t$, the rock's distance after t seconds is the same as it would be if its velocity were the average $-16t$ throughout the fall. Thus the distance is $-16t \times t = -16t^2$.

EXAMPLE 5.11   *A Rock Tossed Upward from a Building*

If we stand atop a 135-foot-high building and toss a rock upward with an initial velocity of 38 feet per second, then the rock will travel upward for a while and then eventually be pulled by gravity down to the ground. Elementary physics can be used to show that the distance up from the ground (in feet) is given by

$$D = -16t^2 + 38t + 135,$$

where t is the time, in seconds, after the rock is thrown. The graph of D versus t will be part of a parabola opening downward. In Figure 5.59 the graph is shown using a horizontal span of $t = 0$ to $t = 5$ and a vertical span of $D = 0$ to $D = 180$.

Figure 5.59: *A rock tossed upward from the top of a building*

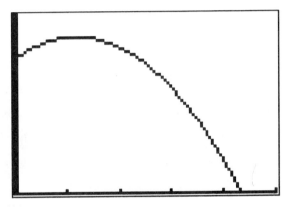

We'll use this function to illustrate the properties of rates of change of quadratic functions. To begin, consider a table of values of D:

Seconds t	0	1	2	3	4
Distance D	135	157	147	105	31

and then a table of first-order differences in D:

Change in t	0 to 1	1 to 2	2 to 3	3 to 4
First-order change in D	$157 - 135 = 22$	$147 - 157 = -10$	$105 - 147 = -42$	$31 - 105 = -74$

To see that this rate of change is linear, compute the second-order differences in D, which are the same as the first-order differences in the rate of change of D. The first-order differences are 22, -10, -42, and -74, and so each term differs from its predecessor by -32. Thus the second-order differences are all -32. We see that the quadratic function $D = -16t^2 + 38t + 135$

has constant second-order differences or, equivalently, a linear rate of change; moreover, here are important values we computed:

Initial value of D	135
Initial first-order difference	22
Second-order difference	-32

We can reconstruct the formula for D from this information. Since all second-order differences are the same, D is quadratic and so may be written in the form $at^2 + bt + c$. The relation turns out to be

$$a = \frac{1}{2} \times \text{second-order difference}$$
$$b = \text{initial first-order difference} - a$$
$$c = \text{initial value.}$$

In our case, $a = \frac{1}{2} \times -32 = -16$, $b = 22 - a = 22 - (-16) = 38$, and $c = 135$, as we expected. In fact, the relationship above gives the formula for quadratic functions in general, if we have data for $t = 0, t = 1,$ and $t = 2$.

EXAMPLE 5.12   *Five-Gallon Water Jug*

In general, if a large cylinder is filled with a fluid and then drained through a small hole, the depth of the fluid will decrease as a quadratic function of time. For example, consider a standard five-gallon water jug with a spigot in it. Certainly, it fills a cup faster when it is full than when it is close to empty. A completely full five-gallon water jug has a water depth of 15.5 inches (above the spigot level). Here is the depth as a function of time:

Minutes t the spigot is open	0	1	2	3	4
Depth D above the spigot, in inches	15.50	11.71	8.45	5.72	3.52

1. Show that D can be modeled as a quadratic function of t.

2. Write the formula for D as a quadratic function of t.

3. How long will it take for the jug to be completely drained?

4. How long does it take for the depth to drop the first $\frac{1}{2}$ inch?

5. How long does it take for the depth to drop the final $\frac{1}{2}$ inch?

Solution to Part 1: To show that D can be modeled by a quadratic function, it suffices to show that the second-order differences are constant:

t	0	1	2	3	4
D	15.50	11.71	8.45	5.72	3.52
First-order difference	-3.79	-3.26	-2.73	-2.20	
Second-order difference	0.53	0.53	0.53		

Since the second-order differences are all 0.53, we see that D can be modeled by a quadratic function.

Solution to Part 2: To get the formula for the quadratic function, we use the standard form $D = at^2 + bt + c$ and the relations

$$a = \frac{1}{2} \times \text{second-order difference}$$
$$b = \text{initial first-order difference} - a$$
$$c = \text{initial value}.$$

Thus $a = \frac{1}{2} \times 0.53 = 0.265$, $b = -3.79 - a = -3.79 - 0.265 = -4.055$, and $c = 15.50$, so $D = 0.265t^2 - 4.055t + 15.50$ is the formula.

Solution to Part 3: The water jug is completely drained when the water depth above the spigot is 0, so we want to solve $D = 0.265t^2 - 4.055t + 15.50 = 0$ for t. There are many ways to do this. In Figure 5.60 we have graphed D with a horizontal span from $t = 0$ to $t = 10$ and a vertical span from $D = 0$ to $D = 16$ and then solved by finding the zero at $t = 7.44$ minutes.

Figure 5.60: *Finding the zero of the water depth function*

Figure 5.61: *Finding when the water depth is 15 inches*

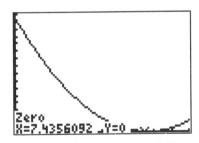

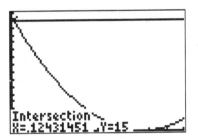

Solution to Part 4: To find how long it takes to drop the first $\frac{1}{2}$ inch, which is from 15.5 inches down to 15 inches, we solve $0.265t^2 - 4.055t + 15.50 = 15$ for t. One way is illustrated in Figure 5.61 with an added line at $D = 15$, which shows that it takes 0.124 minute, or about 7.44 seconds, for the water level to drop the first $\frac{1}{2}$ inch.

Solution to Part 5: To find how long it takes to drop the last $\frac{1}{2}$ inch, we solve $D = 0.5$, or $0.265t^2 - 4.055t + 15.50 = 0.5$, for t. Adding a line at $D = 0.5$ to the graph in Figure 5.60 and finding the intersection in Figure 5.62, we find that $D = 0.5$ when $t = 6.26$ minutes. Since the jug is drained when $t = 7.44$ minutes, it takes $7.44 - 6.26 = 1.18$ minutes, or about 70.8 seconds, for the water level to drop the last $\frac{1}{2}$ inch.

Figure 5.62: *Finding when the water depth is 1/2 inch*

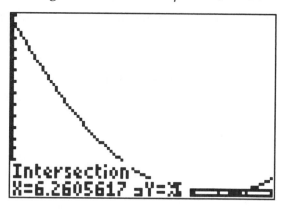

EXAMPLE 5.13   *Cost Per Student of Education*

In the study "Economies of scale in high school operation" by J. Riew,[28] the author studied data from the early 1960's on expenditures for high schools ranging from 150 to 2400 enrollment. The data he observed was similar to that in the table below.

Enrollment	200	600	1000	1400	1800
Cost per student, in dollars	667.4	545.0	461.0	415.4	408.2

1. Show that the cost per student C can be modeled as a quadratic function of the number n of students.

2. J. Riew used the quadratic[29] model $C = 743 - 0.402n + 0.00012n^2$ dollars per student. According to this model, what size enrollment is most efficient in terms of cost per student?

Solution to Part 1: In order to show the data is quadratic in nature, we need to show that the second-order differences are constant. We do this in two steps. To get the first-order differences, we simply subtract each current entry from its preceding entry:

Enrollment	200	600	1000	1400
First-order difference	-122.4	-84.0	-45.6	-7.2

To get the second-order differences, we simply repeat the process, taking differences of the first-order differences:

First-order difference	-122.4	-84.0	-45.6
Second-order difference	38.4	38.4	38.4

---

[28]Published in *Review of Economics and Statistics* **48** (August 1966), 280-287. In our presentation of Riew's results we have suppressed variables such as teacher salaries.

[29]Once data has been recognized as approximately quadratic in nature, the appropriate function can be found using *quadratic regression*, which is available on some calculators. We will not discuss the procedure here.

We see that the second-order differences are constant, and we conclude that the original data can be modeled with a quadratic function.

Solution to Part 2: We want to minimize the function C. To do this, we use the calculator to graph and find the minimum. We choose a horizontal span of 0 to 2400 and a vertical span of 0 to 750. From Figure 5.63 we see that the most efficient enrollment is about 1675 students.

Figure 5.63: *Minimum cost per student*

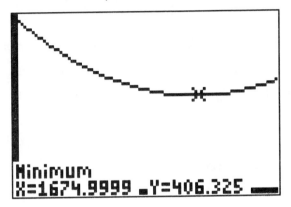

To illustrate the use of the quadratic formula, we look once more at the function C from

The quadratic formula

We have taken the point of view that linear functions should be solved by hand calculation and that others should be solved using the calculator. However, there is a well-known formula, the *quadratic formula*, which allows the solution of quadratic equations by hand calculation, and some instructors may prefer that their students use it. In general, a quadratic equation $ax^2 + bx + c = 0$ has two solutions, which differ only by a sign in one place:

$$x = \frac{-b + \sqrt{b^2 - 4ac}}{2a}$$

and

$$x = \frac{-b - \sqrt{b^2 - 4ac}}{2a}.$$

To illustrate the use of the quadratic formula, we look once more at the function C from Example 5.13. Suppose that we want to know what enrollment will produce a cost per student of $500 per student. That is, we wish to solve the equation

$$0.00012n^2 - 0.402n + 743 = 500.$$

To solve this using the quadratic formula, we first subtract 500 from each side:

$$0.00012n^2 - 0.402n + 243 = 0.$$

Next we identify the values of a, b, and c:

$$a = 0.00012$$

$$b = -0.402$$

$$c = 243.$$

Plugging these values into the quadratic formula, we get

$$n = \frac{-b + \sqrt{b^2 - 4ac}}{2a} = \frac{0.402 + \sqrt{(-.402)^2 - 4 \times 0.00012 \times 243}}{2 \times 0.00012} = 2558.53$$

$$n = \frac{-b - \sqrt{b^2 - 4ac}}{2a} = \frac{0.402 - \sqrt{(-.402)^2 - 4 \times 0.00012 \times 243}}{2 \times 0.00012} = 791.47 \ .$$

Thus we see that enrollments of either 791 or 2559 will produce a cost of about $500 per student. You may wish to check to see that you get the same answers using the crossing graphs method with the calculator. Which procedure do you find easier? Which makes more sense?

Higher-degree polynomials and their roots

Higher-degree polynomials are named for the highest power which occurs in the polynomials. For example, *cubic* polynomials have a highest term of degree three, *quartic* polynomials have a highest term of degree four, and so on. These polynomials often have interesting graphs, such as the cubic graph in Figure 5.64 and the quartic graph in Figure 5.65.

Figure 5.64: The cubic function $Y = X^3 - 4X^2 + X - 2$

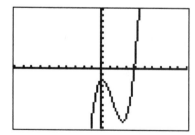

Figure 5.65: The quartic function $Y = -0.1X^4 + 0.2X^3 + X^2$

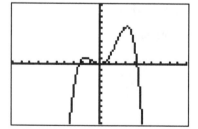

Just as for quadratic functions, the rates of change of cubic functions have a special property: the rate of change of a cubic function is a quadratic function. In particular, data may be modeled exactly by a cubic function precisely if, with evenly spaced values for the variable, the third-order differences in the function values are all the same. In general, data with evenly spaced values for the variable may be modeled exactly by a polynomial of degree n if the nth-order differences are constant, that is, all the same. Just as in the quadratic case, if you know the initial value and all the initial differences (first-order, second-order, and so on), then you can write out the formula for the polynomial; however, this gets progressively more complicated.

For cubic equations, just as for quadratic equations, there is a *cubic formula*, but it is considerably more complicated than the quadratic formula. For example, in the special case $x^3 + cx = d$, the formula gives one root as

$$x = \sqrt[3]{\sqrt{\left(\frac{d}{2}\right)^2 + \left(\frac{c}{3}\right)^3} + \frac{d}{2}} - \sqrt[3]{\sqrt{\left(\frac{d}{2}\right)^2 + \left(\frac{c}{3}\right)^3} - \frac{d}{2}}.$$

The general cubic formula is more complicated. There is also a quartic formula, but it is quite difficult to use. This means that there are formulas to solve any polynomial equation of degree four or less. Amazingly, not only is there no known formula of this type for solving polynomial equations of degree five, but it can be shown that there *cannot* be a formula of this type for solving polynomial equations of degree five! This remarkable fact was proved in the mid-1820's by Niels Henrik Abel when he was in his early twenties.

Rational functions

Rational functions are functions which can be written as a ratio of polynomial functions, that is, as a fraction whose numerator and denominator are both polynomials. Simple examples include any polynomial, $\frac{1}{x}$, $\frac{x^{12} - x}{0.7x^6 + 3}$, and $x + \frac{1}{x^2}$. By way of contrast, e^x, $\sin x$ and $\sqrt{x}$ are not rational functions. The graphs of rational functions can be quite interesting and often exhibit behavior not found in other functions.

One interesting rational function is $\frac{8}{x^2 + 1}$, whose graph is displayed in Figure 5.66. Unlike polynomial functions, for larger values of x, the function gets closer to zero, which is similar to the behavior of decreasing exponential functions. Unlike exponential functions, this function gets close to zero both for large positive and for large (in size) negative values of x.

Figure 5.66: A graph of the rational function $y = \dfrac{8}{x^2 + 1}$

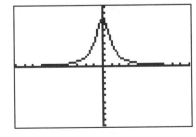

Figure 5.67: A graph of the rational function $y = \dfrac{x}{x^2 - 1}$

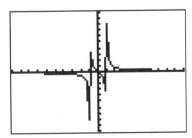

A more interesting rational function is $\frac{x}{x^2 - 1}$, whose graph is displayed in Figure 5.67. The graph looks quite ragged and it is unclear what is happening near $x = -1$ and $x = 1$. The reason for this becomes more clear when we consider that when $x = -1$ or 1, then $x^2 - 1 = 0$; thus the function is trying to evaluate $\dfrac{1}{0}$, which is not possible. In Figure 5.68, we graph $y = \dfrac{x}{x^2 - 1}$, looking more closely near $x = 1$. In Figure 5.69, we stretch out the vertical axis to view the behavior near $x = 1$.

Figure 5.68: A graph of $y = \dfrac{x}{x^2 - 1}$ with horizontal span 0 to 2 and vertical span -10 to 10

Figure 5.69: Same graph as Figure 5.68, but with vertical span -25 to 25

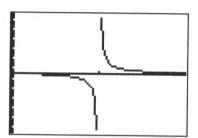

The function $\dfrac{x}{x^2 - 1}$ has no value at $x = 1$. Note that, for x to the left of 1, as x tends towards 1 the function values are negative and become larger and larger in size; by contrast, as x tends towards 1 from the right, the function values are positive and become larger and larger positive numbers. In general a function has a *pole* at $x = a$ if the function is not defined at $x = a$ and the values of the function become larger and larger in size as x gets near a. Not all rational functions have poles, as we saw earlier in considering $y = \dfrac{8}{x^2 + 1}$. In general, poles occur when the denominator of a rational function has a zero, but the numerator does not have the same zero. [30] If a function has a pole at the point $x = a$, then we say that the line $x = a$ is a *vertical asymptote* of the graph of the function, since the graph gets closer and closer to this line as x tends towards a. For example, the graph of the function $\dfrac{x}{x^2 - 1}$ in Figure 5.69 has the line $x = 1$ as a vertical asymptote.

In physical applications, rational functions are far less common than the other functions discussed in the text. When they have poles, the poles usually occur at numbers to which the physical situation does not apply. For example, Ohm's Law says that when electric current is flowing across a resistor, then $i = \dfrac{v}{R}$, where i is current, v is voltage, and R is resistance. For fixed v, the function i is a rational function of R with a pole when $R = 0$. Note that, if $R = 0$, then there is no resistance, so we are no longer in the situation of current flowing across a resistor.

---

[30]We are assuming that factors common to the numerator and denominator have been canceled.

Exercise Set 5.4

1. **Quadratic data:** Show that the following data can be modeled by a quadratic function.

x	0	1	2	3	4
$P(x)$	6	5	8	15	26

2. **Data that is not quadratic:** Show that the following data cannot be modeled by a quadratic function.

x	0	1	2	3	4
$P(x)$	5	8	17	38	77

3. **Quadratic model:** Show that the following data can be modeled by a quadratic function, and find a formula for a quadratic model.

x	0	1	2	3	4
$Q(x)$	5	6	13	26	45

4. **Linear and quadratic data:** One of the two tables below shows data that can be modeled by a linear function, and the other shows data that can be modeled by a quadratic function. Identify which is which, and find a formula for each model.

Table A

x	0	1	2	3	4
$f(x)$	10	17	26	37	50

Table B

x	0	1	2	3	4
$g(x)$	10	17	24	31	38

5. **Leaking can:** The side of a cylindrical can full of water springs a leak, and the water begins to stream out. The depth H (in inches) of water remaining is a function of the distance D in inches (measured from the base of the can) at which the stream of water strikes the ground. Here is a table of values of D and H:

Distance D in inches	0	1	2	3	4
Depth H in inches	1.00	1.25	2.00	3.25	5.00

(a) Show that H can be modeled as a quadratic function of D.

(b) Find the formula for H as a quadratic function of D.

(c) When the depth is 4 inches, how far from the base of the can will the water stream strike the ground?

(d) When the water stream strikes the ground 5 inches from the base of the can, what is the depth of water in the can?

6. **Falling rock:** A rock is thrown downward, and the distance D (in feet) it falls in t seconds is given by $D = 16t^2 + 3t$. Find how long it takes for the rock to fall 400 feet by using

   (a) the quadratic formula

   (b) the crossing graphs method.

7. **Water jug revisited:** Solve Part 3 of Example 5.12 using the quadratic formula.

8. **Natural gas prices:** A cubic polynomial can be used to model the average price of natural gas in the U.S. for residential use.[31] One such model for the years 1980 through 1990 is

$$P = 0.0124t^3 - 0.242t^2 + 1.41t + 3.4.$$

   Here t is the number of years since 1980 and P is the price in dollars per 1000 cubic feet of natural gas.

   (a) Make a graph of P versus t.

   (b) Your graph should have both a maximum and an inflection point. Find the location of both and explain in practical terms what they mean.

9. **Forming a pen:** You want to form a rectangular pen of area 80 square feet. One side of the pen is to be formed by an existing building and the other three sides by a fence. If w is the length (in feet) of the sides of the rectangle perpendicular to the building, then the length of the side parallel to the building is $\dfrac{80}{w}$, and so the total amount $F = F(w)$ (in feet) of fence required is the rational function

$$F = 2w + \frac{80}{w}.$$

   (a) Make a graph of F versus w.

   (b) Explain in practical terms the behavior of the graph near the pole at $w = 0$.

   (c) Determine the dimensions of the rectangle which requires a minimum amount of fence.

---

[31]See D. R. LaTorre, J. W. Kenelly, *et al.*, *Calculus Concepts*, 1998, Houghton Mifflin, Boston.

10. **Inventory:** The yearly inventory expense E (in dollars) of a car dealer is a function of the number Q of automobiles ordered at a time from the manufacturer. A dealer who orders only a few automobiles at a time will have the expense of placing several orders, while if the order sizes are large, then the dealer will have a large inventory of unsold automobiles. For one dealer the formula is

$$E = \frac{425Q^2 + 8000}{Q} \, ,$$

and so E is a *rational* function of Q.

(a) Make a graph of E versus Q covering order sizes up to 10.

(b) Explain in practical terms the behavior of the graph near the pole at $Q = 0$. (Hint: Keep in mind that there is a fixed cost of processing each order, regardless of the size of the order.)

5.5 CHAPTER SUMMARY

While linear and exponential functions are among the most common in mathematical applications, other functions occur often as well. These include *power functions*, *periodic functions*, *quadratic* and other *polynomial functions*, and *rational functions*.

Power Functions

An exponential function is of the form Pa^x, where a is fixed and x is the variable. For *power functions*, these roles are reversed. Power functions are of the form $f(x) = cx^k$. The number k is the *power*, and c is $f(1)$. For a power function with c and x positive:

- If k is positive, then f is increasing.

- If k is negative, then f decreases toward zero.

One common occurrence of power functions is the distance D, in feet, a rock will fall in t seconds if air resistance is ignored: $D = 16t^2$.

A key property of power functions which is often important in applications is the *homogeneity property*. It says that for a power function $f = cx^k$, if x is increased by a factor of t, then f is increased by a factor of t^k. For example, the volume V of a sphere is a power function of its radius: $V = \frac{4\pi}{3}r^3$. The homogeneity property tells us that if the radius of a sphere is doubled, the volume will be increased by a factor of 2^3. That is, the volume will be 8 times larger.

While power functions with positive power increase, in the long term they do not increase as rapidly as exponential functions. The classic dominance of exponential functions over power functions is sometimes key to the choice of an appropriate model. Exponential functions should be used when eventual dramatic growth is anticipated.

Modeling Data with Power Functions

An adaptation of the way the logarithm was used to make exponential models is available for making power function models as well. The key difference is that for exponential data, we take the logarithm of the function values only, leaving the variable values as they are. To make a power model, we take the logarithm of both function value and variable value. The procedure is illustrated below.

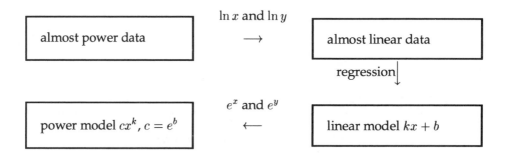

We note that this procedure is sometimes accomplished by plotting the data on logarithmic graphing paper. It is perhaps a bit more tedious than using the calculator, but the practice remains common.

Periodic Functions and Right Triangle Trigonometry

Many natural phenomena such as tidal activity and radio waves display a repetitive or *periodic* behavior, and we use a special function, the *sine* function, to model them. The *period* of a function is the time required for it to repeat. For example, if we are modeling tides which repeat daily, and which range from 3 feet above sea level to 3 feet below sea level, the period is one day. The height the function rises and falls about its middle level is known as the *amplitude*. For these tides, the amplitude is 3 feet.

If we want to model a periodic activity with a given amplitude, we can do so using

$$\text{amplitude} \ \times \sin\left(\frac{360}{\text{period}}\,x\right).$$

The sine function and some of its relatives, the *cosine* and *tangent* functions, can be used to measure right triangles, and hence they become quite important in applications to surveying. If θ is an acute angle of a right triangle, then there is a side *adjacent* to θ, a side *opposite* to θ, and the *hypotenuse*, which is opposite to the right angle. The formulas are

$$\sin \theta = \frac{\text{opposite}}{\text{hypotenuse}}$$

$$\cos \theta = \frac{\text{adjacent}}{\text{hypotenuse}}$$

$$\tan \theta = \frac{\text{opposite}}{\text{adjacent}}.$$

For example, if we stand 300 horizontal feet from the base of a building and find that our transit must be elevated 21 degrees to point toward the top of the building, we can use the tangent function to figure out how tall the building is. The key is that there is a right triangle involved. The adjacent side is the horizontal distance to the building, and the opposite side is the vertical face of the building:

$$\tan 21 = \frac{\text{opposite}}{\text{adjacent}}$$

$$\tan 21 = \frac{\text{height}}{300}$$

$$300 \tan 21 = \text{height}$$

$$115.16 = \text{height}.$$

Polynomial and Rational Functions

A *polynomial* function is one whose formula can be written as a sum of power functions where each of the powers is a nonnegative whole number. Polynomial functions are useful when a single power function does not adequately describe the situation. For example, if a rock is thrown downward with an initial velocity of 5 feet per second, the distance D in feet it travels in t seconds is given by the polynomial $D = 16t^2 + 5t$.

One of the most important kinds of polynomials is the *quadratic function*. It is a polynomial where the highest power occurring is 2. The formula for a quadratic function is $ax^2 + bx + c$. The graph of a quadratic function is a *parabola*, an important geometric shape because it is the path followed by any object ejected near the surface of the Earth: rifle bullet, golf ball, arrow, baseball, etc.

Quadratic functions can be characterized by the fact that their rate of change is linear. For data, this is exhibited by the fact that the *second-order differences* are constant. This can be used as a test to determine if data should be modeled by a quadratic function.

In this text, we have made a practice of solving nonlinear equations using the graphing calculator. But there is a formula, known as the *quadratic formula*, that allows for hand solution of quadratic equations. For the quadratic equation $ax^2 + bx + c = 0$, the solution is given by

$$x = \frac{-b \pm \sqrt{b^2 - 4ac}}{2a}.$$

Higher-degree[32] polynomials are interesting and sometimes useful. Their graphs may have many peaks and valleys, and they are useful in modeling phenomena with more than one maximum or minimum or with an inflection point. Polynomial data for polynomials of degree n is characterized by the fact that n^{th}-order differences are constant.

Rational functions are quotients of polynomials. They occur, for example, in describing the gravitational attraction between planets or other objects.

---

[32]The degree is the largest power that occurs in the polynomial.

CHAPTER 6 *Rates of change*

A key idea in analyzing natural phenomena as well as the functions that may describe them is the *rate of change*, and it is a familiar idea from everyday experience. When you are driving your car, the rate of change in your location is *velocity*. Discussions of cars or airplanes commonly involve velocity because it is virtually impossible to convey key ideas about motion without reference to velocity. The rate of change is no less descriptive for other events. If you step on the gas pedal to pass a slowly moving truck, then your velocity changes, and the rate of change in velocity is *acceleration*. Rates of change occur in other contexts as well. In fact the idea is pervasive in mathematics, science, engineering, social science, and daily life because it is such a powerful tool for description and analysis.

6.1 VELOCITY

We look at velocity first because it is a familiar rate of change. Consider a rock tossed upward from ground level. As the rock rises and then falls back to Earth, we locate its position as the distance up from the ground. In Figure 6.1 we have sketched a possible graph of distance up versus time. It shows the height of the rock increasing as it rises and then decreasing after it reaches its peak and begins to fall.

Getting velocity from directed distance

The *velocity* of the rock is the rate of change in distance up from the ground. That is, at any point in the flight of the rock, the velocity measures how fast the rock is rising or falling. The rock gets some initial velocity at the moment of the toss, but the effect of gravity makes it slow down as it rises toward its peak. After the rock reaches its peak, gravity causes it to accelerate toward the ground, and its *speed* increases. In everyday language, speed and velocity are often used interchangeably, but there is an important, if subtle, difference. Speed is always a positive number, the number you might read on the speedometer of your car, for example. But velocity has an additional component; it has a sign attached which indicates the direction of movement. The key to understanding this is to remember that velocity is the rate of change in *directed distance*. As the rock moves upward, its distance up from the ground is increasing. Thus the rate of change in distance up, the velocity, is positive. But when the rock starts to fall back to Earth, its distance up from the Earth is decreasing. Thus the rate of change in distance up, the velocity, is negative.

Figure 6.1: Distance up of a rock versus time

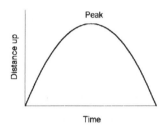

Figure 6.2: Velocity of a rock versus time

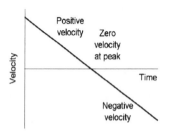

The relationships between the graph of distance up and the graph of velocity are crucial to understanding velocity and indeed rates of change in general. In Figure 6.2 we have sketched a graph of the velocity versus time for the rock. During the period when the rock is moving upward, the graph of distance up is increasing. Since distance up is increasing, velocity is positive. This is shown in Figure 6.2 by the fact that the graph of velocity is above the horizontal axis until the rock reaches the peak of its flight. We also note that during this period the graph of velocity is decreasing toward zero, indicating that the rock is slowing as it rises. At the point where the rock reaches its peak, the graph of velocity crosses the horizontal axis, indicating that the velocity is momentarily zero. When the rock is moving downward, the graph of distance up is decreasing. During this period, velocity is negative, and this is shown in Figure 6.2 by the fact that the graph of velocity is below the horizontal axis. Notice that the *speed* of the rock is increasing as it falls, and so its velocity (the negative of its speed) is decreasing. Velocity continues to decrease until the rock strikes the ground.

In Figure 6.2 we have represented the graph of velocity as a straight line. This is not an apparent consequence of the graph in Figure 6.1, but as we shall see later (see Section 6.3 and also Exercise 9 of Section 6.2), it is a consequence of the fact that (near the surface of the Earth), gravity imparts a constant acceleration. Finally, we note that the sign of velocity depends on the perspective chosen to measure position. In this case, we located the rock using its distance up from the surface of the Earth. In Exercise 2 at the end of this section, you will be asked to analyze the velocity of the rock if the event is viewed from the top of a tall building and the rock's position is considered as its distance down from the top of the building.

The relationships we observed between the graphs of distance up for the rock and its velocity are fundamental, and they are true in a general setting. When directed distance is increasing, velocity is positive. When directed distance is decreasing, velocity is negative. When directed distance is not changing, even momentarily, velocity is zero. In particular, when directed distance reaches a peak (maximum) or valley (minimum), velocity is zero.

> ### KEY IDEA 6.1: VELOCITY AND DIRECTED DISTANCE: THE FUNDAMENTAL RELATIONSHIP
>
> 1. Velocity is the rate of change in directed distance.
>
> 2. **When directed distance is increasing, velocity is positive.** (The graph of velocity is above the horizontal axis.)
>
> 3. **When directed distance is decreasing, velocity is negative.** (The graph of velocity is below the horizontal axis.)
>
> 4. **When directed distance is not changing, velocity is zero.** (The graph of velocity is on the horizontal axis.)

Constant velocity means linear directed distance

In the case of the rock we just looked at, velocity is always changing, but in many situations there are periods when the velocity does not change. A familiar example is that of a car which might accelerate from a yield sign onto a freeway and travel at the same speed for a time before exiting the freeway and parking at its destination. If the car is traveling westward on the freeway, and we locate its position as distance west from the yield sign, a possible graph of its velocity, the rate of change in distance west, is shown in Figure 6.3. We want to consider what this means for the function $L = L(t)$ which gives the location of the car at time t as distance west of the yield sign. Our interest is in determining the nature of L during the period when the car is traveling at a constant velocity on the freeway, say at 65 miles per hour. During this period, each hour the car moves 65 additional miles. This means that the distance L increases by 65 miles each hour, and so it is a linear function with slope 65.

In more general terms, since velocity of the car is the rate of change in L, during this period the rate of change in L is constant. But we know that a function with a constant rate of change is linear, and this is the key observation we wish to make. Whenever velocity, the rate of change in directed distance, is constant, directed distance must be a linear function. Furthermore, the slope of L is the velocity. The linearity of L is reflected in Figure 6.4 by the fact that, for the portion of the graph that represents the time when the car was on the freeway, we drew the graph of directed distance as a straight line.

Figure 6.3: *The velocity of a car traveling on a freeway*

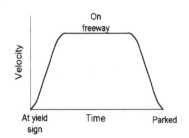

Figure 6.4: *Constant velocity means linear directed distance*

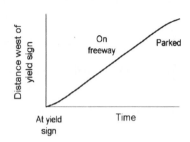

KEY IDEA 6.2: DIRECTED DISTANCE FOR CONSTANT VELOCITY

When velocity is constant, the rate of change in directed distance is constant. Thus directed distance is a linear function with slope equal to the constant velocity, and so its graph is a straight line.

EXAMPLE 6.1   *From New York to Miami*

An airplane leaves Kennedy Airport in New York and flies to Miami, where it is serviced and receives new passengers before returning to New York. Assume that the trip is uneventful and that after each takeoff the airplane accelerates to its standard cruising speed, which it maintains until deceleration prior to landing.

1. Describe what the graph of distance south of New York looks like during the period when the airplane is maintaining its standard cruising speed on the way to Miami.

2. Say we locate the airplane in terms of its distance south of New York. Make possible graphs of its distance south of New York versus time and of the velocity of the airplane versus time.

3. Say we locate the airplane in terms of its distance north of Miami. Make possible graphs of its distance north of Miami versus time and of the velocity of the airplane versus time.

Solution to Part 1: To say that the airplane is maintaining its standard cruising speed means that its velocity is not changing. In other words, the rate of change in distance south is constant. As we noted above, any function with a constant rate of change is linear. We conclude that during this period the graph of distance south is a straight line whose slope is the standard cruising speed.

Solution to Part 2: We will make the graph of distance south in several steps, showing a template for solving problems of this type.

Step 1: Locate and mark the places on the graph where directed distance is zero and where it reaches its extremes. The distance south of New York is zero at the beginning and end of the trip. These points will lie on the horizontal axis, and we have marked and labeled them in Figure 6.5. The graph of distance south will be at its maximum while the airplane is being serviced in Miami. This is also marked and labeled in Figure 6.5.

Figure 6.5: *Zeros and extremities of distance south*

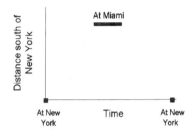

Figure 6.6: *Regions of increase and decrease for distance south*

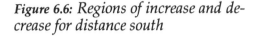
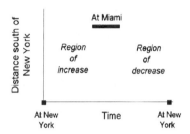

Step 2: Label on the graph the regions where directed distance is increasing and the regions where it is decreasing. Directed distance is increasing on the trip from New York to Miami and is decreasing on the return leg of the trip. These regions are labeled in italics in Figure 6.6.

Step 3: Complete the graph, incorporating any additional information known about directed distance. In this case, the important additional information is that for most of the trip the airplane is flying at its standard cruising speed. As we observed in Part 1, this means that the graph of distance during these periods is a straight line. Our graph of distance south of New York for the airplane is in Figure 6.7. Notice that in the completed graph, we have labeled the horizontal and vertical axes as well as other important features. You are encouraged to follow this practice.

Now let's consider velocity for the same trip. To make the graph we use steps similar to those used to graph distance south.

Step 1: Locate and label the points on the graph where the velocity is zero. The velocity is zero when the airplane is on the ground at the beginning and end of the trip and during the period it is being serviced in Miami. These times must lie on the horizontal axis and are so marked and labeled in Figure 6.8. For the graph of velocity, we put the horizontal axis in the middle since we expect to graph both positive and negative values.

Figure 6.7: *The completed graph of distance south from New York*

Figure 6.8: *Times when the velocity is zero*

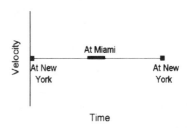

Step 2: Label the regions where velocity is positive and where it is negative. On the flight from New York to Miami, distance south of New York is increasing, so its rate of change, or velocity, is positive. On the return leg, distance south of New York is decreasing, so its rate of change, or velocity, is negative. These are labeled in italics in Figure 6.9. At this point, it is important to check that the places in Figure 6.6 that we marked as *increasing* match the places in Figure 6.9 that we marked as *positive*. Similarly, the *decreasing* labels in Figure 6.6 must match the *negative* labels in Figure 6.9.

Step 3: Complete the graph, incorporating any other known features of the graph. For velocity, the important additional feature is that during most of both legs of the trip the airplane is maintaining a constant cruising speed. That means that the graph of velocity must be horizontal in these regions. This is shown in our completed graph in Figure 6.10. Note once again that we have labeled the axes as well as other important features of the graph.

Figure 6.9: *Labeling regions of positive and negative velocity*

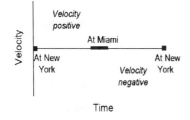

Figure 6.10: *The completed graph of velocity*

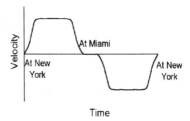

We want to emphasize that Figure 6.7 and Figure 6.10 show the fundamental relationship between graphs of directed distance and velocity that we noted earlier. Observe in particular that during the flight from New York to Miami the graph of directed distance in Figure 6.7 is increasing, and the corresponding part of the graph of velocity in Figure

6.10 is above the horizontal axis. Also, during the return leg of the trip the graph of directed distance in Figure 6.7 is decreasing, and the corresponding graph of velocity in Figure 6.10 is below the horizontal axis.

Solution to Part 3: We want to look now at the airplane flight from the perspective of a Miami resident. We will follow the same steps as we did in Part 2 to get the graphs. If you are waiting at the Miami airport, the distance north to the airplane is a large positive number at the beginning and end of the trip, when the plane is in New York. These points are marked and labeled in Figure 6.11. The distance north is zero while the airplane is in Miami, and this region is also marked in Figure 6.11.

As the plane flies from New York toward Miami, the distance north from Miami decreases, and this is noted in italics in Figure 6.11. On the return leg, distance north increases, and this is also noted in italics in Figure 6.11. Following the notes in Figure 6.11, we complete the graph in Figure 6.12, showing as before the regions where the graph is a straight line.

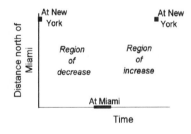

Figure 6.11: Interesting points and regions of increase and decrease for distance north

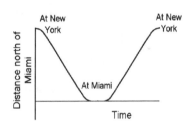

Figure 6.12: The completed graph of location from the Miami perspective

We proceed similarly to get the graph of velocity. We first mark the places, New York and Miami, where the velocity is zero. Our change in perspective does not affect this, so the appropriate picture is the same as Figure 6.8. But the change in perspective does affect the sign of velocity. On the first leg of the flight, distance north of Miami is decreasing, so the velocity is negative. This is marked in italics in Figure 6.13, and we note that it corresponds to the *decreasing* label in Figure 6.11. Similarly, on the return trip, distance north of Miami is increasing, so the velocity is positive. This is marked in italics in Figure 6.13, and we note as before that it corresponds to the *increasing* label in Figure 6.11. We complete the picture in Figure 6.14, being careful to incorporate constant cruising speed.

Figure 6.13: *Labeling velocity features from the Miami perspective*

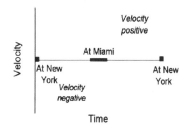

Figure 6.14: *The completed graph of velocity from the Miami perspective*

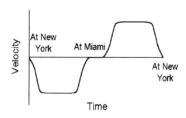

EXAMPLE 6.2   *A Forgotten Wallet*

A man leaves home driving west on a straight road. We locate the position of the car as the distance west from home. Suppose the man begins his trip and sets his cruise control, but after a few minutes he notices that he has forgotten his wallet. He returns home, gets his wallet, and then resumes his journey. He figures that he may be late, so he sets the cruise control a little higher than before. Make graphs of distance west versus time and velocity versus time for the forgetful man's car.

Solution: We first make the graph showing the location of the car following the steps outlined in Example 6.1. The distance west from home is zero when the trip starts and when the man is back home retrieving his wallet. In Figure 6.15 we have marked these places, as well as the point where the forgotten wallet was remembered and the car turned around and headed home. Distance west is increasing when the car is moving away from home. This occurs at the beginning of the trip and also later, after the wallet has been retrieved. The only period when distance west is decreasing is when the man is returning home to get the wallet. These regions are marked in italics in Figure 6.15.

To get the completed graph in Figure 6.16, we made use of the notes from Figure 6.15 and also incorporated the information we have about cruise control settings: On the first leg of the trip, the cruise control is set, so the velocity is constant. This means that, for this period, distance west is a linear function, and so we have drawn the graph there as a straight line. We have also made the graph for the trip back home a straight line, though that information is not provided in the problem. For the final leg of the trip, we know once again that the graph is a straight line, but we also know a little more. The cruise control is set a little higher than it was on the first leg of the trip, and therefore the velocity is greater. Thus we must draw a line with a larger slope (a steeper line) than for the first part of the trip.

Figure 6.15: *Important features in the location of a car*

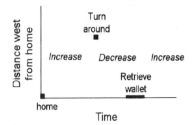

Figure 6.16: *The completed graph for a car trip*

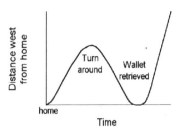

We proceed with a similar analysis for velocity. We know that the velocity is zero when the trip starts and when the wallet is being retrieved. But the velocity is also zero when the car turns around. At this point, just as we saw at the peak of the rock's flight, the rate of change in distance west is momentarily zero. We have marked and labeled all of these places in Figure 6.17.

We can get the right regions for positive and negative velocity from the increasing and decreasing labels in Figure 6.15. Each *increase* label in Figure 6.15 corresponds to a *positive velocity* label in Figure 6.17, and the *decrease* label corresponds to the *negative velocity* label in Figure 6.17. We use the notes in Figure 6.17 to get the completed graph of velocity in Figure 6.18, keeping in mind that since the cruise control is set for most of the trip, the velocity is constant most of the time. This means that the graph of velocity is horizontal in these regions. Finally, we note that since the car went faster on the last part of the trip, the graph of velocity is higher there.

Figure 6.17: *Important features for velocity*

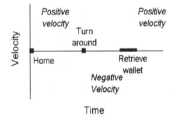

Figure 6.18: *The completed graph for the velocity of the car*

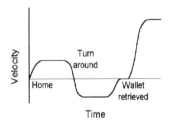

When distance is given by a formula

Many times, directed distance or velocity is given by a formula. For example, if we stand atop a building 30 feet high and toss a rock upward with an initial velocity of 18 feet per second, then elementary physics can be used to show that the distance up $D = D(t)$ from the ground of the rock t seconds after it is tossed is given by

$$D = 30 + 18t - 16t^2 \text{ feet.}$$

We are assuming that, when the rock comes back down, it does not hit the top of the building where we are standing but falls all the way to the ground. Let's begin our analysis by making a graph of D versus t. We can use our everyday experience to choose a window setting. The rock surely won't go over 50 feet high, and it will take only a few seconds for the rock to hit the ground. Thus in Figure 6.19, which shows the flight of the rock, we used $\boxed{6.1}$ a horizontal span of $t = 0$ to $t = 5$ and a vertical span of $D = 0$ to $D = 50$.

How long did it take the rock to hit the ground? That happens when the distance up is zero. Thus we want to solve for t in the equation

$$30 + 18t - 16t^2 = 0.$$

We have used the single graph method $\boxed{6.2}$ to do that in Figure 6.20, and we see that it takes about 2.04 seconds for the rock to complete its flight.

Figure 6.19: *A rock tossed upward from the top of a building*

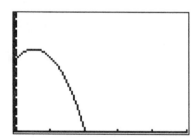

Figure 6.20: *When the rock strikes the ground*

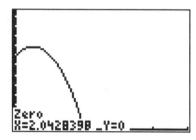

We might also want to know how high the rock went and when it reached its peak. In Figure 6.21 we got that information by using the calculator to locate $\boxed{6.3}$ the maximum. We see that the rock reached its highest point of 35.06 feet just over half a second after it was tossed.

Figure 6.21: *The peak of the rock's flight*

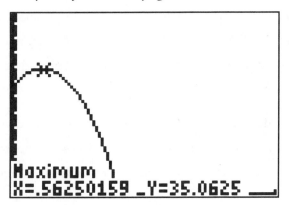

Let's look now at the velocity $V = V(t)$ of the rock. We proceed as in earlier examples, first marking important features, and noting in particular where velocity is positive, where it is negative, and where it is zero. This is shown in Figure 6.22. We use this information to complete the graph of velocity in Figure 6.23.

Figure 6.22: *Interesting features of velocity*

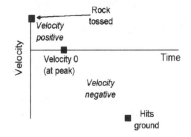

Figure 6.23: *The graph of velocity versus time*

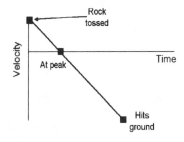

Exercise Set 6.1

1. **From New York to Miami again:** The city of Richmond, Virginia, is about halfway between New York and Miami. A Richmond resident might locate the airplane in Example 6.1 using distance north of Richmond. Make the graphs of location and velocity of the airplane from this perspective.

2. **The rock with a changed reference point:** Make graphs of position and velocity for a rock tossed upward from ground level as it might be viewed by someone standing atop a tall building. Thus the location of the rock is measured by its distance down from the top of the building.

3. **The rock with a formula:** If from ground level we toss a rock upward with a velocity of 30 feet per second, we can use elementary physics to show that the height in feet of the rock above the ground t seconds after the toss is given by $S = 30t - 16t^2$.

   (a) Use your calculator to plot the graph of S versus t.

   (b) How high does the rock go?

   (c) When does it strike the ground?

   (d) Sketch the graph of the velocity of the rock versus time.

4. **Getting velocity from a formula:** When a man jumps from an airplane with an opening parachute, the distance $S = S(t)$ in feet that he falls in t seconds is given by

   $$S = 20 \left(t + \frac{e^{-1.6t} - 1}{1.6} \right) .$$

   (a) Use your calculator to make a graph of S versus t for the first 5 seconds of the fall.

   (b) Sketch a graph of velocity for the first 5 seconds of the fall.

5. **A rubber ball:** A rubber ball is dropped from the top of a building. The ball lands on concrete and bounces once before coming to rest on the grass. Measure the location of the ball as its distance up from the ground. Make graphs of the location and velocity of the ball.

6. **Gravity on Earth and on Mars:** The acceleration due to gravity near the surface of a planet depends on the mass of the planet; larger planets impart greater acceleration than do smaller ones. Mars is much smaller than Earth. A rock is dropped from the top of a cliff on each planet. Give its location as the distance down from the top of the cliff.

   (a) On the same coordinate axes, make a graph of distance down for each of the rocks.

(b) On the same coordinate axes, make a graph of velocity for each of the rocks.

7. **Traveling in a car:** Make graphs of location and velocity for each of the following driving events. In each case, assume that the car leaves from home moving west down a straight road and that position is given as the distance west from home base.

    (a) *A vacation*: Being anxious to begin your overdue vacation, you set your cruise control and drive faster than you should to the airport. You park your car there and get on an airplane to Spain. When you fly back two weeks later, you are tired and drive at a leisurely pace back home. (<u>Note</u>: Here we are talking about location of your car, not the airplane.)

    (b) *On a country road*: A car driving down a country road encounters a deer. The driver slams on the brakes and the deer runs away. The journey is cautiously resumed.

    (c) *At the movies*: In a movie chase scene, our hero is driving his car rapidly toward the bad guys. When the danger is spotted, he does a Hollywood 180 degree turn and speeds off in the opposite direction.

8. **Making up a story about a car trip:** You begin from home on a car trip. Initially your velocity is a small positive number. Shortly after you leave, velocity decreases momentarily to zero. Then it increases rapidly to a large positive number and remains constant for this part of the trip. After a time, velocity decreases to zero and then changes to a large negative number.

    (a) Make a graph of velocity for this trip.

    (b) Discuss your distance from home during this driving event and make a graph.

    (c) Make up a driving story that matches this description.

9. **Car trips with given graphs:**

   (a) The graph in Figure 6.24 shows your distance west of home on a car trip. Make a graph of velocity.

   (b) The graph in Figure 6.25 shows your velocity on a different car trip. Assuming you start at home, make a graph of distance west of home.

Figure 6.24: *A graph of distance west of home*

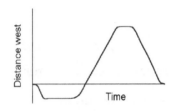

Figure 6.25: *A graph of velocity for a different car trip*

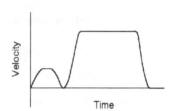

10. **A car moving in an unusual way:** In making graphs for location and velocity, the authors of this text have tried to avoid sharp corners. This problem is designed to show you why. Suppose a car's distance west of home is given by the graph in Figure 6.26.

Figure 6.26: *Position for an unusual driving event*

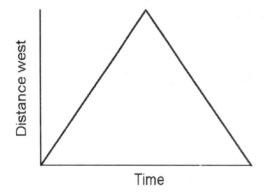

   (a) Make a graph of velocity versus time. In making your graph, take extra care near the peak shown in Figure 6.26.

   (b) Carefully describe the motion of the car near the peak in Figure 6.26. Do you think it is possible to drive in a way that matches the graph in Figure 6.26?

11. **Sporting events:** Analyze the following sporting activities according to the instructions.

    (a) *Skiing*: A skier is going down a ski slope at Park City, Utah. The hill is initially gentle but gets steep about halfway down before flattening out at the bottom. Locate position on the slope as the distance from the top and make graphs of location and velocity.

    (b) *Practicing for the NCAA basketball tournament*: You are bouncing a basketball. Locate the basketball by its distance up from the floor. Make graphs of location and velocity.

    (c) *Hiking*: The graph of the distance west of base camp for a hiker in Colorado is given in Figure 6.27. Make a graph of velocity versus time and then make up a story that matches this description.

Figure 6.27: A hiking trip

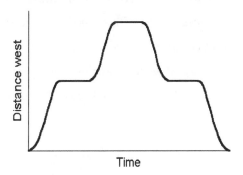

6.2 RATES OF CHANGE FOR OTHER FUNCTIONS

Velocity is the rate of change in directed distance, and this same idea applies to any function. For a function $f = f(x)$, we can look at the rate of change in f with respect to x, which tells how f changes for a given change in x. Rates of change are so pervasive in mathematics, science, and engineering that they are given a special notation, most commonly $\dfrac{df}{dx}$ or $f'(x)$, and a special name, the *derivative of f with respect to x*. We will use the notation $\dfrac{df}{dx}$ but will consistently refer to it as the *rate of change in f with respect to x*. We should note that some applications texts use the notation $\dfrac{\Delta f}{\Delta x}$ for the rate of change.[1]

Examples of rates of change

The notation $\dfrac{df}{dx}$ means the rate of change in f with respect to x. Specifically, $\dfrac{df}{dx}$ tells how much f is expected to change if x increases by one unit. We note that $\dfrac{df}{dx}$ is a function of x; in general, the rate of change varies as x varies. Some examples may help clarify things.

1. If $S = S(t)$ gives directed distance for an object as a function of time t, then $\dfrac{dS}{dt}$ is the rate of change in directed distance with respect to time. This is *velocity*. It tells the additional distance we expect to travel in one unit of time. For example, if we are currently located $S = 100$ miles south of Dallas, Texas, and if we are traveling south with a velocity $\dfrac{dS}{dt}$ of 50 miles per hour, then in one additional hour we would expect to be 150 miles south of Dallas.

2. If $V = V(t)$ is the velocity of an object as a function of time t, then $\dfrac{dV}{dt}$ is the rate of change in velocity with respect to time. This is *acceleration*. It tells the additional velocity we expect to attain in one unit of time. For example, if we are traveling with a velocity $V = 50$ miles per hour, and if we start to pass a truck, our acceleration $\dfrac{dV}{dt}$ might be 2 miles per hour each second. Then one second in the future, we would expect our velocity to be 52 miles per hour.

3. If $T = T(D)$ denotes the amount of income tax (in dollars) you pay on an income of D dollars, then $\dfrac{dT}{dD}$ is the rate of change in tax with respect to the money you earn. It is known as the *marginal tax rate*. If you have already accumulated D dollars, then $\dfrac{dT}{dD}$ is the additional tax you expect to pay if you earn one additional dollar.   For example,

---

[1] This is seen most often in business and agricultural applications but in other places as well. There is a technical difference between the meanings of $\dfrac{df}{dx}$ and $\dfrac{\Delta f}{\Delta x}$. The notation $\dfrac{\Delta f}{\Delta x}$ means an *average rate of change over a given change in x* while $\dfrac{df}{dx}$ is the *instantaneous rate of change*. The distinction is important in advanced mathematics, but less so in many applications.

suppose we have a tax liability of \$3000 and our marginal tax rate $\dfrac{dT}{dD}$ is 0.2, or 20 cents per dollar. Then if we earn an additional \$1, we would expect our tax liability to increase by 20 cents to a total of \$3000.20; if instead we earn an additional \$100, we would expect our tax liability to increase by \$20 to a total of \$3020. The marginal tax rate is a crucial bit of information for financial planning.

4. If $P = P(i)$ is the profit (in dollars) you expect to earn on an investment of i dollars, then $\dfrac{dP}{di}$ is the rate of change in profit with respect to dollars invested. It tells how much additional profit is to be expected if one additional dollar is invested. In economics this is known as *marginal profit*. For example, if our current investment in a project gives a profit of \$1000, and if our marginal profit $\dfrac{dP}{di}$ is 0.2, or 20 cents per dollar, then we would expect that an additional investment of \$100 would yield an additional profit of \$20 for a total profit of $P = 1020$ dollars.

Common properties of all rates of change

The list could be extended indefinitely, but fortunately all rates of change have common properties which we have already seen in our study of velocity. If $S = S(t)$ denotes directed distance at time t, then we know that when S is increasing, then the velocity $\dfrac{dS}{dt}$ is positive; when S is decreasing, then the velocity $\dfrac{dS}{dt}$ is negative; and when S is not changing, then the velocity $\dfrac{dS}{dt}$ is zero. This fundamental relationship holds true for all rates of change.

KEY IDEA 6.3: FUNDAMENTAL PROPERTIES OF RATES OF CHANGE

For a function $f = f(x)$ we will use the notation $\dfrac{df}{dx}$ to denote the rate of change in f with respect to x.

1. The expression $\dfrac{df}{dx}$ tells how f changes in relation to x. It gives the additional value that is expected to be added to f if x increases by one unit.

2. When f is increasing, then $\dfrac{df}{dx}$ is positive.

3. When f is decreasing, then $\dfrac{df}{dx}$ is negative.

4. When f is not changing, then $\dfrac{df}{dx}$ is zero.

EXAMPLE 6.3 *Passing a Truck*

You are driving with your cruise control set when you encounter a slow-moving truck. You speed up to pass the truck. When you have overtaken the truck, you slow down and resume your previous speed. Let $V = V(t)$ denote your velocity during this event as a function of time t.

1. Explain in practical terms the meaning of $\dfrac{dV}{dt}$.

2. Make a graph of $V = V(t)$, marking important points on the graph.

3. Make a graph of acceleration $A = \dfrac{dV}{dt}$.

Solution to Part 1: The function $\dfrac{dV}{dt}$ is the rate of change in velocity. This is acceleration. It tells the additional velocity you expect to attain over one unit of time.

Solution to Part 2: Since the cruise control is set, velocity is positive and constant at the beginning. That means the graph of V versus t starts above the horizontal axis and is horizontal. When we start around the truck we speed up, making the graph of $V(t)$ go up. After we overtake the truck, we slow down and resume our original velocity. This makes the graph of velocity go back down to its original level and flatten out again. Our graph is shown in Figure 6.28.

Figure 6.28: *Velocity when passing a truck*

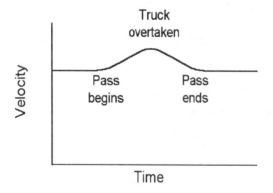

Solution to Part 3: We will get the graph of acceleration directly from the graph of velocity in Figure 6.28 and then verify that it makes sense. The basic tools we use are the fundamental properties of rates of change: When V is increasing then $\dfrac{dV}{dt}$ is positive, when V is decreasing then $\dfrac{dV}{dt}$ is negative, and when V is not changing then $\dfrac{dV}{dt}$ is zero.

Before and after passing, the graph of velocity in Figure 6.28 is horizontal and so V is not changing at all. This means that the rate of change in velocity $A = \dfrac{dV}{dt}$ is zero during these periods. Thus the graph of A lies on the horizontal axis, as we have shown in Figure 6.29. From the time the pass begins until the truck is overtaken, the graph of velocity in Figure 6.28 is increasing. This means that the rate of change in velocity $\dfrac{dV}{dt}$ is positive, and we have marked that in italics in Figure 6.29. At the peak of the graph in Figure 6.28, the acceleration is zero; but from there to the time the pass is completed, velocity is decreasing, so its rate of change, namely the acceleration, is negative, as is marked in italics in Figure 6.29.

We used the information from Figure 6.29 to draw the completed graph of $\dfrac{dV}{dt}$ in Figure 6.30. We note that the graph makes sense. At the beginning and end of the graph, the

Figure 6.29: *Important features for acceleration*

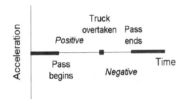

Figure 6.30: *Completed graph of acceleration when passing a truck*

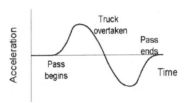

cruise control is set, so velocity is not changing. That is, acceleration is zero. As we begin the pass, we accelerate. Thus the graph of acceleration is above the horizontal axis for this period, as is shown in Figure 6.30. Once the truck is overtaken, we let off the gas pedal and the car *decelerates* back to the original cruising speed. Deceleration is the same as negative acceleration, and so from the time we overtake the truck until the pass is completed, the graph of acceleration is below the horizontal axis, as is represented in Figure 6.30. Thus the graph we made using the fundamental properties of rates of change agrees with our intuitive analysis.

When the rate of change is zero

In Example 6.3 we used the graph of velocity V in Figure 6.28 to get the graph of acceleration $\dfrac{dV}{dt}$ in Figure 6.30. In the process we noted two situations where the acceleration was zero: when velocity is constant while the cruise control is set (here acceleration is zero over a span of time), and when velocity reaches its maximum value at the point when the truck is overtaken (here acceleration is zero for one instant in time). Once again, these observations remain true for rates of change in general. When f is constant, its rate of change $\dfrac{df}{dx}$ is zero over a span of x values. This is intuitively clear, since the statement that $\frac{df}{dx}$ is zero is the same as saying that f is not changing and therefore remains at a constant value. At a peak or valley of f, the rate of change $\frac{df}{dx}$ will be zero for that single value of x. These important facts are illustrated in Figure 6.31 and Figure 6.32.

Figure 6.31: Constant function has persistent zero rate of change

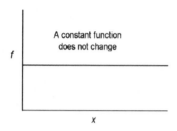

Figure 6.32: At extrema the rate of change is momentarily zero

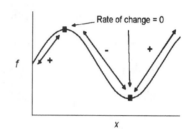

Let's look at an example to see how we use the fact that the rate of change is zero at extrema.

EXAMPLE 6.4   Marginal Profit for a Tire Company

The CEO of a tire company has kept records of the profit $P = P(n)$ the company makes when it produces n tires per day.

1. The rate of change $\frac{dP}{dn}$ is known as the *marginal profit*. Explain in practical terms the meaning of marginal profit.

2. What action should the CEO take if the marginal profit is positive?

3. What action should the CEO take if the marginal profit is negative?

4. The CEO has used his records to make the graph of marginal profit shown in Figure 6.33. According to this graph, how many tires per day should the company be producing?

Solution to Part 1: The marginal profit $\frac{dP}{dn}$ is the rate of change in profit with respect to the number of items produced. It is the change in profit the CEO can expect to earn by producing one more tire.

Solution to Part 2: Marginal profit tells the CEO how much profit can be expected to change if one additional tire is produced. Since marginal profit is positive, additional profit can be made by increasing production. In other words, since the rate of change $\frac{dP}{dn}$ is positive, then the profit P is an increasing function of the production level n. This means that the CEO should increase production, since then the profit will increase. This situation might occur if, for example, the demand for tires were larger than current production.

Figure 6.33: Marginal profit for a tire company

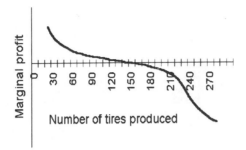

Solution to Part 3: In this case the marginal profit is negative, meaning that if production is increased, profits will change by a negative amount. That is, profits will decrease. In other words, since the rate of change $\frac{dP}{dn}$ is negative, then the profit P is a *decreasing* function of n. In this scenario, it would be wise for the CEO to cut back on production. This might happen if the number of tires currently being produced exceeded consumer demand.

Solution to Part 4: In Part 2 we noted that if the marginal profit is positive, then the profit P is increasing with n, and production should be increased. In Part 3, we saw that when the marginal profit is negative, then the profit P is decreasing with n, and production should be decreased. Thus it makes sense that for the CEO, the ideal value of marginal profit is zero. In other words, to get a maximum profit P, we want to find the point where P switches from increasing to decreasing (since at this point the profit is at its highest value), and that is where the marginal profit $\frac{dP}{dn}$ switches from positive to negative. This is where the graph in Figure 6.33 crosses the horizontal axis, i.e., where the marginal profit is zero. The crossing point occurs at $n = 160$, so the company should be producing 160 tires per day.

When the rate of change is constant

We observed in Section 6.1 that when velocity $\frac{dS}{dt}$ is constant, we can conclude that directed distance S is a linear function whose slope is the constant velocity. Virtually all of the observations we made about the relationships between velocity and directed distance are true for rates of change in general, and this one is no exception: If the rate of change in f, namely $\frac{df}{dx}$, is constant, then f is a linear function with slope $\frac{df}{dx}$. We would emphasize that this is not really a new observation. The characterization of linear functions as those with constant rate of change has been used repeatedly in this course. The only thing new about our expression of this fact here is the use of $\dfrac{df}{dx}$ to denote the rate of change.

EXAMPLE 6.5   A Leaky Balloon

A balloon is initially full of air but springs a leak. Let $B = B(t)$ represent the volume in liters of air in the balloon at time t measured in minutes.

1. Explain in practical terms what $\dfrac{dB}{dt}$ means. As air is leaking out of the balloon, is $\dfrac{dB}{dt}$ positive or negative?

2. Assume air is leaking from the balloon at a constant rate of 0.5 liter per minute.[2] What does this information tell you about $\dfrac{dB}{dt}$? What does it tell you about B?

3. A little later the leak is patched, a pump which outputs 2 liters of air per minute is attached to the balloon, and it is inflated back to its original size. As the balloon inflates, what can we conclude about $\dfrac{dB}{dt}$? What can we conclude about B?

4. Make a graph of B versus t as the balloon leaks air and then is reinflated to its original size.

5. Make a graph of $\dfrac{dB}{dt}$ versus t as the balloon leaks air and then is reinflated to its original size.

Solution to Part 1: The function $\dfrac{dB}{dt}$ is the rate of change in B with respect to t. This is the rate of change in the volume of air in the balloon with respect to time. In practical terms, $\dfrac{dB}{dt}$ is the change in the volume of air in the balloon that we expect in one minute of time. Since the volume B of air in the balloon is decreasing, $\dfrac{dB}{dt}$ is negative.

Solution to Part 2: To say that air is leaking from the balloon at 0.5 liter per minute means that the volume B of air in the balloon is decreasing by 0.5 liter per minute. In other words, B is changing by -0.5 liter per minute. We conclude that $\dfrac{dB}{dt} = -0.5$.

Since B is a function with a constant rate of change of -0.5, we know that it is a linear function with slope -0.5. We were not asked to do so, but with this information we have almost everything we need to write a formula for B, namely $B = -0.5t + b$. In this formula, what is the practical meaning of b?

---

[2]This is not in fact how we would expect a balloon to leak. Rather, we would expect it to leak rapidly when it is almost full and more slowly when it is nearly empty. We will examine this more realistic description of a leaky balloon in the next section.

Solution to Part 3: As the balloon is inflated, the volume B of air in the balloon is increasing by 2 liters per minute. That is, $\dfrac{dB}{dt} = 2$ liters per minute. Since B has a constant rate of change of 2, we conclude that B is a linear function with slope 2 during the period of reinflation.

As in Part 2, we were not asked to write a formula for B, but we have almost everything we need to do that. The formula is $B = 2t + c$. Explain the practical meaning of c in this formula.

Solution to Part 4: We know that while the balloon is losing air, B is linear with slope -0.5, and that after the pump is attached, B is linear with slope 2. Thus, the graph of B should start as a straight line with slope -0.5 and then change to a straight line with slope 2. These features are reflected in Figure 6.34.

Note in Figure 6.34 that we have rounded the graph at the bottom where the pump is attached. Alternatively, one might draw this graph with a sharp corner at the bottom. Which do you think is correct?

Figure 6.34: *The graph of B versus t*

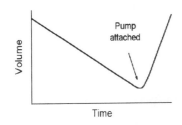

Figure 6.35: *The graph of* $\dfrac{dB}{dt}$ *versus t*

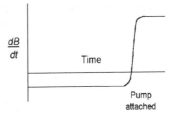

Solution to Part 5: We know that while the balloon is leaking, $\dfrac{dB}{dt}$ has a constant value of -0.5. Thus its graph during this period is a horizontal line located 0.5 unit below the horizontal axis. After the pump is attached, $\dfrac{dB}{dt}$ has a constant value of 2. Thus during this period, its graph is a horizontal line 2 units above the horizontal axis. How the two line segments are connected near the time the pump is attached depends on how we drew the graph of B near that point. If you think there should be a sharp corner there, how would this affect the graph of $\dfrac{dB}{dt}$ we have drawn in Figure 6.35?

Exercise Set 6.2

1. **Estimating rates of change:** Use your calculator to make the graph of $f(x) = x^3 - 5x$.

   (a) Is $\dfrac{df}{dx}$ positive or negative at $x = 2$?

   (b) Identify a point on the graph of f where $\dfrac{df}{dx}$ is negative.

2. **The spread of AIDS:** The following table shows the cumulative number $N = N(t)$ of AIDS cases in the United States that have been reported to the Centers for Disease Control by end of the year given. (The source for this data, the U.S. Centers for Disease Control and Prevention in Atlanta, cautions that this data is subject to retrospective change.)

t = year	1986	1987	1988	1989	1990	1991	1992
N = total cases reported	28,711	49,799	80,518	114,113	155,766	199,467	244,939

   (a) What in practical terms does $\dfrac{dN}{dt}$ mean?

   (b) From 1986 to 1992 was $\dfrac{dN}{dt}$ ever negative?

3. **Mileage for an old car:** The gas mileage M you get on your car depends on the age t (in years) of the car.

   (a) Explain in practical terms the meaning of $\dfrac{dM}{dt}$.

   (b) As your car ages and performance degrades, do you expect $\dfrac{dM}{dt}$ to be positive or negative?

4. **Investing in the stock market:** You are considering buying three stocks whose prices at time t are given by $P_1(t)$, $P_2(t)$, and $P_3(t)$. You know that $\dfrac{dP_1}{dt}$ is a large positive number, $\dfrac{dP_2}{dt}$ is near zero, and $\dfrac{dP_3}{dt}$ is a large negative number. Which stock will you buy? Explain your answer.

5. **Hiking:** You are hiking in a hilly region and $E = E(t)$ is your elevation at time t.

   (a) Explain in practical terms the meaning of $\dfrac{dE}{dt}$.

   (b) Where might you be when $\dfrac{dE}{dt}$ is a large positive number?

   (c) You reach a point where $\dfrac{dE}{dt}$ is briefly zero. Where might you be?

   (d) Where might you be when $\dfrac{dE}{dt}$ is a large negative number?

6. **Marginal profit:** A small firm produces at most 15 widgets in a week. Its profit P (in dollars) is a function of n, the number of widgets manufactured in a week. The marginal profit $\frac{dP}{dn}$ for the firm is given by the formula

$$\frac{dP}{dn} = 72 + 6n - n^2 .$$

(a) Use your calculator to make a graph of $\frac{dP}{dn}$ versus n.

(b) For what values of n is the profit P decreasing?

(c) How many widgets should the firm produce in a week to maximize profit P?

7. **Health plan:** The manager of an employee health plan for a firm has studied the balance B (in millions of dollars) in the plan account as a function of t, the number of years since the plan was instituted. He has determined that the rate of change $\frac{dB}{dt}$ in the account balance is given by the formula

$$\frac{dB}{dt} = 12 - 10e^{0.1t} .$$

(a) Use your calculator to make a graph of $\frac{dB}{dt}$ versus t over the first 5 years of the plan.

(b) During what period is the account balance B increasing?

(c) At what time is the account balance B at its maximum?

8. **A race:** A man enters a race that involves running and swimming. He lines up at the starting gate and begins running very fast, hoping to take the initial lead before settling in to a constant slower pace for a distance run. To complete the race he must swim across a river and back before running back to the starting gate. He swims at a constant rate but cannot swim as fast as he runs. Let $S = S(t)$ denote the distance of the contestant from the starting gate.

(a) Make a graph of S versus t. Note on your graph the places where you know that S is linear.

(b) Make a graph of the contestant's velocity.

(c) Make a graph of the contestant's acceleration.

9. **The acceleration due to gravity:** From the time of Galileo, physicists have known that near the surface of the Earth, gravity imparts a constant acceleration of 32 feet per second per second. If air resistance is ignored, explain how this shows that velocity for a falling object is a linear function of time.

10. **Water in a tank:** Water is leaking out of a tank. The amount of water in the tank t minutes after it springs a leak is given by $W(t)$ gallons.

   (a) Explain in practical terms what $\dfrac{dW}{dt}$ means.

   (b) As water leaks out of the tank, is $\dfrac{dW}{dt}$ positive or negative?

   (c) For the first 10 minutes, water is leaking from the tank at a rate of 5 gallons per minute. What do you conclude about the nature of the function W during this period?

   (d) After about 10 minutes, the hole in the tank suddenly gets larger, and water begins to leak out of the tank at 12 gallons per minute.

      i. Make a graph of W versus t. Be sure to incorporate linearity where it is appropriate.

      ii. Make a graph of $\dfrac{dW}{dt}$ versus t.

11. **A population of bighorn sheep:** There is an effort in Colorado to restore the population of bighorn sheep. Let $N = N(t)$ denote the number of sheep in a certain protected area at time t.

   (a) Explain in practical terms the meaning of $\dfrac{dN}{dt}$.

   (b) A small breeding population of bighorn sheep is initially introduced into the protected area. Food is plentiful and conditions are generally favorable for bighorn sheep. What would you expect to be true about the sign of $\dfrac{dN}{dt}$ during this period?

   (c) This summer a number of dead sheep were discovered, and all were infected with a disease that is known to spread rapidly among bighorn sheep and is almost always fatal. How would you expect an unchecked spread of this disease to affect $\dfrac{dN}{dt}$?

   (d) If the re-introduction program goes well, then the population of bighorn sheep will grow to the size the available food supply can support and remain at about that same level. What would you expect to be true about $\dfrac{dN}{dt}$ when this happens?

12. **Eagles:** In an effort to restore the population of bald eagles, a breeding group is introduced into a protected area. Let $N = N(t)$ denote the population of bald eagles at time t. Over time you observe the following information about $\dfrac{dN}{dt}$.

   - Initially $\dfrac{dN}{dt}$ is a small positive number.

   - A few years later $\dfrac{dN}{dt}$ is a much larger positive number.

   - Many years later $\dfrac{dN}{dt}$ is positive but near zero.

Make a possible graph of $N(t)$.

13. **Visiting a friend:** I live in the suburbs and my friend lives out in the country. The speed limit between my home and a stop sign one mile away is 35 miles per hour. After that the speed limit is 65 miles per hour. Assume I drove out to visit my friend. As I pulled into his driveway, I saw he wasn't at home and so I drove back to my house. Assume I obeyed all the traffic laws and had an otherwise uneventful trip. Make graphs of $S = S(t)$, my location in relation to my home; the velocity $V = V(t)$; and the acceleration $A = A(t)$. Be sure your graphs show linearity where it is appropriate.

14. **A car trip:** You are moving away from home in your car, and throughout this story you do not turn back. Your acceleration is initially a large positive number, but it decreases slowly to zero, where it remains for a while. Suddenly your acceleration decreases rapidly to a large negative number before returning slowly to zero.

    (a) Make a graph of acceleration for this event.

    (b) Make a graph of velocity for this event.

    (c) Make a graph of the location of the car.

    (d) Make up a driving story which matches this description.

15. **An airplane** leaves Atlanta flying to Dallas. Due to heavy air traffic, it circles the airport at Dallas for a time before landing. Let $D(t)$ be the distance west from Atlanta to the airplane. Make a graph of $D(t)$. Thinking of velocity $V(t)$ as the rate of change in $D(t)$, make a graph of the velocity of the airplane.

16. **Growth in height:** The following graph gives for a certain man his height $H = H(t)$ in inches at age t in years.

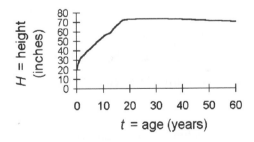

    (a) Explain in practical terms what $\dfrac{dH}{dt}$ means.

    (b) Sketch a graph of $\dfrac{dH}{dt}$ versus t.

6.3 EQUATIONS OF CHANGE: LINEAR AND EXPONENTIAL FUNCTIONS

Many natural phenomena can be understood by first analyzing their rates of change, and this can often be used to construct a mathematical model. For example, in 1798, Malthus considered the problem of poverty in terms of the rate of change of population as opposed to the rate of change of food production: "Population, when unchecked, increases in a geometrical ratio. Subsistence increases only in an arithmetic ratio. A slight acquaintance with numbers will shew the immensity of the first power in comparison with the second." [3] Even earlier Galileo observed that near the surface of the Earth downward acceleration due to gravity is constant. Physicists use the letter g to denote this constant, and its value is about 32 feet per second per second. For a freely falling object, the observation that acceleration is constant is actually an observation about $\frac{dV}{dt}$, the rate of change in velocity with respect to time. (We measure downward velocity V in feet per second and time t in seconds.) Furthermore, we can express this in an equation:

$$\text{Acceleration} = \text{Constant}$$
$$\frac{dV}{dt} = g.$$

We will refer to equations of this type as *equations of change* since the equation expresses how the function changes. These equations are a special case of what, in advanced mathematics, science, and engineering, are known more formally as *differential equations*.

Equation of change for linear functions

The key mathematical observation that we need to make about the equation of change $\frac{dV}{dt} = g$ is that it tells us that the rate of change in V is the constant g. But we already know that functions with constant rate of change are linear functions. We conclude that the velocity V is a linear function of time with slope $g = 32$.

Since V is a linear function with slope 32, we know that we can write V as $V = 32t + b$. In order to complete the formula for V, we need only know the initial value b of velocity. For example, if we throw a rock downward with an initial velocity of 10 feet per second, then $b = 10$, and $V = 32t + 10$.

The same analysis holds true for any function with constant rate of change.

---

[3] From page 14 of *First Essay on Population, 1798,* by Thomas Robert Malthus, 1926 reprint by Macmillan & Co. Ltd., London.

KEY IDEA 6.4: EQUATIONS OF CHANGE AND LINEAR FUNCTIONS

The equation of change $\dfrac{df}{dx} = m$, where m is a constant, says that f has a constant rate of change m and hence that f is a linear function with slope m. That is, $f = mx + b$. An initial condition is needed to determine the value of b.

We want to emphasize that Key Idea 6.4 is not a new idea. It is just a restatement, in terms of equations of change, of a fundamental property of linear functions which we have already studied in some detail.

EXAMPLE 6.6 *Dropping a Rock on Mars*

On Mars, a falling object satisfies the equation of change

$$\frac{dV}{dt} = 12.16 \, ,$$

where V is downward velocity in feet per second and t is time in seconds.

1. What is the value of acceleration due to gravity on Mars?

2. Suppose an astronaut stands atop a cliff on Mars and throws a rock downward with an initial velocity of 8 feet per second. What is the velocity of the rock 3 seconds after release?

Solution to Part 1: The equation of change $\dfrac{dV}{dt} = 12.16$ tells us that the rate of change in velocity for an object falling near the surface of Mars is 12.16 feet per second per second. Since the rate of change in velocity is acceleration, we conclude that acceleration is 12.16 feet per second per second.

Solution to Part 2: The equation of change $\dfrac{dV}{dt} = 12.16$ tells us that the rate of change in V has a constant value of 12.16. We conclude that V is a linear function with slope 12.16. Since the rock was tossed downward with a velocity of 8 feet per second, the initial value of V is 8. We now have the information we need to write a formula for V:

$$V = 12.16t + 8 \, .$$

We want the velocity of the rock when $t = 3$:

$$V = 12.16 \times 3 + 8 = 44.48 \text{ feet per second.}$$

Instantaneous rates of change

Linear functions have a constant rate of change, but for other functions this rate varies, and we need to think more carefully about how to interpret the rate of change. When we study equations of change, we are thinking of instantaneous rates of change, not average rates of change.

For example, a car's speedometer reading gives the *instantaneous* rate of change in distance. To understand this, consider the following question: How do we interpret a reading of 60 miles per hour? It would be inaccurate to interpret this reading as saying that we will actually travel 60 miles in the next hour, since the car could slow down or speed up through the hour. It would be more accurate to say that we expect to go one mile in the next minute; this is the same average rate of change, but it is computed over a shorter time interval, so there should be less variation in speed. Still more accurate would be the interpretation that we will go one-sixtieth of a mile (88 feet) in a second. In all three interpretations the average rate of change is 60 miles per hour, but our common sense tells us that the last interpretation is the most accurate, because computing over a time interval of one second is the closest to instantaneous change.

We stress that, even though our rates of change are instantaneous, the units used are the same as for average rates of change. For example, our speedometer readings are measured in miles per hour, not in "miles per instant."

Equation of change for exponential functions

We know that the equation of change for a linear function is $\frac{df}{dx} = m$ because that equation tells us that the rate of change is constant. Let's look at the balance in a savings account as a familiar example of an exponential function and find out its equation of change. Suppose we open an account by investing $100 with a financial institution that advertises an APR of 3%, or (as a decimal) 0.03, with interest compounded *continuously*. Our balance $B = B(t)$ (with B in dollars and time t in years since opening the account) is an exponential function since it has a constant percentage rate of change. We want to express the fact that B has a constant percentage rate of change using an equation of change. All that is required is to look more closely at what the description of B tells us. The rate of change $\frac{dB}{dt}$ is the amount by which we expect our account balance to increase in a unit of time. This is just the interest. On an annual basis, the interest is 3% of the balance, or $0.03 \times B$. We find that the equation of change is $\frac{dB}{dt} = 0.03B$.

One question that arises is why we use the APR of 0.03 here and not the EAR (effective annual rate), which takes into account the compounding. The reason is that the rate of change is instantaneous, and during an instant we can ignore compounding. We compute on an annual basis, though, because the unit of measurement is a year, not an instant.

Which exponential function does the equation of change $\dfrac{dB}{dt} = 0.03B$ represent? To answer this we need to find the yearly growth factor for B. We saw in Section 4 of Chapter 4 that when interest is compounded continuously we find the yearly growth factor by exponentiating the APR. Then the growth factor for our exponential function B is $e^{0.03}$. Since the initial value is 100 dollars (our initial investment), we have the formula

$$B = 100 \times \left(e^{0.03}\right)^t,$$

or, since $e^{0.03}$ is about 1.0305, the approximate formula

$$B = 100 \times 1.0305^t.$$

As we saw in Section 4 of Chapter 4, an alternative form of B avoids this last step of computing the exponential of the APR: We just write

$$B = 100e^{0.03t}.$$

This description of the account balance is typical of all exponential functions. If the rate of change in f is a constant multiple of f, that is, if $\dfrac{df}{dx} = rf$, then f has constant proportional rate of change and hence is an exponential function with exponential growth rate r. A formula for f in the alternative form is then $f = Pe^{rx}$, where P is the initial value of f. To find the standard form, we first get the growth (or decay) factor by exponentiating:

$$\text{Growth (or decay) factor} = e^r.$$

The formula for f is then $f = P \times (e^r)^x$. Note that finding the alternative form requires one less step, and it is more accurate since it avoids the rounding error in calculating the growth factor.

KEY IDEA 6.5: EQUATIONS OF CHANGE AND EXPONENTIAL FUNCTIONS

The equation of change $\dfrac{df}{dx} = rf$, where r is a constant, says that f has constant proportional (and hence percentage) rate of change and is therefore an exponential function. The exponential growth rate for f is r, so the growth (or decay) factor is e^r. That is,

$$f = Pe^{rx},$$

or

$$f = P \times (e^r)^x,$$

where P is the initial value of f.

As with equations of change for linear functions, we want to emphasize that Key Idea 6.5 is not really new. It is simply a restatement, in terms of equations of change, of the fundamental properties of exponential functions.

EXAMPLE 6.7   *Newton's Law of Cooling*

Several exercises in Chapters 1 and 4 involved the way objects heat or cool. The formulas used in those exercises were all derived from Isaac Newton's *law of cooling*. Newton observed that if a hot object is placed in the open air to cool, then the way it cools depends on the difference between the temperature of the object and the temperature of the air. Specifically, if $D = D(t)$ is the difference between the temperature of the object and the temperature of the air, then the rate of change in D is proportional to D. The value of the constant of proportionality depends on the nature of the cooling object and is normally determined experimentally. For a hot cup of coffee placed on the table to cool, if we measure our time t in minutes since the coffee was poured, the value of the cooling constant is about -0.06 per minute. Assume that coffee poured from the pot is 190 degrees and that room temperature is 72 degrees.

1. Write an equation of change that shows how the coffee cools.

2. Describe the function D and give a formula for it.

3. Find a formula for the temperature of the coffee at time t.

4. What will be the temperature of the coffee after 5 minutes? When will the temperature of the coffee be 130 degrees?

Solution to Part 1: The function D is the temperature difference, so the rate of change in temperature difference is $\dfrac{dD}{dt}$. Newton's law of cooling tells us that the rate of change in temperature difference is -0.06 times D:

$$\text{Rate of change in temperature difference} \;=\; -0.06 \times \text{Temperature difference}$$
$$\frac{dD}{dt} \;=\; -0.06 \times D \;.$$

Solution to Part 2: The equation of change we got is that of an exponential function because it gives the rate of change in D as a constant multiple of D.

The constant of proportionality, or cooling constant, is -0.06. Therefore D is an exponential function with exponential growth rate -0.06. To write a formula for D we need to know its initial value. When the coffee was poured, its temperature was 190 degrees. Room temperature is 72 degrees. Thus the initial temperature difference, the initial value for D, is $190 - 72 = 118$ degrees, so

$$D = 118e^{-0.06t}.$$

This gives a formula for D in the alternative form. To find the standard form, we first calculate the decay factor: $e^{-0.06}$ is about 0.94. The standard form is then $D = 118 \times 0.94^t$. In the rest of the solution we will use the alternative form since it is more accurate.

Solution to Part 3: Let $T = T(t)$ denote the temperature of the coffee at time t. We already know a formula for the difference D between coffee temperature and air temperature. To get a formula for T, we need to add the temperature of the air:

$$\text{Temperature difference} = T - \text{Air temperature}$$
$$T = \text{Air temperature} + \text{Temperature difference}$$
$$T = 72 + D$$
$$T = 72 + 118e^{-0.06t} .$$

Solution to Part 4: To get the temperature of the coffee after 5 minutes, we put 5 in place of t in the formula we found in Part 3:

$$T(5) = 72 + 118e^{-0.06 \times 5} = 159.4 \text{ degrees.}$$

To answer the second question, we want to know when the temperature of the coffee is 130 degrees. That is, we want to solve for t the equation

$$72 + 118e^{-0.06t} = 130 .$$

We do this using the crossing graphs method. First we enter $\boxed{6.4}$ the function and the target temperature and record the appropriate correspondences:

$$Y_1 = T, \text{ temperature on vertical axis}$$
$$Y_2 = 130, \text{ target temperature}$$
$$X = t, \text{ minutes on horizontal axis.}$$

Next we make the table of values shown in Figure 6.36 to determine how to set the graphing window. The table shows that we can expect the temperature difference to reach 130 degrees between 10 and 15 minutes after the coffee cup is left to cool. Thus we set up the graphing window using a horizontal span of $t = 0$ to $t = 15$ and a vertical span of $T = 100$ to $T = 250$. In Figure 6.37 we made the graphs and then used the calculator to find $\boxed{6.5}$ the intersection point. We see that the coffee reaches a temperature of 130 degrees after about 12 minutes of cooling.

Figure 6.36: A table of values for a cooling cup of coffee

X	Y₁	Y₂
0	190	130
5	159.42	130
10	136.76	130
15	119.98	130
20	107.54	130
25	98.329	130
30	91.505	130

X=0

Figure 6.37: When the temperature reaches 130 degrees

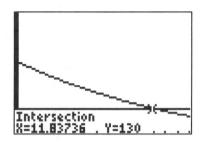

Intersection
X=11.83736 . Y=130 . . . .

Why equations of change?

The equation of change we encountered in Example 6.7 is of the form $\frac{df}{dx} = rf$, and it is a remarkable fact that this same equation actually describes a diverse set of natural phenomena. As you will see in Exercise 4 at the end of this section, this equation describes how a balloon leaks air. It is not apparent that leaky balloons and cooling coffee cups have anything at all in common, but the temperature of the coffee and the volume of air in the balloon actually behave in a very similar fashion. One important power of mathematics is its ability to distill ideas from many areas into a common one. This is a good case in point. The following table shows a number of phenomena which all obey an equation of change of the form $\frac{df}{dx} = rf$.

Phenomenon	Equation of change	Comments
Newton's law of cooling	$\frac{dD}{dt} = rD$	D is temperature difference between cooling object and ambient air; r depends on the object.
Exponential population growth	$\frac{dN}{dt} = rN$	N is population size; r is the exponential growth rate.
Compound interest	$\frac{dB}{dt} = rB$	B is the account balance; r is the APR with continuous compounding, t in years.
Radioactive decay	$\frac{dA}{dt} = rA$	A is the amount of radioactive substance remaining; r depends on the radioactive substance.

These and other phenomena can be modeled by exponential functions because they all obey equations of change of the form $\frac{df}{dx} = rf$. These phenomena also have another important feature, which we will see in the next section: In making a mathematical model, it is easiest first to understand the rate of change and then to make an equation of change to model that. In this section, we have exploited the equation of change to get a formula to serve as our model. In the next section, we will see that there are many other ways to get information from equations of change.

Exercise Set 6.3

1. **Looking up:** The constant $g = 32$ feet per second per second is the *downward* acceleration due to gravity near the surface of the Earth. If we stand on the surface of the Earth and locate objects using their distance up from the ground, then the positive direction is up, and so down is the negative direction. With this perspective, the equation of change in velocity for a freely falling object would be expressed as

$$\frac{dV}{dt} = -g \,.$$

(We measure upward velocity V in feet per second and time t in seconds.) Consider a rock tossed upward from the surface of the Earth with an initial velocity of 40 feet per second upward.

(a) Use a formula to express the velocity $V = V(t)$ as a linear function. (<u>Hint</u>: You get the slope of V from the equation of change. The vertical intercept is the initial value.)

(b) How many seconds after the toss does the rock reach the peak of its flight? (<u>Hint</u>: What is the velocity of the rock when it reaches its peak?)

(c) How many seconds after the toss does the rock strike the ground? (<u>Hint</u>: How does the time it takes for the rock to rise to its peak compare with the time it takes it to fall back to the ground?)

2. **Baking muffins:** Chocolate muffins are baking in a 350 degree oven. Let $M = M(t)$ be the temperature of the muffins t minutes after they are placed in the oven, and let $D = D(t)$ be the difference between the temperature of the oven and the temperature of the muffins. Then D satisfies Newton's law of cooling, and its equation of change is

$$\frac{dD}{dt} = -0.04D \,.$$

(a) If the initial temperature of the muffins is room temperature, 73 degrees, find a formula for D.

(b) Find a formula for M.

(c) The muffins will be done when they reach a temperature of 225 degrees. When should we take the muffins out of the oven?

3. **Borrowing money:** Suppose you borrow $10,000 at 7% APR and that interest is compounded continuously. The equation of change for your account balance $B = B(t)$ is

$$\frac{dB}{dt} = 0.07B \; .$$

Here t is the number of years since the account was opened, and B is measured in dollars.

(a) Explain why B is an exponential function.

(b) Find a formula for B using the alternative form for exponential functions.

(c) Find a formula for B using the standard form for exponential functions. (Round the growth factor to 3 decimal places.)

(d) Assuming no payments are made, use your formula from Part (b) to determine how long it would take for your account balance to double.

4. **A leaky balloon:** In Example 6.5 we used a linear function to describe air leaking from a balloon. We indicated there that this is not in fact an accurate description. Let's look more carefully. Suppose a balloon initially holds 2 liters of air. Let $B = B(t)$ denote the volume, in liters, of air in the balloon t seconds after it starts leaking. When a balloon is almost full, the air pressure inside is large, and so it leaks rapidly. When it is almost empty, it will leak more slowly. More formally, the rate of change in B is proportional to B. Assume that for this particular balloon, the constant of proportionality is -0.05.

(a) Write an equation of change for B.

(b) Find a formula for B.

(c) How long will it take for half of the air to leak out of the balloon?

5. **A population of bighorn sheep:** A certain group of bighorn sheep live in an area where food is plentiful and conditions are generally favorable to bighorn sheep. Consequently the population is thriving. There are initially 30 sheep in this group. Let $N = N(t)$ be the population t years later. The population changes each year due to births and deaths. The rate of change in the population is proportional to the number of sheep currently in the population. For this particular group of sheep, the constant of proportionality is 0.04.

(a) Express the sentence "The rate of change in the population is proportional to the number of sheep currently in the population" as an equation of change. (Incorporate in your answer the fact that the constant of proportionality is 0.04 .)

(b) Find a formula for N.

(c) How long will it take this group of sheep to grow to a level of 50 individuals?

6. **Water flow:** Water is flowing into a tank at a steady rate of 2 gallons per minute. Let t be the time (in minutes) since the process began, and let V be the volume (in gallons) of water in the tank.

   (a) Explain why V a linear function of t, and write an equation of change for V.

   (b) If initially there were 4 gallons of water in the tank, find a formula for V.

   (c) The tank can hold 20 gallons. How long will it take to fill the tank?

7. **Growing child:** A certain girl grew steadily between the ages of 3 and 12 years, gaining five and a half pounds each year. Let W be the girl's weight (in pounds), as a function of her age t in years, between the ages of $t = 3$ and $t = 12$.

   (a) Is W a linear function or an exponential function? Be sure to explain your reasoning.

   (b) Write an equation of change for W.

   (c) Given that the girl weighed 30 pounds at age 3, find a formula for W.

8. **An investment:** You initially invest $500 with a financial institution which offers an APR of 4.5%, with interest compounded continuously. Let B be your account balance (in dollars), as a function of the time t in years since you opened the account.

   (a) Write an equation of change for B.

   (b) Find a formula for B.

   (c) If you had invested your money with a competing financial institution, the equation of change for your balance M would have been $\dfrac{dM}{dt} = 0.04M$. If this competing institution compounded interest continuously, what APR would they offer?

9. **Radioactive decay:** The amount remaining A (in grams) of a radioactive substance is a function of time t, measured in days since the experiment began. The equation of change for A is

$$\frac{dA}{dt} = -0.05A.$$

   (a) What is the exponential growth rate for A?

   (b) If initially there are 3 grams of the substance, find a formula for A.

   (c) What is the half-life of this radioactive substance?

10. **Magazines:** Two magazines, *Alpha* and *Beta*, started at the same time with the same circulation of 100. The circulation of *Alpha* is given by the function A, which has the equation of change

$$\frac{dA}{dt} = 0.10A \ .$$

The circulation of *Beta* is given by the function B, which has the equation of change

$$\frac{dB}{dt} = 10 \ .$$

Here t is the time in years since the magazines started.

(a) One of these functions is growing in a linear way, while the other is growing exponentially. Identify which is which, and find formulas for both functions.

(b) Which magazine is growing more quickly in circulation? Be sure to explain your reasoning.

6.4 EQUATIONS OF CHANGE: GRAPHICAL SOLUTIONS

When an equation of change is of the form $\frac{df}{dx} = c$, we know that f is a linear function with slope c. When it is of the form $\frac{df}{dx} = cf$, we know that f is an exponential function with growth (or decay) factor e^c. Determining which function goes with more complicated equations of change involves calculus and is beyond the scope of this text. Indeed, for many equations of change, even calculus is not powerful enough to allow recovery of the function. Nevertheless, even without being able to find the exact function, we are often able to sketch a graph of it and to find useful information about the phenomena being modeled. The key tool which we will use is the fundamental relationship between a function and its rate of change. When the rate of change is positive, the function is increasing. When the rate of change is negative, the function is decreasing. When the rate of change is momentarily 0, the function may be at a maximum or minimum. When the rate of change is zero over a period of time, the function is not changing and so is constant over that period.

Equilibrium solutions

Equilibrium or *steady-state* solutions for equations of change are solutions which never change and hence are constant. Thus, they occur where the rate of change is 0 over a period of time. Let's look at an example to see what we mean. The formulas we have used to study falling objects subject to air resistance actually came from an equation of change. When objects fall, they are accelerated downward by gravity. On the other hand, air resistance slows down the fall, which is why a feather falls so slowly. In general the downward acceleration of a falling object subject to air resistance is given by

Acceleration = Acceleration due to gravity − Retardation from air resistance. (6.1)

This drag due to air resistance is in practice difficult to determine, and engineers use wind tunnels and other high-tech devices to help in the design of more aerodynamic cars, airplanes, or rockets. The simplest model for air resistance assumes that the drag is proportional to the velocity. That is,

Retardation due to drag $= rV$,

where V is the downward velocity, and r is a constant known as the *drag coefficient*. We will show in Exercise 3 one way in which its value may be experimentally determined. Downward acceleration is the rate of change in velocity with respect to time, $\dfrac{dV}{dt}$. We put these bits of information into Equation (6.1) above to get the equation of change:

Acceleration $\dfrac{dV}{dt}$	=	Acceleration due to gravity g	−	Retardation from air resistance rV

Thus the velocity of an object falling subject to air resistance follows the equation of change $\dfrac{dV}{dt} = g - rV$. The acceleration due to gravity is $g = 32$ feet per second per second, and for an average size man, the drag coefficient has been determined to be approximately $r = 0.1818$ per second. Hence a skydiver falls subject to the equation of change

$$\frac{dV}{dt} \;=\; 32 - 0.1818V \,. \qquad (6.2)$$

(We measure velocity V in feet per second and time t in seconds.)

Let's find the equilibrium solutions of Equation (6.2) and investigate their physical meaning. Equilibrium solutions are those which never change, that is, those for which the rate of change is 0. To find them, we should put 0 in for $\dfrac{dV}{dt}$ and solve the resulting equation for V:

$$0 = 32 - 0.1818V \,.$$

This is a linear equation, which we proceed to solve. (You may if you wish solve it with the calculator.) We get

$$
\begin{aligned}
0 &= 32 - 0.1818V \\
0.1818V &= 32 \\
V &= \frac{32}{0.1818} = 176 \text{ feet per second.}
\end{aligned}
$$

Thus $V = 176$ feet per second is an equilibrium or steady-state solution. When $V = 176$, then $\dfrac{dV}{dt}$ is zero. Consequently, if the skydiver reaches a velocity of 176 feet per second, he will continue to fall at that same speed until the parachute opens. This is the velocity at which the downward force of gravity matches retardation due to air resistance: the terminal velocity of the skydiver. To emphasize this, we have graphed the steady-state solution $V = 176$ in Figure 6.38. The fact that the graph of velocity versus time is a horizontal line emphasizes the fact that velocity is not changing. Such horizontal lines are typical of equilibrium solutions.

Figure 6.38: *The equilibrium solution* $V = 176$ *for the equation of change of a skydiver*

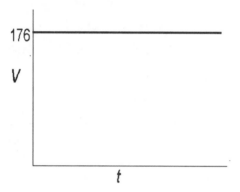

There are two important things to emphasize here. One is that the equilibrium solution was easy to find from the equation of change. The second is that the equilibrium solution had an important physical interpretation. It is the velocity we would expect the sky diver to have in the long term. This is typical of steady-state solutions: They yield important information, and that is the reason we use equations of change to find them.

KEY IDEA 6.6: EQUILIBRIUM OR STEADY-STATE SOLUTIONS

Equilibrium or steady-state solutions of an equation of change occur where the rate of change is 0. We get them by setting the rate of change equal to 0 and solving. In many cases they show long-term behavior.

On several occasions earlier in this text we have looked at logistic population growth. Logistic growth is one of many growth models which assume that there is an upper limit, the *carrying capacity*, beyond which the population cannot grow. In essentially all of these models the population grows rapidly at first, but the growth rate slows as the population approaches the carrying capacity. The logistic model is perhaps the simplest model for limited growth and was first proposed in 1838 by P.F. Verhulst. In its original formulation it was presented as an equation of change.

If a population exhibits logistic growth in an area with a carrying capacity K, then the logistic equation for the population $N = N(t)$ can be written

$$\frac{dN}{dt} = rN\left(1 - \frac{N}{K}\right) .$$

The constants K and r depend on the species and environmental factors, and their values are normally determined experimentally. The number N is sometimes the actual number of individuals, but more commonly it is the combined weight of the population (the *biomass*). In several examples and exercises in this text we have presented logistic growth with a formula. That formula was obtained by using calculus to solve the equation of change above, a feat we will not reproduce here. Rather, we will gain information directly from the equation of change itself. The logistic model of growth is useful in the study of some species, and we illustrate it in the case of Pacific sardines.

EXAMPLE 6.8   *Logistic Population Growth: Sardines*

In California in the 1930's and 1940's, a large part of the fishing-related industry was based on the catch of Pacific sardines. Studies have shown that the sardine population grows approximately logistically; moreover, the numbers K and r have been determined experimentally to be $K = 2.4$ million tons and $r = 0.338$ per year.[4] For Pacific sardines, our logistic equation is therefore

$$\frac{dN}{dt} = 0.338N\left(1 - \frac{N}{2.4}\right) , \tag{6.3}$$

where N is measured in millions of tons of fish and t is measured in years.

Find the equilibrium solutions of Equation (6.3) and give a physical interpretation of their meaning.

Solution: We get the equilibrium solutions where $\dfrac{dN}{dt}$ is zero. Thus we want to solve the equation

$$0.338N\left(1 - \frac{N}{2.4}\right) = 0 .$$

We use the single graph method to do that. To make and understand the graph, it is crucial that we identify correspondences among the variables. The importance of carefully employing this practice with all graphs made in this section cannot be overemphasized. In this case we have

$$Y_1 = \frac{dN}{dt}, \text{ rate of change in biomass on vertical axis}$$

$$X = N, \text{ biomass on horizontal axis.}$$

---

[4]From a study by G. I. Murphy, in "Vital statistics of the Pacific sardine (*Sardinops caerulea*) and the population consequences," *Ecology* **48** (1967), 731–736.

Now we enter $\boxed{6.6}$ the formula for the function $\frac{dN}{dt}$ given in Equation (6.3) and look at the table of values in Figure 6.39 to see how to set up the graphing window. The table shows that $N = 0$ is one equilibrium solution and that another occurs between $N = 2$ and $N = 3$. Allowing a little extra room, we use a window setup with a horizontal span of $N = 0$ to $N = 3$ and a vertical span of $\frac{dN}{dt} = -0.2$ to $\frac{dN}{dt} = 0.3$. This gives the graph in Figure 6.40. Using the calculator we get $\boxed{6.7}$ the second equilibrium solution $N = 2.4$ as shown in Figure 6.40.

Figure 6.39: *A table of values to set up a graphing window*

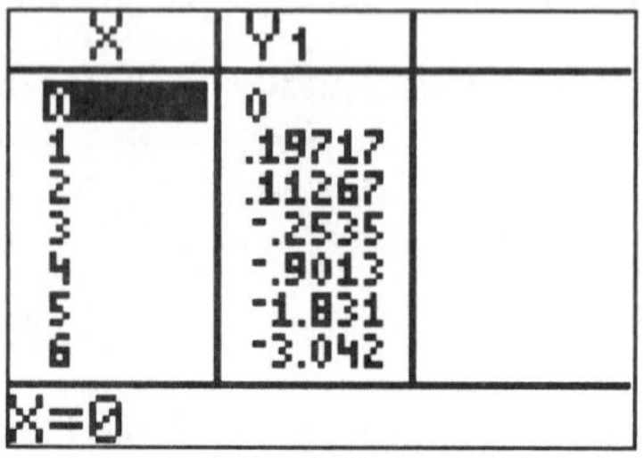

Figure 6.40: *Getting an equilibrium solution from the graph of* $\frac{dN}{dt}$ *versus N*

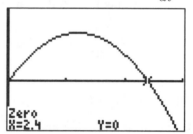

We have two equilibrium solutions, $N = 0$ and $N = 2.4$ million tons. If the biomass ever reaches either of these levels, it will remain there and not change. This is certainly obvious when $N = 0$: If there are no fish present, there will be no new ones born. The solution $N = 2.4$ million tons is the environmental carrying capacity for sardines in this region. If the biomass ever reaches 2.4 million tons of fish, then environmental limitations match growth tendencies, and the population stays the same.

We emphasize once more that Figure 6.40 does not show the graph of N versus t. It does not directly give the size of the fish biomass. Rather it is the graph of $\frac{dN}{dt}$ versus N. It shows the growth rate that can be expected for a given size of biomass. Our next topic is how to use this graph to sketch a graph of the biomass itself.

Sketching graphs

Equilibrium solutions for equations of change can yield important information about the physical situation that they describe, but the fundamental relationship between a function and its rate of change can tell us even more. To see this, let's continue our analysis of the equation of change $\frac{dN}{dt} = 0.338N\left(1 - \frac{N}{2.4}\right)$ for Pacific sardines. We know that if the biomass today is either 0 or 2.4 million tons, then in the future, the biomass will not change. Can we say what is to be expected to happen in the future if there are today 0.4 million tons of Pacific sardines? The key is the graph of $\frac{dN}{dt}$ versus N that we made in Figure 6.40. Table 6.1 was made by looking at where the graph is above the horizontal axis and where it is below.

Table 6.1: *The sign of* $\dfrac{dN}{dt}$

Range for N	from 0 to 2.4	greater than 2.4
Sign of $\dfrac{dN}{dt}$	positive	negative

The key to understanding what happens to the population N is the fundamental property of rates of change: When $\dfrac{dN}{dt}$ is positive, then N is increasing, and when $\dfrac{dN}{dt}$ is negative, then N is decreasing. In Table 6.2, we have added another row to Table 6.1 to include this information.

Table 6.2: *The information we get from the graph of* $\dfrac{dN}{dt}$ *versus* N

Range for N	from 0 to 2.4	greater than 2.4
Sign of $\dfrac{dN}{dt}$	positive	negative
Effect on N	increasing	decreasing

The information we get from this table is that whenever N is between 0 and 2.4 million tons, the graph of N versus t will increase; that is, the biomass will grow. When N is greater than 2.4 million tons, the graph of N versus t will decrease; that is, the biomass will decrease over time. Since $N = 0.4$ is between 0 and 2.4, we know that the graph of N versus t will start out increasing. For further evidence of this, in Figure 6.41 we have put the graphing cursor at $N = 0.4$, and we see that the value of $\dfrac{dN}{dt}$ is 0.11, a positive number. Since $\dfrac{dN}{dt}$ is positive, N is increasing. Thus if we start with a biomass of 0.4 million tons, we can expect the biomass to increase.

We want to start with this bit of information and make a hand-drawn sketch of the graph of N versus t. We started our picture in Figure 6.42 by first drawing the equilibrium solutions $N = 0$ and $N = 2.4$ and indicating what we already know about how the graph starts out.

To complete the graph of N versus t, we see from Figure 6.40 (or from Table 6.2) that $\dfrac{dN}{dt}$ remains positive as long as N is less than the equilibrium solution of $N = 2.4$ million tons. This tells us that the graph of N continues to increase up toward the equilibrium solution, as we have drawn in Figure 6.43. Notice that the horizontal lines representing equilibrium solutions provided us with important guides in sketching the graph of N versus t. Whenever you are asked to sketch a graph from an equation of change, you should first look to see if there are equilibrium solutions. If so, draw them in before you start.

Figure 6.41: When $N = 0.4$, the rate of change $\dfrac{dN}{dt}$ is positive

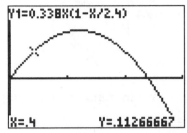

Figure 6.42: Equilibrium solutions and beginning the graph of N versus t

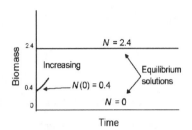

Figure 6.43: Completing the graph of biomass versus time

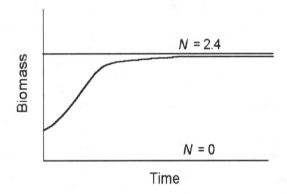

As we expected, the graph in Figure 6.43 looks like the classic logistic curve. But we would emphasize that it is a free-hand drawing and that graphs made in this way cannot be expected to show accurate function values. What is important is to use properly information about the rate of change to get the shape of the graph as nearly correct as is possible.

Let's analyze Pacific sardines in one final situation. What will the biomass be in the future if today it is 2.8 million tons? We proceed as before. In Figure 6.44, on the graph of $\dfrac{dN}{dt}$ versus N we have put the cursor at $N = 2.8$ million tons. We see that when the biomass is at this level, then $\dfrac{dN}{dt} = -0.16$, a negative number. Thus, the graph of N versus t will be decreasing. In Figure 6.45 we have started our graph just as we did in the first case.

We see further from Figure 6.40 (or from Table 6.2) that as long as N is greater than the equilibrium solution $N = 2.4$, then $\dfrac{dN}{dt}$ will remain negative. Thus the graph of N versus t continues to decrease toward the horizontal line representing the equilibrium solution. Our completed graph is in Figure 6.46. The graph in Figure 6.46 is reasonable because the biomass started out greater than the environmental carrying capacity of 2.4 million tons. The environment cannot support that many fish, and so we expect the biomass to decrease.

Figure 6.44: When biomass is 2.8 million tons, $\dfrac{dN}{dt}$ is negative

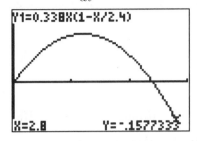

Figure 6.45: *Starting the graph of* N *versus* t *when* $N(0) = 2.8$

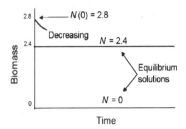

Figure 6.46: When population is greater than carrying capacity

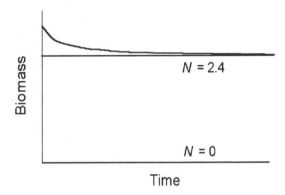

KEY IDEA 6.7: USING THE EQUATION OF CHANGE TO GRAPH FUNCTIONS

To make a hand-drawn sketch of the graph of f versus x from the equation of change for f, first draw horizontal lines representing any obvious equilibrium solutions. Next, use the calculator to graph $\dfrac{df}{dx}$ versus f. (If the equation of change is of the form

$$\frac{df}{dx} = \text{right-hand side}$$

you graph the right-hand side.)

1. Where the graph of $\dfrac{df}{dx}$ versus f crosses the horizontal axis, we have $\dfrac{df}{dx} = 0$, and so f does not change. This is an equilibrium solution for f.

2. When the graph of $\dfrac{df}{dx}$ versus f is above the horizontal axis, the graph of f versus x is increasing.

3. When the graph of $\dfrac{df}{dx}$ versus f is below the horizontal axis, the graph of f versus x is decreasing.

EXAMPLE 6.9 *Further Analysis of Falling Objects*

Earlier we found the equilibrium solution $V = 176$ feet per second for the equation of change

$$\frac{dV}{dt} = 32 - 0.1818V$$

which governs the velocity of a skydiver.

1. Sketch the graph of the skydiver's velocity if the initial downward velocity is 0. (This might occur if the skydiver jumped from an airplane which was flying level.)

2. Sketch the graph of the skydiver's velocity if the initial downward velocity is 200 feet per second. (This might occur if a pilot ejected from an airplane which was diving out of control.)

Solution to Part 1: We enter $\boxed{6.8}$ the right-hand side of the equation of change in the calculator and record the important correspondences:

$$\mathsf{Y_1} = \frac{dV}{dt}, \text{ acceleration on vertical axis}$$
$$\mathsf{X} = V, \text{ velocity on horizontal axis.}$$

To get the graph of $\frac{dV}{dt}$ versus V shown in Figure 6.47, we wanted to be sure the value $V = 200$, which will be needed in Part 2, was included. We used $\boxed{6.9}$ a horizontal span of $V = 0$ to $V = 250$, and we got the vertical span $\frac{dV}{dt} = -20$ to $\frac{dV}{dt} = 35$ from a table of values. We know already that the crossing point $V = 176$ shown in Figure 6.47 corresponds to the equilibrium solution that represents terminal velocity.

Figure 6.47: *The graph of* $\frac{dV}{dt}$, *acceleration, versus velocity* V

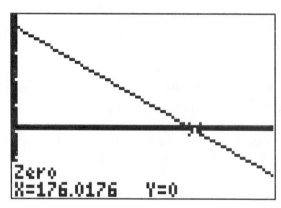

We have recorded in Table 6.3 the information about V we get from Figure 6.47.

Table 6.3: *The information we get from the graph of $\dfrac{dV}{dt}$ versus V*

Range for V	less than 176	greater than 176
Sign of $\dfrac{dV}{dt}$	positive	negative
Effect on V	increasing	decreasing

Since the initial velocity for this part is 0, we know from Table 6.3 that the graph of V starts out increasing. We make the graph of V versus t in two steps. In Figure 6.48 we have graphed the equilibrium solution and started our increasing graph. Consulting Table 6.3 once again, we see that velocity will continue to increase toward the terminal velocity of 176 feet per second. Thus the graph will continue to increase toward the horizontal line representing this equilibrium solution. Our graph is in Figure 6.49.

Figure 6.48: *Starting the graph of V versus t*

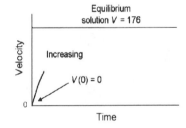

Figure 6.49: *Completing the graph of V versus t*

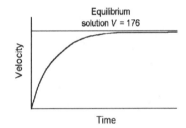

Solution to Part 2: Since the initial velocity is 200 feet per second, we see from Table 6.3 that velocity is decreasing. This bit of information is incorporated into the start of our graph in Figure 6.50. But Table 6.3 also shows that velocity will continue to decrease toward the equilibrium solution of 176 feet per second. Thus, in Figure 6.51 we completed the graph by making it decrease toward the horizontal line corresponding to terminal velocity. We note that the graph in Figure 6.51 makes sense because, at the skydiver's large initial velocity, the retardation due to air resistance is greater than the downward acceleration due to gravity. Thus, his speed will decrease toward terminal velocity.

Figure 6.50: *Starting the graph when* $V(0) = 200$

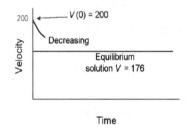

Figure 6.51: *The completed graph of* V *versus* t *for* $V(0) = 200$

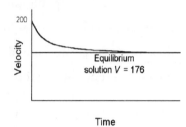

Exercise Set 6.4

1. **Equation of change for logistic growth:** The logistic growth formula $N = \dfrac{6.21}{0.035 + 0.45^t}$ we used in Chapter 2 for deer on the George Reserve actually came from the following equation of change:

$$\frac{dN}{dt} = 0.8N\left(1 - \frac{N}{177}\right).$$

   (a) Plot the graph of $\dfrac{dN}{dt}$ versus N, and use it to find the equilibrium solutions. Explain their physical significance.

   (b) On the same graph, sketch the equilibrium solutions and graphs of N versus t in each of the two cases $N(0) = 10$ and $N(0) = 225$.

   (c) To what starting value for N does the solution $N = \dfrac{6.21}{0.035 + 0.45^t}$ correspond?

2. **A catfish farm:** Catfish in a commercial pond can be expected to exhibit logistic population growth. Consider a pond with a carrying capacity of $K = 4000$ catfish. Take the r value for catfish in this pond to be $r = 0.06$.

   (a) Write the equation of change for logistic growth of the catfish population. (<u>Hint</u>: If you have difficulty here, refer to Example 6.8.)

   (b) Make a graph of $\dfrac{dN}{dt}$ versus N.

   (c) For what values of N would the catfish population be expected to increase?

   (d) For what values of N would the catfish population be expected to decrease?

   (e) There are several strategies which may be applied to populations which are regularly harvested. The *maximum sustainable yield* model says that a renewable resource should be maintained at a level where its growth rate is at a maximum, since this allows the population to replenish itself quickly. According to this model, at what level should the catfish population be maintained?

3. **Experimental determination of the drag coefficient:** When retardation due to air resistance is proportional to downward velocity V (in feet per second), falling objects obey the equation of change

$$\frac{dV}{dt} = 32 - rV,$$

where r is known as the *drag coefficient*. One way to measure the drag coefficient is to measure and record terminal velocity.

(a) We know that an average size man has a terminal velocity of 176 feet per second. Use this to show that the value of the drag coefficient is $r = 0.1818$ per second. (Hint: To say that the terminal velocity is 176 feet per second means that when the velocity V is 176, velocity will not change. That is, $\frac{dV}{dt} = 0$. Put these bits of information into the equation of change and solve for r.)

(b) An ordinary coffee filter has a terminal velocity of about 4 feet per second. What is the drag coefficient for a coffee filter?

4. **Other models of drag due to air resistance:** When some objects move at high speeds, air resistance has a more pronounced effect. For such objects, retardation due to air resistance is often modeled as being proportional to the square or even higher powers of velocity.

(a) A rifle bullet fired downward has an initial velocity of 2100 feet per second. If we use the model that gives air resistance as the square of velocity, then the equation of change for the downward velocity V (in feet per second) of a rifle bullet is

$$\frac{dV}{dt} = 32 - rV^2 .$$

If the terminal velocity for a bullet is 1600 feet per second, find the drag coefficient r.

(b) A meteor may enter the Earth's atmosphere at a velocity as high as 90,000 feet per second. If the downward velocity V (in feet per second) of a meteor in the Earth's atmosphere is governed by the equation of change

$$\frac{dV}{dt} = 32 - 2 \times 10^{-18} V^4 ,$$

how fast will it be traveling when it strikes the ground, assuming it has reached terminal velocity?

5. **Fishing for sardines:** This is a continuation of Example 6.8. If we take into account an annual fish harvest of F million tons of fish, then the equation of change for Pacific sardines becomes

$$\frac{dN}{dt} = 0.338N\left(1 - \frac{N}{2.4}\right) - F\,.$$

(a) Suppose that there are currently 1.8 million tons of Pacific sardines off the California coast and that you are in charge of the commercial fishing fleet. It is your goal to leave the Pacific population of sardines as you found it. That is, you wish to set the fishing level F so that the biomass of Pacific sardines remains stable. What value of F will accomplish this? (Hint: You want to choose F so that the current biomass level of 1.8 million tons is an equilibrium solution.)

(b) For the remainder of this exercise take the value of F to be 0.1 million tons per year. That is, assume the catch is 100,000 tons per year.

    i. Make a graph of $\frac{dN}{dt}$ versus N, and use it to find the equilibrium solutions.

    ii. For what values of N will the biomass be increasing? For what values will it be decreasing?

    iii. On the same graph, sketch all equilibrium solutions, and the graphs of N versus t for each of the initial populations $N(0) = 0.3$ million tons, $N(0) = 1.0$ million tons, and $N(0) = 2.3$ million tons.

    iv. Explain in practical terms what the picture you made in Part iii tells you. Include in your explanation the significance of the equilibrium solutions.

6. **Sprinkler irrigation in Nebraska:** Logistic growth can be used to model not only population growth but also economic and other types of growth. For example, the total number of acres $A = A(t)$ (in millions) in Nebraska which are being irrigated by modern sprinkler systems has shown approximate logistic growth since 1955, closely following the equation of change

$$\frac{dA}{dt} = 0.15A\left(1 - \frac{A}{3}\right)\,.$$

Here time t is measured in years.

(a) According to this model, how many total acres in Nebraska can be expected eventually to be irrigated by sprinkler systems? (Hint: This corresponds to the carrying capacity in the logistic model for population growth.)

(b) How many acres of land were under sprinkler irrigation when sprinkler irrigation was expanding at its most rapid rate?

7. **Logistic growth with a threshold:** Most species have a survival *threshold* level, and populations of fewer individuals than the threshold cannot sustain themselves. If the carrying capacity is K and the threshold level is S, then the logistic equation of change for the population $N = N(t)$ is

$$\frac{dN}{dt} = -rN \left(1 - \frac{N}{S}\right)\left(1 - \frac{N}{K}\right) .$$

For Pacific sardines, we may use $K = 2.4$ million tons and $r = 0.338$ per year, as in Example 6.8. Suppose we also know that the survival threshold level for the sardines is $S = 0.8$ million tons.

(a) Write the equation of change for Pacific sardines under these conditions.

(b) Make a graph of $\frac{dN}{dt}$ versus N and use it to find the equilibrium solutions. How do the equilibrium solutions correspond with S and K?

(c) For what values of N is the graph of N versus t increasing and for what values is it decreasing?

(d) Explain what can be expected to happen to a population of 0.7 million tons of sardines.

(e) At what population level will the population be growing at its fastest?

8. **Chemical reactions:** In a *second-order reaction*, one molecule of a substance A collides with one molecule of a substance B to produce a new substance, the *product*. If t denotes time and $x = x(t)$ denotes the concentration of the product, then its rate of change $\frac{dx}{dt}$ is called the *rate of reaction*. Suppose the initial concentration of A is a and the initial concentration of B is b. Then, assuming a constant temperature, x satisfies the equation of change

$$\frac{dx}{dt} = k(a - x)(b - x)$$

for some constant k. This is since the rate of reaction is proportional both to the amount of A that remains untransformed and to the amount of B that remains untransformed. Here we study a reaction between isobutyl bromide and sodium ethoxide in which $k = 0.0055$, $a = 51$, and $b = 76$; the concentrations are in moles per cubic meter, and time is in seconds.[5]

(a) Write the equation of change for the reaction between isobutyl bromide and sodium ethoxide.

(b) Make a graph of $\frac{dx}{dt}$ versus x. Include a span of $x = 0$ to $x = 100$.

---

[5]From a study by I. Dostrovsky and E. D. Hughes, as described by Gordon M. Barrow in *Physical Chemistry*, 4th edition, 1979, McGraw-Hill, New York.

(c) Explain what can be expected to happen to the concentration of the product if the initial concentration of the product is 0.

9. **Competition between bacteria:** Suppose there are two types of bacteria, type A and type B, in a place with limited resources. If the bacteria had unlimited resources, they would both grow exponentially, with exponential growth rates a for type A and b for type B. Let P be the *proportion* of type A bacteria, so P is a number between 0 and 1. For example, if $P = 0$, then there are no type A bacteria and all are of type B. If $P = 0.5$, then half of the bacteria are of type A and half are of type B. Each population of bacteria grows in competition for the limited resources, and so the proportion P changes over time. The function P is subject to the equation of change

$$\frac{dP}{dt} = (a - b)P(1 - P) \,.$$

Suppose $a = 2.3$ and $b = 1.7$.

(a) Does P have a logistic equation of change?

(b) What happens to the populations if initially $P = 0$?

(c) What happens to P in the long run if $P(0)$ is positive (but not zero)? In this case, does it matter what the exact value of $P(0)$ is?

10. **Growth of fish:** Let $w = w(t)$ denote the weight of a fish as a function its age t. For the North Sea cod, the equation of change

$$\frac{dw}{dt} = 2.1w^{\frac{2}{3}} - 0.6w$$

holds. Here w is measured in pounds and t in years.

(a) Explain in practical terms what $\dfrac{dw}{dt}$ means.

(b) Make a graph of $\dfrac{dw}{dt}$ against w. Include weights up to 45 pounds.

(c) What is the weight of the cod when it is growing at the fastest rate?

(d) To what weight does the cod grow?

11. **Grazing sheep:** The amount C of food consumed in a day by a merino sheep is a function of the amount V of vegetation available. The equation of change for C is

$$\frac{dC}{dV} = 0.01(2.8 - C) \ .$$

Here C is measured in pounds and V in pounds per acre.

(a) Explain in practical terms what $\dfrac{dC}{dV}$ means.

(b) Make a graph of $\dfrac{dC}{dV}$ versus C. Include consumption levels up to 3 pounds.

(c) What is the most you would expect a merino sheep to consume in a day?

6.5 ESTIMATING RATES OF CHANGE

Up to now we have looked at rates of change qualitatively, emphasizing how the sign of a rate of change affects the function. But exact values or estimates for rates of change can provide additional important information. It is easiest to make such estimates for functions given by tables.

Rates of change for tabulated data

The following brief table shows the location S (measured as distance in miles east of Los Angeles) of an airplane flying toward Denver.

Time	1:00 p.m.	1:30 p.m.
Distance from L.A.	360 miles	612 miles

We want to know the velocity $\dfrac{dS}{dt}$ at 1:00 p.m. We do this using a familiar formula:

$$\text{Average velocity} = \frac{\text{Distance traveled}}{\text{Elapsed time}} \ .$$

Between 1:00 and 1:30 the airplane traveled $612 - 360 = 252$ miles, and it took half an hour to travel that far:

$$\text{Average velocity from 1:00 to 1:30} = \frac{252}{0.5} = 504 \text{ miles per hour} \ .$$

It is important to point out that this calculation can be expected to give the exact velocity $\dfrac{dS}{dt}$ at 1:00 p.m. only if the airplane is traveling at the same speed over the entire 30 minute time interval. If the airplane was speeding up between 1:00 and 1:30, then 504 miles per hour is the average velocity over the 30 minute time interval, but its exact velocity at 1:00 would have been somewhat less than 504 miles per hour. If, on the other hand, the airplane was slowing down, then its exact velocity at 1:00 would have been more than 504 miles per hour. In general, when velocity is calculated from a table of values as we did here, it should be regarded as an approximation of the exact velocity. Furthermore, it is clear that shorter time intervals yield better approximations because the moving object doesn't have as much time to change velocity. (Recall the discussion in Section 3 of this chapter.) Suppose, for example, that instead of the table above, we had been given the following table.

Time	1:00 p.m.	1 second after 1:00 p.m.
Distance from L.A.	360 miles	360.140 miles

If we use this table to calculate velocity, we see that the airplane traveled 0.140 mile in one second. Since there are 3600 seconds in an hour, we get

$$\text{Average velocity from 1:00 p.m. to 1 second later} = \frac{0.140}{\frac{1}{3600}} = 504 \text{ miles per hour.}$$

Since the velocity of the airplane can change very little over a one second time interval, we would now feel confident reporting its velocity at 1:00 p.m. as 504 miles per hour.

Exactly the same idea can be used to approximate the rate of change for any function given by tabulated data. For example, the following table shows the percentage $B = B(t)$ of babies in the U.S. born to unmarried women in 1980 and 1990.

Year	1980	1990
Born to unmarried women	18%	28%

The function $\frac{dB}{dt}$ is the rate of change in B. If we were able to calculate it exactly for 1980, it would tell us how fast the percentage of births to unmarried women was increasing in 1980. That would say how much growth in this percentage could be expected in one year's time. As with the example of the airplane, the table does not give enough information to allow for an exact calculation of $\frac{dB}{dt}$ for 1980, but we can approximate its value using the average rate of change from 1980 to 1990 just as we did for the velocity of the airplane.

During the period from 1980 to 1990 the change in percentage was $28 - 18 = 10$ percentage points. This occurred over a period of 10 years, so we divide to get

$$\text{Average rate of change from 1980 to 1990} = \frac{10}{10} = 1 \text{ percentage point per year.}$$

We can use this number as an approximation of the value of $\frac{dB}{dt}$.

KEY IDEA 6.8: ESTIMATING RATES OF CHANGE FOR TABULATED DATA

If $f = f(x)$ is a function given by a table of values, then $\dfrac{df}{dx}$, the rate of change of f with respect to x, cannot without further information be calculated exactly. But it can be estimated using

$$\text{Average rate of change in } f \text{ with respect to } x = \frac{\text{Change in } f}{\text{Change in } x}.$$

Calculating rates of change in this way for tabulated data is not a new idea at all, and we have already made and used this type of calculation many times in the text. Look back at Example 1.3 in Section 2 of Chapter 1, where we studied tabulated data for the number $W = W(t)$ of women in the United States employed outside the home. In the solution to Part 3 of that example, we noted that the number increased by 12.3 million from 1970 to 1980. We concluded that during the decade of the 70's the number of women employed outside the home was increasing by an average of $\dfrac{12.3}{10} = 1.23$ million per year, and we used that number to estimate the number of women employed outside the home in 1972. This is exactly the calculation presented in the key idea above. Thus in the language of rates of change, we would say that $\dfrac{dW}{dt}$ is approximately the average rate of change from 1970 to 1980, 1.23 million per year, and that this number tells us the increase in W which would be expected in one year. This is once again an illustration of how mathematics distills ideas from many contexts into a single fundamental concept. Precisely the same mathematical idea, the rate of change, is used to analyze applications from velocity to population growth to numbers of women employed outside the home.

EXAMPLE 6.10   *Water Flowing from a Tank*

Consider the following table, which shows the number of gallons W of water left in a tank t hours after it starts to leak.

t = hours	0	3	6	9	12
W = gallons left	860	725	612	515	433

1. Explain in practical terms the meaning of $\dfrac{dW}{dt}$ and estimate its value at $t = 6$ using the average rate of change from $t = 6$ to $t = 9$.

2. Use your answer from Part 1 to estimate the amount of water in the tank 8 hours after the leak begins.

Solution to Part 1: The function $\dfrac{dW}{dt}$ is the rate of change in water remaining in the tank. This is the change in volume we expect over an hour. Since the amount of water is decreasing, $\dfrac{dW}{dt}$ is negative. Its size is the rate at which water is leaking from the tank.

We emphasize that the calculation we make here is the same one we made for women employed outside the home in Example 1.3 of Chapter 1. We are using a new language but not a new idea. Six hours after the leak began there were 612 gallons in the tank. When $t = 9$ there are 515 gallons left. Thus the water level changed by $515 - 612 = -97$ gallons over the 3 hour period. Hence

$$\text{Average rate of change from 6 hours to 9 hours} = \frac{-97}{3} = -32.33 \text{ gallons per hour.}$$

This is the estimate for $\dfrac{dW}{dt}$ that we were asked to find.

Solution to Part 2: For each hour after 6, we expect the number of gallons in the tank to decrease by 32.33 gallons. From 6 to 8 hours is a 2 hour span, and so we expect $2 \times 32.33 = 64.66$ gallons to leak out. That leaves $612 - 64.66 = 547.34$ gallons in the tank 8 hours after the leak began.

Rates of change for functions given by formulas

For functions given by formulas, it is possible using calculus to find exact formulas for rates of change. But for many applications, a close approximation to the rate of change is sufficient, and that is what we will use in this text. The idea is first to use the formula to generate a brief table of values and then to compute the rate of change just as we did above. Let's look, for example, at a falling rock. If air resistance is ignored, elementary physics can be used to show that the rock falls $S = 16t^2$ feet during t seconds of fall. Let's estimate the downward velocity $\dfrac{dS}{dt}$ of the rock 2.5 seconds into the fall. We will first show the steps involved in making this computation and then show how the graphing calculator offers a shortcut.

We know from the computations we made for the velocity of the airplane that we get more reliable answers if we keep the time interval short. Below we have made a brief table of values for $S = 16t^2$ using only $t = 2.5$ and $t = 2.50001$. You may wish to use your calculator to check that $16 \times 2.5^2 = 100$ and $16 \times 2.50001^2 = 100.0008$.

t	2.5	2.50001
S	100	100.0008

Now the change in S is $100.0008 - 100 = 0.0008$ foot. This is how far the rock falls from 2.5 to 2.50001 seconds into the fall. It takes $2.50001 - 2.5 = 0.00001$ seconds to fall this far, so the rate of change in S with respect to t (in other words, the velocity) is approximated as follows:

$$\frac{dS}{dt} = \frac{\text{Change in } S}{\text{Change in } t} = \frac{0.0008}{0.00001} = 80 \text{ feet per second .}$$

We will do this computation again, but this time we use the automated features provided by the calculator. The first step is to make the graph $\boxed{6.10}$ as shown in Figure 6.52.

We used a window setup with a horizontal span of $t = 0$ to $t = 5$ and a vertical span of $S = 0$ to $S = 300$. Next we use the calculator $\boxed{6.11}$ to get the rate of change $\dfrac{dS}{dt}$ at $t = 2.5$. The result in Figure 6.53 shows that the velocity $\dfrac{dS}{dt}$ at $t = 2.5$ is 80 feet per second, and this agrees with our earlier computation. We would emphasize that no magic has been performed. The calculator has just internally made a table of values and performed a computation similar to the one we did when we made the computation by hand. The calculator just automates the procedure. For the exact keystrokes[6] needed to do this, consult the *Keystroke Guide*.

Figure 6.52: *Graph of distance versus time*

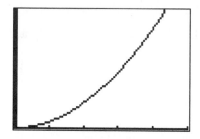

Figure 6.53: *Getting $\dfrac{dS}{dt}$, the velocity*

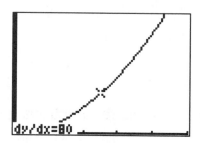

EXAMPLE 6.11   A Cannonball

A cannonball fired from the origin with a muzzle velocity of 300 feet per second follows the path of the graph of

$$h = x - 32 \left(\frac{x}{300} \right)^2 ,$$

where distances are measured in feet. (This simple model ignores air resistance.)

1. Plot the graph of the flight of the cannonball.

2. Use the graph to estimate the height h of the cannonball 734 feet downrange.

3. By looking at the graph of h, do you expect $\dfrac{dh}{dx}$ to be positive or negative at $x = 734$?

4. Calculate $\dfrac{dh}{dx}$ at 734 feet downrange and explain in practical terms what this number means.

Solution to Part 1: To get the graph we first enter $\boxed{6.12}$ the function and record the appropriate variable correspondences:

$$\mathsf{Y_1} \;=\; h, \text{ height on vertical axis}$$

$$\mathsf{X} \;=\; x, \text{ distance downrange on horizontal axis.}$$

---

[6]When the procedure is executed on some calculators, you will sometimes not be able to get dy/dx for the exact x value you want. See the *Keystroke Guide* for additional information.

Consulting the table of values in Figure 6.54, we choose a window setup with a horizontal span of $x = 0$ to $x = 3000$ and a vertical span of $h = 0$ to $h = 750$. The resulting graph is in Figure 6.55.

Figure 6.54: A table of values for the height of the cannonball

X	Y1	
0	0	
500	411.11	
1000	644.44	
1500	700	
2000	577.78	
2500	277.78	
3000	-200	

X=0

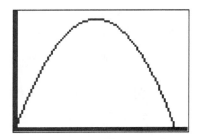

Figure 6.55: The flight of the cannonball

Solution to Part 2: In Figure 6.56 we have put the cursor [6.13] at X=734, and we read from the bottom of the screen that the height of the cannonball 734 feet downrange is $h = 542.44$ feet.

Solution to Part 3: From Figure 6.56 we see that at $x = 734$ the graph of h is increasing. That is, at this distance downrange, the cannonball is still rising. Since h is increasing, we know that $\frac{dh}{dx}$ is positive.

Solution to Part 4: In Figure 6.57 we have used the calculator to get [6.14] the value of $\frac{dh}{dx}$ at $x = 734$ feet downrange. The rate of change, 0.48 foot per foot (rounded to two decimal places), can now be read[7] from the dy/dx= prompt at the bottom of Figure 6.57. This is a measure of how steeply the cannonball is rising. It tells us that if we move one more foot downrange to 735 feet, we can expect the cannonball to rise 0.48 foot from 542.44 to $542.44 + 0.48 = 542.92$ feet.

Figure 6.56: The height of the cannonball 734 feet downrange

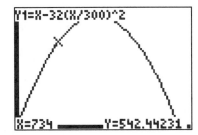

Figure 6.57: The rate of change in height 734 feet downrange

[7]Some calculators will give a slightly different answer since they do not calculate the rate of change exactly at $x = 734$.

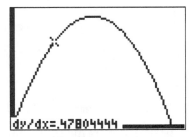

Exercise Set 6.5

1. **Population growth:** The following table[8] shows the population of reindeer on an island as of the given year.

Date	1945	1950	1955	1960
Population	40	165	678	2793

We let t be the number of years since 1945, so that $t = 0$ corresponds to 1945, and we let $N = N(t)$ denote the population size.

(a) Approximate $\dfrac{dN}{dt}$ for 1955 using the average rate of change from 1955 to 1960, and explain in practical terms what this number means.

(b) Use your work from Part (a) to estimate the population in 1957.

(c) The number you calculated in Part (a) is an approximation to the actual rate of change. As you will be asked to show in the next exercise, the reindeer population growth can be closely modeled by an exponential function. With this in mind, do you think your answer in Part (a) is too large or too small? Explain your reasoning.

2. **Further analysis of population growth:** *This is a continuation of Exercise 1. Our goal is to make an exponential model of the data and use it to get a more accurate estimate of the rate of change in population in 1955.*

(a) Plot the logarithm of the data from Exercise 1 and use regression to obtain an exponential model for population growth. (For details on the method here, see Section 4.3.) Your final answer should be an exponential formula for $N = N(t)$.

(b) Use the formula you found in Part (a) to get $\dfrac{dN}{dt}$ in 1955.

(c) How does your answer from Part (b) of this exercise compare with your answer from Part (a) of Exercise 1? Does this agree with your answer in Part (c) of Exercise 1?

3. **Deaths from heart disease:** The following tables show the deaths per 100,000 caused by heart disease in the United States for males and females age 55 to 64 years. The function H_m gives deaths per 100,000 for males, and H_f gives deaths per 100,000 for females.

Heart disease deaths per 100,000 for males age 55 to 64 years

t = year	1970	1980	1985	1990	1991
H_m = deaths per 100,000	987.2	746.8	651.9	537.3	520.8

---

[8]The table is based on the study by D. Klein, "The introduction, increase, and crash of reindeer on St. Matthew Island," *J. Wildlife Management* **32** (1968), 350–367.

Heart disease deaths per 100,000 for females age 55 to 64 years

t = year	1970	1980	1985	1990	1991
H_f = deaths per 100,000	351.6	272.1	250.3	215.7	210

(a) Approximate the value of $\dfrac{dH_m}{dt}$ in 1980 using the average rate of change from 1980 to 1985.

(b) Explain in practical terms the meaning of the number you calculated in Part (a). You should, among other things, tell what the sign means.

(c) Use your answer from Part (a) to estimate the heart disease death rate for males age 55 to 64 years in 1983.

(d) Approximate the value of $\dfrac{dH_f}{dt}$ for 1980 using the average rate of change from 1980 to 1985.

(e) Explain what your calculations from Parts (a) and (d) tell you about comparing heart disease deaths for men and women in 1980.

4. **The cannon with a different muzzle velocity:** If the cannonball from Example 6.11 is fired with a muzzle velocity of 370 feet per second, it will follow the graph of

$$h = x - 32 \left(\frac{x}{370}\right)^2 ,$$

where distances are measured in feet.

(a) Plot the graph of the flight of the cannonball.

(b) Find the height of the cannonball 3000 feet downrange.

(c) By looking at the graph of h, determine if $\dfrac{dh}{dx}$ is positive or negative at 3000 feet downrange.

(d) Calculate $\dfrac{dh}{dx}$ at 3000 feet downrange and explain in practical terms what this number means.

5. **Falling with a parachute:** If an average size man with a parachute jumps from an airplane, he will fall

$$S = 12.5(0.2^t - 1) + 20t$$

feet in t seconds.

(a) Plot the graph of S versus t over at least the first 10 seconds of the fall.

(b) How far does the parachutist fall in 2 seconds?

(c) Calculate $\dfrac{dS}{dt}$ at 2 seconds into the fall and explain in practical terms what the number you calculated means.

6. **Free fall subject to air resistance:** Gravity and air resistance contribute to the characteristics of a falling object. An average size man will fall

$$S = 968(e^{-0.18t} - 1) + 176t$$

feet in t seconds after the fall begins.

(a) Plot the graph of S versus t over the first 5 seconds of the fall.

(b) How far will the man fall in 3 seconds?

(c) Calculate $\dfrac{dS}{dt}$ at 3 seconds into the fall and explain in practical terms what the number you calculated means.

7. **A yam baking in the oven:** A yam is placed in a preheated oven to bake. An application of Newton's law of cooling gives the temperature Y, in degrees, of the yam t minutes after it is placed in the oven as

$$Y = 400 - 325e^{-\frac{t}{50}} .$$

(a) Make a graph of the temperature of the yam at time t over 45 minutes of baking time.

(b) Calculate $\dfrac{dY}{dt}$ at the time 10 minutes after the yam is placed in the oven.

(c) Calculate $\dfrac{dY}{dt}$ at the time 30 minutes after the yam is placed in the oven.

(d) Explain what your answers in Parts (b) and (c) tell you about the way the yam heats over time.

8. **A floating balloon:** A balloon is floating upward and its height in feet above the ground t seconds after it is released is given by a function $S = S(t)$. Suppose you know that $S(3) = 13$ and that $\dfrac{dS}{dt}$ is 4 when t is 3. Estimate the height of the balloon 5 seconds after it is released. Explain how you got your answer.

9. **A pond:** Water is running out of a pond through a drainpipe. The amount of water (in gallons) in the pond t minutes after the water began draining is given by a function $G = G(t)$.

   (a) Explain in practical terms the meaning of $\dfrac{dG}{dt}$.

   (b) While water is running out of the pond, do you expect $\dfrac{dG}{dt}$ to be positive or negative?

   (c) When $t = 30$, water is running out of the drainpipe at a rate of 8000 gallons per minute. What is the value of $\dfrac{dG}{dt}$?

   (d) When $t = 30$, there are 2,000,000 gallons of water in the pond. Using the information from Part (c), estimate the value of $G(35)$.

10. **Marginal profit:** *This refers to Example 6.4 of Section 6.2.* A small firm produces at most 20 widgets in a week. Its profit P (in dollars) is a function of n, the number of widgets manufactured in a week, and the formula is

$$P = 26n - n^2 .$$

   Recall that the rate of change $\dfrac{dP}{dn}$ is called the *marginal profit*.

   (a) Make a graph of the profit as a function of the number of widgets manufactured in a week.

   (b) Determine the marginal profit if the firm produces 10 widgets in a week, and explain in practical terms what your answer means.

   (c) Determine the marginal profit if the firm produces 15 widgets in a week, and explain in practical terms what your answer means.

   (d) How many widgets should be produced in a week to maximize profit?

   (e) Use the calculator to determine the marginal profit at the production level you found in Part (d). How does your answer compare to what Example 6.4 of Section 6.2 indicates for the marginal profit when the profit is maximized?

6.6 CHAPTER SUMMARY

This chapter develops the notion of the rate of change for a function. Earlier in the book, rates of change arose in contexts such as studying tabulated functions using average rates of change, analyzing graphs in terms of concavity, and characterizing linear functions as having a constant rate of change. In the current chapter we see the value of a unified approach to rates of change, and we apply this point of view to a variety of important real-world problems.

Velocity

Velocity is the rate of change in location, or directed distance. There are just a few basic rules that enable us to relate directed distance and velocity:

- When directed distance is increasing, then velocity is positive.

- When directed distance is decreasing, then velocity is negative.

- When directed distance is not changing, even for an instant, then velocity is zero.

- When velocity is constant (for example, when the cruise control on a car is set), then directed distance is a linear function of time, and so its graph is a straight line.

Using these basic rules, from a verbal description of how an object moves we can sketch graphs of directed distance and velocity.

General rates of change

The idea of velocity as the rate of change in directed distance applies to any function. The rate of change of a function $f = f(x)$ is denoted by $\dfrac{df}{dx}$. It tells how much f is expected to change if x increases by one unit. For example, if V is the velocity of an object as a function of time t, then $\dfrac{dV}{dt}$ is the change in velocity we expect in one unit of time. This is the acceleration of the object. For example, a car might gain 2 miles per hour in velocity each second.

As with velocity, there are just a few basic rules that express the fundamental relationship between a function f and its rate of change $\dfrac{df}{dx}$:

- When f is increasing, then $\dfrac{df}{dx}$ is positive.

- When f is decreasing, then $\dfrac{df}{dx}$ is negative.

- When f is not changing, then $\dfrac{df}{dx}$ is zero.

- When $\dfrac{df}{dx}$ is constant, then f is a linear function of x.

One important observation is that we expect the rate of change to be zero at a maximum or minimum of a function. For example, the CEO of a tire company wants to maximize profit by adjusting the production level. In Figure 6.58 is a graph of marginal profit, which is the rate of change in profit as a function of the production level. Using the graph we see that marginal profit is positive if the production level is less than 160 tires per day, so the profit increases up to this level. If the level is more than 160 tires per day, then marginal profit is negative, and the profit is decreasing. The maximum profit occurs where the marginal profit is zero, and that is at a production level of 160 tires per day.

Figure 6.58: *Marginal profit for a tire company*

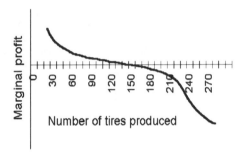

Equations of change

An equation of change for a function $f = f(x)$ is an equation expressing the rate of change $\dfrac{df}{dx}$ in terms of f and x. Equations of change arise in many applications where we are given a formula not for the function itself but for its rate of change.

The equation of change $\dfrac{df}{dx} = m$, where m is a constant, says that f has a constant rate of change m, and so f is a linear function with slope m. Once we know the initial value b of f, we can write down the linear formula $f = mx + b$.

The equation of change $\dfrac{df}{dx} = rf$, where r is a constant, says that f has a constant proportional (or percentage) of change, and so f is an exponential function. The exponential growth rate is r, and so the growth (or decay) factor is e^r. Once we know the initial value P of f, we can write down the exponential formula $f = Pe^{rx}$, or $f = P \times (e^r)^x$.

Many equations of change which arise in practice do not have the simple characterization we found for linear and exponential functions. But we can still get important information about a function by using the fundamental relationship between the function and its rate of change. The first step is to find the equilibrium solutions of an equation of change. These are the constant solutions, and so we find them by setting the rate of change equal to zero. In many cases they show the long-term behavior we can expect. The next step after finding equilibrium solutions is to determine for what values of the function its rate of change is positive or negative, since this will tell us where the function is increasing or decreasing. In many cases of interest these two steps can be accomplished by analyzing a graph of the rate of change versus the function. After these two steps we can sketch a graph of the function for a given initial condition.

For example, the population N of deer in a certain area as a function of time t in years has the logistic equation of change

$$\frac{dN}{dt} = 0.5N \left(1 - \frac{N}{150} \right).$$

A graph of $\frac{dN}{dt}$ versus N is shown in Figure 6.59. (We used a horizontal span of 0 to 160 and a vertical span of -5 to 20.) In addition to the equilibrium solution $N = 0$, from the graph we find the equilibrium solution $N = 150$. This is the first step. For the second step we make Table 6.4 based on the graph in Figure 6.59.

Figure 6.59: *A graph of $\dfrac{dN}{dt}$ versus N*

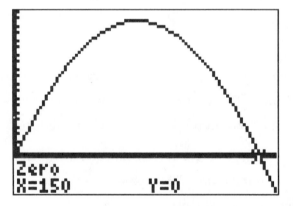

Table 6.4: *The information we get from the graph of $\dfrac{dN}{dt}$ versus N*

Range for N	from 0 to 150	greater than 150
Sign of $\dfrac{dN}{dt}$	positive	negative
Effect on N	increasing	decreasing

The table tells us, for example, that an initial population of 10 deer should increase in time to a size of 150.

Estimating rates of change

The rate of change of a function is again a function, and estimating the rate of change at a point can be of practical importance. The method we use depends on how the function is presented.

For a function $f = f(x)$ given by a table, we can estimate $\dfrac{df}{dx}$ by using the average rate of change given by

$$\frac{\text{Change in } f}{\text{Change in } x}.$$

For a function given by a formula, we can make a graph and then ask the calculator to estimate the rate of change at a point.

CHAPTER 7 *Mathematics of population ecology*

Population ecology is the study of the structure and growth of populations in a particular environment, and mathematical models play an important role in this study. Investigations have been conducted for a wide variety of populations, ranging from yeast cultures to insects to humans. It is noteworthy that certain fundamental principles apply even in the midst of this diversity. This chapter will explore the mathematics associated with some of these principles.

7.1 POPULATION DYNAMICS: EXPONENTIAL GROWTH

Population dynamics is the study of the rate of change of populations over time. The number N of individuals in a population is a function of time t, and the rate of change of the population is then $\frac{dN}{dt}$. If this quantity is positive at a given time, the population is growing then; if it is negative, the population is declining.

The two basic factors affecting the rate of change of a population are the number of births B and the number of deaths D per unit of time.[1] It should be emphasized that B and D are functions of time, and in general they will change over time. The basic equation in population dynamics is

Rate of change of population = Births per unit of time − Deaths per unit of time ,

or, expressed as an equation of change,

$$\frac{dN}{dt} \;=\; B - D \,. \tag{7.1}$$

Clearly the quantities B and D should be interpreted in terms of the size of the population: We expect more births in a population of a million individuals than we do in a population of a hundred, other factors being the same. One way to measure individual contributions to population changes is to introduce the *birth rate* and the *death rate*. The birth rate b is defined as the number of births per unit of time divided by the size of the population: $b = \frac{B}{N}$, or $B = bN$. The death rate d is defined in a similar way as the number of deaths per unit of time

---

[1]Other factors are immigration and emigration, which measure how populations disperse. Ignoring these will simplify the discussion.

divided by the size of the population: $d = \dfrac{D}{N}$, or $D = dN$. Then the basic equation, Equation (7.1), becomes

$$\frac{dN}{dt} \;=\; bN - dN \;,$$

or

$$\frac{dN}{dt} \;=\; (b - d)N \;. \tag{7.2}$$

Exponential growth

The simplest model of population dynamics involves exponential growth. For this we assume that the number of births B and the number of deaths D per unit of time are both proportional to the size of the population or, equivalently, that the birth rate b and death rate d are both constant. Then, by Equation (7.2), the rate of change of the population is proportional to the size of the population, with constant of proportionality r defined by $r = b - d$. With these modifications we can write Equation (7.2) as

$$\frac{dN}{dt} = rN \;. \tag{7.3}$$

We know from Chapter 6, Section 3, that the equation of change given in Equation (7.3) is characteristic of exponential functions, and we conclude that

$$N(t) = N(0)e^{rt} \;, \tag{7.4}$$

where $N(0)$ is the initial size of the population. Note that some ecologists prefer to write the time variable as a subscript, and they would write Equation (7.4) as

$$N_t = N_0 e^{rt} \;.$$

The constant r is the exponential growth rate, a measure of the intrinsic per capita growth rate. It represents the population's capacity for growth in the absence of constraints. It is measured in units such as per year or per day. (Note that even though its units involve discrete time such as a year, r represents an instantaneous rate of change.) Clearly, if the birth rate is larger than the death rate, then $r = b - d$ is positive, and the population is growing exponentially. If the birth rate is smaller than the death rate, then r is negative, and the population is declining exponentially. If the birth rate and death rate are equal, then r is zero, and the population size does not change. We are interested in the case when r is positive. We should also note that the use of the letter r is standard here, and ecologists often refer to the r value of a species in an environment.

To understand the rapid growth implied by this model, we consider an example.

EXAMPLE 7.1 *Aphid Population Growth*

There are initially 100 individuals in an aphid population which has an exponential growth rate of $r = 0.243$ per day.[2]

1. How many aphids will there be after twelve hours? After two weeks?

2. How long will it take an initial population of 100 aphids to double in size?

3. What is the doubling time if instead the initial population is 1000?

Solution to Part 1: With an initial population of 100, we have from Equation (7.4) that

$$N = 100e^{0.243t} ,$$

with t measured in days. Twelve hours make half of a day, and so we compute

$$N(0.5) = 100e^{0.243 \times 0.5} = 113 ,$$

rounded to the nearest integer. After twelve hours there are about 113 aphids.

Two weeks is 14 days, and proceeding as above we find that

$$N(14) = 100e^{0.243 \times 14} = 3002 ,$$

rounded to the nearest integer. There are just over three thousand aphids after two weeks.

Solution to Part 2: To find the doubling time we want to know the time t_d so that $N(t_d) = 200$. That is, we want to solve the equation

$$100e^{0.243t_d} = 200 .$$

We solve it using the crossing graphs method as shown in Figure 7.1. To make the graph we used a table of values to help us choose a horizontal span from 0 to 5 and a vertical span from 50 to 250. From the prompt at the bottom of the screen we see that t_d is about 2.85. So the doubling time is about 2.85 days, or about 2 days and 20 hours.

Solution to Part 3: If the initial population is 1000 instead of 100, then the new population function is

$$N = 1000e^{0.243t} .$$

To find the doubling time in this case we need to solve

$$1000e^{0.243t_d} = 2000 .$$

---

[2]This r value is from the study by R. Root and A. Olson which appeared as "Population increase of the cabbage aphid, *Brevicoryne brassicae*, on different host plants," *Can. Ent.* **101** (1969), 768–773.

In Figure 7.2 we show the solution of this equation using the crossing graphs method. Here we used a horizontal span from 0 to 5 and a vertical span from 500 to 2500. From the bottom of the screen, we see that the answer, $t_d = 2.85$ days, is the same one we got in Part 2!

Figure 7.1: Doubling time for an initial population of 100

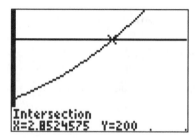

Figure 7.2: Doubling time for an initial population of 1000

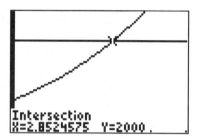

Doubling times

In Example 7.1 we saw that the time required for the aphid population to double is the same whether we start with 100 or 1000 aphids. The doubling time is an interesting measure of how rapidly a population grows, and we want to look at how to calculate it in general. The doubling time for an exponential function $N(0)e^{rt}$ is the time t_d that gives a population of $N(0) \times 2$. That is when e^{rt} is equal to 2. Thus we get the doubling time by solving the equation

$$e^{rt_d} = 2 .$$

In words, this equation says that rt_d is the power of e that gives 2. Thus, in terms of the natural logarithm,

$$rt_d = \ln 2 . \tag{7.5}$$

We divide to complete the solution:

$$t_d = \frac{\ln 2}{r} . \tag{7.6}$$

Check with your calculator that when $r = 0.243$ then Equation (7.6) gives the doubling time $t_d = 2.85$ that we found in Example 7.1. Note that the initial population $N(0)$ does not appear in Equation (7.6), so the doubling time t_d depends only on r, not on the initial population $N(0)$. This is what Example 7.1 suggested, and it is a special feature of exponential growth. One interpretation of this is that it doesn't matter when the clock begins: How long we have to wait before the population doubles in size is independent of the starting time. As you will be asked to show in Exercise 2, similar remarks apply to tripling times, etc.

To further illustrate the relationship between the doubling time and the exponential growth rate, we have graphed $\frac{\ln 2}{r}$ versus r in Figure 7.3. We used a horizontal span from 0 to 1 and a vertical span from 0 to 10. The horizontal axis corresponds to the exponential growth rate r, and the vertical axis corresponds to the doubling time t_d. The graph shows that, as intuition would suggest, doubling time decreases as exponential growth rate increases.

Figure 7.3: *Graph of doubling time versus exponential growth rate*

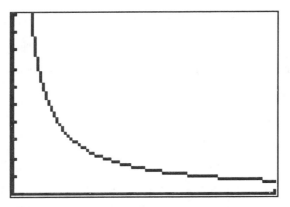

Equation (7.5) could alternatively be solved for r instead of t_d. Such a solution allows us to calculate the exponential growth rate r if we know the doubling time. You will be asked to explore this further in Exercise 3.

Empirical data

The exponential model of population growth is realistic only under special conditions, such as growth of a laboratory culture and colonization of a new area, or over relatively short time spans. In this connection it is quite remarkable that the United States population showed exponential growth for 70 years from 1790 through 1860. The size of the parameter r depends of course on the species. But it can also depend on environmental conditions such as temperature and humidity. It is difficult in practice accurately to measure this parameter, but such measurements are crucial and can provide important comparisons between similar species, or between the same species in different environments. This is illustrated in Exercises 8 and 9. Some known examples of r and the corresponding doubling time are given in Table 7.1, where the basic unit is a year.[3] For example, a population of flour beetles would, according to the exponential growth model, double in size in 0.03 year, or about 11 days.

An approximate value of r can be found from several measurements of population size,

---

[3]This table is adapted from p. 269 of Robert E. Ricklefs, *The Economy of Nature*, 3rd edition, 1993, W. H. Freeman and Company, New York.

Table 7.1: *Value of r and doubling time for some populations*

Population	r	t_d
Water flea	69	0.01
Flour beetle	23	0.03
Field vole	3.18	0.218
Ring-necked pheasant	1.02	0.68
Elephant seal	0.091	7.617

using the technique of fitting exponential data given in Chapter 4, Section 3: We plot the natural logarithm of the population data and calculate the regression line. The slope of this line then gives an estimate of r for the population, since this slope represents the exponential growth rate. Here is an example.

EXAMPLE 7.2 *Finding Exponential Growth Rate for Yeast*

The following table gives the size of a yeast population.[4] Time t is measured in hours.

Time t	0	1	2	3	4	5	6	7
Amount N of yeast	9.6	18.3	29.0	47.2	71.1	119.1	174.6	257.3

Plot the natural logarithm of the data points to show that an exponential model is appropriate for this population. Then use this information to estimate the exponential growth rate r and the doubling time t_d.

Solution: Since we expect to model this data with an exponential function, we approach the problem exactly as we did in Section 4.3. That is, we proceed to find the regression line for the natural logarithm of the data. The properly entered data ⌐7.1⌐ is in Figure 7.4. The plot ⌐7.2⌐ of the logarithm of the data is shown in Figure 7.5. It appears that the

Figure 7.4: *Entering the logarithm of yeast data*

Figure 7.5: *Plotting the logarithm of yeast data*

L1	L2	L3	2
0	2.2616	9.6	
1	2.9069	18.3	
2	3.3673	29	
3	3.8544	47.2	
4	4.2641	71.1	
5	4.78	119.1	
6	5.1625	174.6	

L2(1)=2.261763098…

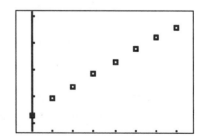

logarithm of N is a straight line, and so it is reasonable to model population growth

---

[4] The data is taken from an early study of yeast population growth by Tor Carlson, as described by R. Pearl in "The Growth of Populations," *Quart. Rev. Biol.* **2** (1927), 532–548.

using an exponential function. We calculate $\boxed{7.3}$ the regression line, with the result shown in Figure 7.6. From this we see that the slope of the regression line is about 0.46. We conclude that the exponential growth rate for yeast is about $r = 0.46$ per hour. In Figure 7.7 we have added the graph of the regression line to the plot of the logarithm of the data.

Figure 7.6: *Constants for the regression line of* $\ln N$

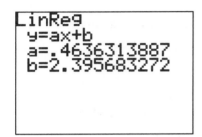

Figure 7.7: *Graph of regression line added to plot of the logarithm of yeast data*

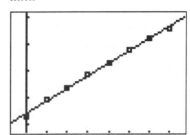

To get the doubling time we use Equation (7.6), putting 0.46 in place of r:

$$t_d = \frac{\ln 2}{0.46} \text{ , about 1.5 hours.}$$

Discrete generations

The above model for exponential growth is reasonable for species with overlapping generations and extended breeding seasons. For many species, however, breeding occurs only during certain seasons, and so population growth should be measured over discrete intervals of time. The exponential growth model has a discrete counterpart, sometimes called *geometric growth*, and in many ways the mathematics is unchanged. For geometric growth we assume that a unit of time represents the time between breeding seasons. For example, with annual breeding one unit represents a year. Then the basic assumption is that the population at time $t + 1$ is proportional to the population at time t. This constant of proportionality is usually denoted by the Greek letter λ. Thus we are assuming that

$$N(t + 1) = \lambda N(t)$$

for all t. It follows that

$$N(1) = \lambda N(0)$$
$$N(2) = \lambda N(1) = \lambda^2 N(0)$$
$$N(3) = \lambda N(2) = \lambda^3 N(0) .$$

This gives rise to the formula

$$N(t) = N(0)\lambda^t .$$

We note that this is exactly the same as an exponential function with initial value $N(0)$ and growth factor λ. But for geometric growth the model is used a little differently, since for species with discrete breeding times the formula is reasonable only when t is a whole number of units. Between breeding seasons there are no births, and the population actually declines due to deaths.

EXAMPLE 7.3   *Population of an Annual Plant*

The population of a species of an annual plant grows geometrically. If initially there are 10 plants, and after one year there are 200, how many plants are there after three years? Make a graph of the population function.

Solution: We started with 10 plants and one year later there were 200. Since the function is exponential, we know how to get the growth factor by dividing:

$$\frac{200}{10} = 20 .$$

Thus the growth factor λ is 20. To get the population in succeeding years, we multiply each year by 20. So after two years there are $200 \times 20 = 4000$ plants present. After three years there are $4000 \times 20 = 80,000$ plants. We can see that if the same geometric growth continues, the spread of the plant will be quite dramatic in succeeding years.

Since we know that the initial value is 10 and the growth factor is 20, we can write the general formula for N:

$$N = 10 \times 20^t .$$

Using this we could have computed the number of plants after three years as $N(3) = 10 \times 20^3 = 80,000$. The graph of N versus t is shown in Figure 7.8, in which we have used a horizontal span from 0 to 4 and a vertical span from 0 to 1,600,000. The steepness of the graph is startling, and we know that such a rate of growth cannot be maintained

---

Figure 7.8: A picture of rapid geometric growth

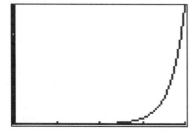

Figure 7.9: A discrete picture of geometric growth

indefinitely. We should also note that since the geometric growth formula only applies for integral values of t, it might have been more appropriate to plot the graph as discrete data points. We have done this in Figure 7.9.

Exercise Set 7.1

1. **Norway rat:** A population of the Norway rat undergoes exponential growth with exponential growth rate $r = 3.91$ per year.

   (a) If the initial population size is 6, find the population size after six months, and after two years.

   (b) Suppose again that the initial population size is 6. How long will it take for the population to double to 12? How long will it take to reach a size of 24? (Hint: Think before you calculate!)

   (c) If the initial population size is not given, but it is known that the population has reached 150,000 after 2 years, determine the initial population size.

2. **Tripling times:** We noted that for exponential population growth, the doubling time can be calculated, if the exponential growth rate is known, by using the formula

$$t_d = \frac{\ln 2}{r}.$$

   In this exercise, we seek a similar formula that gives the time it takes such a population to triple. The population is modeled by the function

$$N = N(0)e^{rt} .$$

   (a) What value of e^{rt} will make the population triple?

   (b) Write an equation whose solution gives the tripling time.

   (c) Use logarithms to solve the equation in Part (b).

   (d) Use the formula you obtained in Part (c) to obtain the tripling time for a population with exponential growth rate $r = 0.15$ per year.

3. **Exponential growth rate from doubling time:** To get the doubling time in terms of r, we solved the equation

$$rt = \ln 2$$

   for t. What do you get if you solve the equation instead for r? Use the formula you get to find the exponential growth rate for a population that takes 7 years to double in size.

4. **A species of mammal with known doubling time:** *This is a continuation of Exercise 3.* A certain species of mammal grows at an exponential rate, doubling its population size every 56 days.

   (a) Find the exponential growth rate r.

   (b) If initially there are 10 individuals, how many will there be after 1 year?

5. **Caribou:** The following table describes the growth of a population of caribou introduced in 1962 on Brunette Island of Newfoundland, Canada.[5] Time t is measured in years since 1962.

Time t	0	1	2	3	4	5
Population	17	27	40	54	78	100

   Use this table to calculate r for Newfoundland caribou.

6. **Insects:** The following table gives the size of an insect population which is believed to be growing exponentially. Time t is measured in days.

Time t	0	2	4	6	8
Population N	10	14	20	29	42

   (a) Use this information to estimate the parameter r and the doubling time t_d. Be sure to state the units of each. (Suggestion: You may use regression to solve this problem. Alternatively, you may be able to get a pretty close estimate of the doubling time by looking at the data. Then see Exercise 3.)

   (b) Estimate the number of insects present after 10 days.

7. **Field mice:** The following table gives the size of a population of field mice. Time t is measured in years.

Time t	0	0.5	1	1.5	2
Population N	4	38	360	3416	32,412

   (a) Plot the graph of the logarithm of the data and determine if it is reasonable to use an exponential model for this population.

   (b) What are the values of r and the doubling time t_d? Be sure to state the units of each.

   (c) Find an exponential model for the data.

   (d) Estimate the number of mice present after 9 months.

---

[5] The data is taken from A. Bergerud, "The Population Dynamics of Newfoundland Caribou," *Wildlife Monographs* **25** (1971), 1–55.

8. **Aphid growth on broccoli:** This problem and the following one are based on the results of a study by R. Root and A. Olson of population growth for aphids on various host plants.[6] The value of $r = 0.243$ per day used in Example 6.1 was for aphids reared on broccoli plants outdoors. The following table describes the growth of an aphid population reared on broccoli plants *indoors* (in a controlled environment). Time t is measured in days.

Time t	0	1	2	3	4
Population	25	30	37	44	54

Calculate r for this table. Did the aphids fare better indoors or outdoors?

9. **Aphid growth on different plants:** *This is a continuation of Exercise 8.* In the study described in Exercise 8, aphids were also reared indoors on collard and on yellow rocket (a decorative flower). Use the following tables to determine whether aphids fare better on collard or on yellow rocket. Time t is measured in days.

Collard

Time t	0	1	2	3	4
Population	28	33	40	48	57

Yellow rocket

Time t	0	1	2	3	4
Population	48	53	58	64	70

10. **Field voles:** Two regions are being colonized with field voles. The first colony starts with 10 individuals and grows exponentially. The second colony starts with 100 individuals and grows only by immigration: 125 individuals are added to the colony each year.

   (a) Use a formula to express the population of the first colony as a function of time t. Be sure to state the units of t. (Use Table 7.1 to find r.)

   (b) Use a formula to express the population of the second colony as a function of time t.

   (c) After how long will the first colony overtake the second in size?

---

[6] "Population increase of the cabbage aphid, *Brevicoryne brassicae*, on different host plants," *Can. Ent.* **101** (1969), 768–773. We have used the r values they obtained and their initial population sizes to construct our hypothetical population growth data.

11. **Microbes:** A population of microbes grows geometrically, increasing threefold every day. Initially there are 2 microbes. Plot the number of microbes as a function of time over a 5 day period, and determine how many microbes there are at the end of this period.

12. **Microbes which grow more rapidly:** A population of microbes grows geometrically, increasing tenfold every 2 days. Initially there are 2 microbes.

    (a) How many microbes will there be after 4 days?

    (b) Find the parameter λ.

    (c) How many microbes will there be after 5 days?

13. **How the size of λ affects geometric growth:** If the parameter λ for geometric growth is less than 1 (but positive), what implications does this have for the population? What if $\lambda = 1$?

7.2   POPULATION DYNAMICS: LOGISTIC GROWTH

Although the exponential growth model is simple, in reality a population cannot undergo such rapid growth indefinitely. The species will begin to exhaust local resources, and we expect that, instead of remaining constant, the birth rate will begin to decrease while the death rate begins to increase. Thus there will be a reduced rate of growth as the population increases in size. This regulation of growth is described by the logistic equation.

Logistic growth model

For an example of such population growth we consider a yeast culture. In Example 7.2 we looked at the first 7 hours of growth and saw that it was approximately exponential. In Figure 7.10 we see what happens over a longer period, namely 18 hours. The dots represent the population data. Note the conformity to the pattern of growth described above: At first growth is rapid; then it tapers off. In Figure 7.11 a curve has been fitted to the data. This S-shaped curve (sometimes called a sigmoid shape) is the trademark of logistic growth. Note that the graph resembles that of an exponential function early on. But as time goes by, the slope of the graph (representing the rate of change of the population) begins to decrease; eventually the graph flattens out, indicating that the population size is stabilizing.

In many circumstances the logistic model is a more reasonable description of population growth than is the exponential model. It was introduced by P.F. Verhulst in 1838 to describe the growth of human populations. In 1920, R. Pearl and L. J. Reed derived the same

Figure 7.10: *Population growth data for yeast*

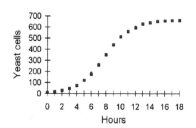

Figure 7.11: *Yeast data fitted with a logistic curve*

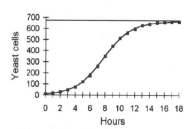

model to describe the growth of the population of the United States since 1790 and to attempt predictions of that population at future times.[7] We will return to their predictions at the end of this section.

The logistic model implies that there is an upper limit beyond which a population cannot grow. The smallest such limit is called the *carrying capacity* of the environment in which the population lives; it is denoted K. It can be thought of as a saturation level for the population. According to this model the population will actually get closer and closer to the carrying capacity as time goes on. For the yeast study above, the approximate value of K is 665. The horizontal line corresponding to this value has been added to the graph in Figure 7.11, and the relationship between this line and that of the logistic growth curve described above is apparent there.

Our discussion of the logistic model so far has been qualitative—we have given no formulas. But even at this level it yields interesting results, as the following application shows.

Application to harvesting renewable resources

In the management of a harvested population (as in a marine fishery), an important problem is to determine at what level to maintain the population so as to sustain a maximum harvest. This is the problem of optimum yield. The logistic model for population growth gives a theoretical basis for solving this problem, leading to the theory of *maximum sustainable yield*.

If we harvest when the population is small, then the population level will be reduced so far that it will take a long time to recover and, in extreme cases, may be driven to extinction. In Figure 7.12 we have illustrated harvesting at the lower part of the curve for a species undergoing logistic growth. The unbroken curve shows how the population would grow if left alone, and the broken graph shows the result of periodic harvesting. It shows that if we

---

[7]"On the rate of growth of the population of the United States since 1790 and its mathematical representation," *Proc. Nat. Acad. Sci.* **6** (1920), 275–288.

harvest when the population reaches a level of about 125, we will be able to harvest about 100 individuals each third day. At the other extreme, we could maintain the population near its maximum level. But if the harvested population grows according to the logistic model, this means maintaining the population near the carrying capacity of the environment, where the rate of population growth is *slow* since the graph flattens out there. After a small harvest we will have to wait a relatively long time for the population to recover. This is illustrated in Figure 7.13, and we see that in the case shown, we are able to harvest about 100 individuals each two days.

Figure 7.12: *Harvesting a logistic population at the lower part of the curve*

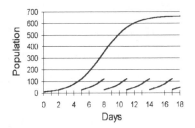

Figure 7.13: *Harvesting a logistic population near the carrying capacity*

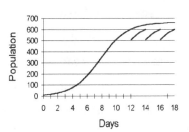

To get the best harvest, it makes sense to determine the population size at which the rate of growth is the largest, since the population will recover most quickly near that size. This scenario is illustrated in Figure 7.14, and we see that under these conditions, we are able to harvest 100 individuals each day. That population size, known as the *optimum yield level*, corresponds to the steepest point on the graph, which is near the middle of the curve. This suggests that the optimum yield level is half of the carrying capacity. It is marked in Figure 7.15. Before the population reaches this steepest point, the rate of growth is increasing; after this point, the rate of growth is decreasing. In other words, the optimum yield level corresponds to the inflection point on the graph. Our conclusion is that optimum yields come from populations maintained not at maximum size but at maximum growth rates, and that this occurs at half of the carrying capacity.

Note: The importance of wise management is demonstrated by the history of the Peruvian anchovy fishery, which flourished off the coast of Peru from the mid-1950's until the early 1970's. In 1970 it accounted for 18% of the total world harvest of fish. But in the early 1970's, a change in environmental conditions, combined with overfishing, caused the collapse of the fishery. This had a dramatic impact on food prices worldwide. The fishery has never recovered from this collapse.

Figure 7.14: *Harvesting near the middle of the logistic curve*

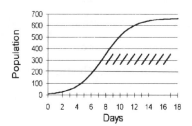

Figure 7.15: *The level of optimum yield is half the carrying capacity*

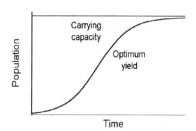

Logistic equation of change

Now we turn to the formula for $\frac{dN}{dt}$ which gives the logistic model. The idea is simple: For exponential growth we assume a constant per capita growth rate r. For logistic growth we assume instead that each individual added to the population decreases the per capita growth rate by an equal amount, until this rate is zero at the carrying capacity K. The effect of this is to multiply r by a factor which is a linear function of the population size N, namely the function $1 - \frac{N}{K}$. From this we get the logistic equation of change, sometimes called the logistic equation in its *differential form*:

$$\frac{dN}{dt} = rN\left(1 - \frac{N}{K}\right) .$$ (7.7)

Let's examine this equation of change more closely to see how the shape of the logistic curve arises from it. The linear factor $1 - \frac{N}{K}$ is 1 when $N = 0$, so for small initial populations Equation (7.7) is approximately the same as the equation of change for exponential growth: $\frac{dN}{dt} = rN$. This is reflected in the fact that the logistic curve resembles the graph of an exponential function for small population sizes. The linear factor $1 - \frac{N}{K}$ is 0 when $N = K$, so the rate change of N, as given by Equation (7.7), decreases as the population size N approaches the saturation level K. This explains why the curve flattens out over time. In summary, Equation (7.7) does indeed describe the S-shaped logistic curve: approximate exponential growth early on, followed by a gradual leveling off of growth.

In light of the preceding application involving optimum yield, it is illuminating to examine a graph of $\frac{dN}{dt}$ against N. According to Equation (7.7), this is a graph of $rN(1 - \frac{N}{K})$ as a function of N. The graph is the *parabola* shown in Figure 7.16. This again illustrates the basic form of logistic growth: The rate of growth of the population at first increases with increasing population and then, at a certain level, begins to decrease. By symmetry it is clear that the maximum growth rate, the maximum of $\frac{dN}{dt}$ as a function of N, is attained at the population level $N = \frac{1}{2}K$. In connection with the application to harvesting renewable resources, this confirms that the population level for optimum yield is half of the carrying capacity K.

Figure 7.16: *Population growth rate as a function of population size*

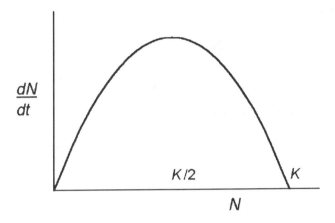

In our discussion we have assumed that the initial population level is below the carrying capacity K. If instead this level is above K, then initially $1 - \frac{N}{K}$ is negative, and according to Equation (7.7) the population will decrease in size. In fact, the population will decrease to K in the limit. This is what intuition would suggest. Thus the logistic model implies that, no matter what the initial population size is, in the long run the population will stabilize at the carrying capacity K of the environment. In mathematical terms, K is a *stable equilibrium* for N.

Empirical data: We can estimate r and K for a given population using a sample of population sizes. One way of doing this is to graph the (varying) per capita growth rate as a function of population size, i.e., to graph $\frac{1}{N}\frac{dN}{dt}$ against N. According to Equation (7.7), this is a graph of $r(1 - \frac{N}{K})$ as a function of N. Since

$$r\left(1 - \frac{N}{K}\right) = r - \frac{r}{K}N ,$$

this is a line with slope $-\frac{r}{K}$, vertical intercept r, and horizontal intercept K. The graph is shown in Figure 7.17.

To determine r and K, then, we use regression to find the best linear fit to the data for $\frac{1}{N}\frac{dN}{dt}$ versus N. The vertical intercept of this line is r, and if we call its slope m, then we solve the equation $m = -\frac{r}{K}$ for K to get $K = -\frac{r}{m}$. Alternatively, we can find K by graphing the line to determine its horizontal intercept, as is shown in the next example.

Figure 7.17: *Per capita growth rate as a function of population size*

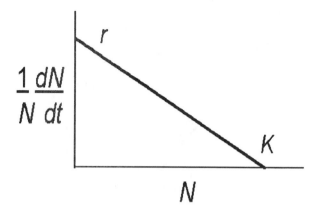

EXAMPLE 7.4 *Logistic Growth for the Water Flea*

The following table gives population sizes and corresponding per capita growth rates for the water flea.[8] Time t is measured in days, and N is measured as number per cubic centimeter.

N	1	2	4	8	16
$\frac{1}{N}\frac{dN}{dt}$	0.230	0.216	0.208	0.145	0.057

1. Plot the given data points to show that a linear model for $\dfrac{1}{N}\dfrac{dN}{dt}$ versus N is appropriate for this population.

2. Use the data to estimate r and K for the water flea in this environment.

Solution to Part 1: The plot $\boxed{7.4}$ is shown in Figure 7.18. It appears that a linear model is appropriate.

Solution to Part 2: We calculate $\boxed{7.5}$ the regression line, with the result shown in Figure 7.19. From this we see that the vertical intercept of the regression line is about 0.24, and this is our estimate for r. Also, the slope of the regression line is about -0.012, so by the preceding discussion this gives the estimate $K = -\frac{0.24}{-0.012} = 20$. We conclude that, for the water flea in this environment, r is about 0.24 per day, and the carrying capacity K is about 20 individuals per cubic centimeter.

---

[8]The data is taken from a study by P. W. Frank, C. D. Boll, and R. W. Kelly, in "Vital statistics of laboratory cultures of *Daphnia pulex* DeGeer as related to density," *Physiol. Zool.* **30** (1957), 287–305.

Figure 7.18: *Data for per capita growth rate as a function of population size*

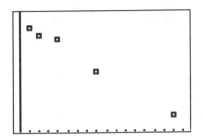

Figure 7.19: *Equation of the regression line*

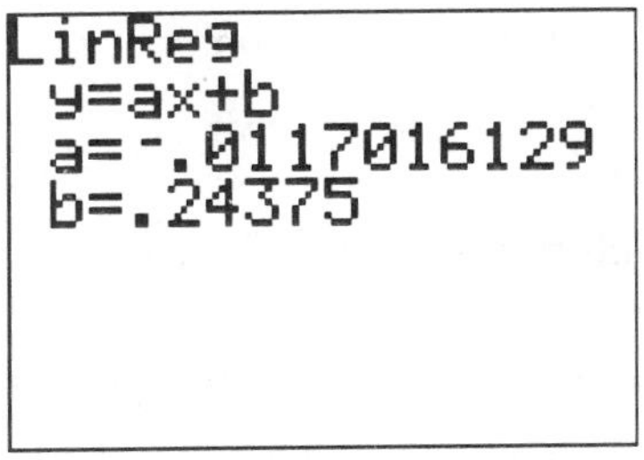

```
LinReg
y=ax+b
a=-.0117016129
b=.24375
```

Note: The procedure described above requires that we know the per capita growth rates for a sample of population sizes. In practice these rates can be difficult to obtain. There are more sophisticated ways to fit logistic functions to data without using per capita growth rates.

Integral form of the logistic equation

Using more advanced techniques, Equation (7.7) can be solved to obtain an explicit formula for N as a function of t. This is the integral form of the logistic equation:

$$N = \frac{K}{1 + be^{-rt}} , \qquad (7.8)$$

where b is the constant $\frac{K}{N(0)} - 1$.

To illustrate this, we return to the example of the Pacific sardine population.

EXAMPLE 7.5 *The Pacific Sardine*

Studies to fit a logistic model to the Pacific sardine population have yielded

$$N = \frac{2.4}{1 + 239e^{-0.338t}} ,$$

where t is measured in years and N is measured in millions of tons of fish.[9]

1. What is r for the Pacific sardine?

2. What is the environmental carrying capacity K?

3. What is the optimum yield level?

4. Make a graph of N.

5. At what time t should the population be harvested?

---

[9]From a study by G. I. Murphy, in "Vital statistics of the Pacific sardine (*Sardinops caerulea*) and the population consequences," *Ecology* **48** (1967), 731–736.

6. What portion of the graph is concave up? Concave down?

Solution to Part 1: The formula for N is written in the standard form of Equation (7.8) for a logistic function. Since the coefficient of t in the exponential is -0.338, it must be that $r = 0.338$ per year.

Solution to Part 2: The constant in the numerator is 2.4, so $K = 2.4$ million tons of fish.

Solution to Part 3: The optimum yield level is half of the carrying capacity, and by Part 2 this is $\frac{1}{2} \times 2.4 = 1.2$ million tons of fish.

Solution to Part 4: Using a table of values we determine a horizontal span from 0 to 40 and a vertical span from 0 to 2.5. In Figure 7.20 we have made the graph.

Solution to Part 5: The population should be harvested at the optimum yield level, and according to Part 3 this is at the population level $N = 1.2$, so we want to solve the equation $N(t) = 1.2$ for t. We do this using the crossing graphs method, as shown in Figure 7.21. We find that the time t is about 16.2 years.

Figure 7.20: *Population growth for the Pacific sardine*

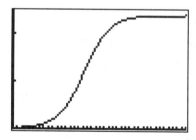

Figure 7.21: *Time of optimum yield*

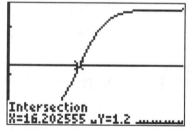

Intersection
X=16.202555 Y=1.2

Solution to Part 6: By examining Figure 7.20 we see that the graph is concave up until the population reaches the level for optimum yield and is concave down after that. Using Part 5 we see that the graph is concave up over the first 16.2 years and concave down thereafter.

Note: The Pacific sardine fishery along the California coast expanded rapidly from about 1920 until the 1940's, with an annual catch of around 800 thousand tons at its height. In the late 1940's to early 1950's the fishery collapsed, and yield levels have remained very low since then. The economic consequences of the collapse have been severe. Major factors contributing to the collapse were the stress inflicted on the population by heavy fishing and by environmental changes which seem to have favored a competing fish, the anchovy.[10]   This example serves to illustrate the point that the model of maximum sustainable yield which we have developed is only a first step in the study of managing renewable resources, and that a variety of factors must be considered.

The value of the logistic model

The logistic model for population growth, like the exponential, has limitations. As we mentioned above, the original article by Pearl and Reed in 1920 used this model to predict population growth in the United States. According to their data, the population was to stabilize at about 197 million. Of course, this level has been far surpassed. At a more fundamental level, the basic assumptions of the logistic model have been called into question. Various other models have been suggested, but they involve more complicated mathematics.

The main value of the models we have examined lies in their qualitative form. They enable us to discuss, in general terms, population trends and the reasons for such trends. Rather than memorizing the complicated formula in Equation (7.8), you should concentrate on understanding how the S-shaped graph reflects the underlying assumptions of this model.

---

[10]For more information on the Pacific sardine fishery, see the account by Michael Culley in *The Pilchard*, 1971, Pergamon Press, Oxford, England.

Exercise Set 7.2

1. **Estimating optimum yield:** In Figure 7.22 a logistic growth curve is sketched. Estimate the optimum yield level and the time when this population should be harvested.

Figure 7.22: A logistic growth curve

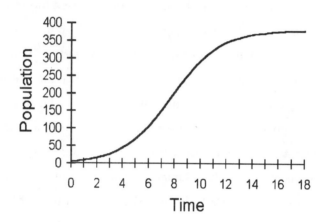

2. **Estimating carrying capacity:** In Figure 7.23 a portion of a logistic   growth curve is sketched. Estimate the optimum yield level and the environmental carrying capacity.

Figure 7.23: A portion of a logistic growth curve

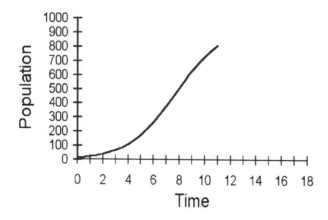

3. **Northern Yellowstone elk:** The northern Yellowstone elk winter in the northern range of Yellowstone National Park.[11] A moratorium on elk hunting was imposed in 1969, and after that the growth of the elk population was approximately logistic. Use the following table to estimate r and K for this elk population. (The unit for the per capita growth rate $\frac{1}{N}\frac{dN}{dt}$ is per year. The date is included in the table only for the sake of interest.)

Year	1968	1969	1970	1971	1972	1973	1974
N	3,172	4,305	5,543	7,281	8,215	9,981	10,529
$\frac{1}{N}\frac{dN}{dt}$	0.305	0.253	0.273	0.121	0.195	0.053	0.180

Note: At one time the gray wolf was a leading predator of the elk, but it was not a factor during this study period. The level at which the elk population stabilized suggests that food supply (and not just predators) can effectively regulate population size in this setting.

4. **Negative growth rates:** The original water flea study described in Example 7.4 also included growth from population sizes of $N = 24$ and $N = 32$.

   (a) Use the regression line computed in Example 7.4 to estimate the per capita growth rates for these two values of N.

   (b) Explain in terms of the carrying capacity why the growth rates from Part (a) should both be negative.

5. **More on the Pacific sardine:** In this problem we explore further the Pacific sardine population using the model in Example 7.5.

   (a) If the current level of the Pacific sardine population is 50,000 tons, how long will it take for the population to recover to the optimum growth level of 1.2 million tons?

   (b) The value of r used in Example 7.5 ignores the effects of fishing. If fishing mortality is taken into account, then r drops to 0.215 per year (with the carrying capacity still at 2.4 million tons). Answer the question in Part (a) using this lower value of r.

   Note: The population estimate of 50,000 tons and the adjusted value of r are given in the paper by Murphy, *op. cit.* Murphy points out that factoring in the growth of the competing anchovy population makes the recovery times even longer, and he states: "It is disconcerting to realize how slowly the population will recover to its level of maximum productivity . . . even if fishing stops."

---

[11] This problem is based on the study by Douglas B. Houston in *The Northern Yellowstone Elk*, 1982, Macmillan Publishing Co., New York. Houston uses a slightly different method to fit a logistic model.

6. **Eastern Pacific yellowfin tuna:** Studies to fit a logistic model to the Eastern Pacific yellowfin tuna population have yielded

$$N = \frac{148}{1 + 3.6e^{-2.61t}} \, ,$$

where t is measured in years and N is measured in thousands of tons of fish.[12]

  (a) What is r for the Eastern Pacific yellowfin tuna?

  (b) What is the environmental carrying capacity K for the Eastern Pacific yellowfin tuna?

  (c) What is the optimum yield level?

  (d) Use your calculator to graph N against t.

  (e) At what time was the population growing the fastest?

7. **Maximum growth rate for tuna:** *This is a continuation of Exercise 6.*

  (a) With your calculator make a graph of $\dfrac{dN}{dt}$ against t, using

$$\frac{dN}{dt} = rN\left(1 - \frac{N}{K}\right)$$

  and the formula for N given in Exercise 6.

  (b) Use your graph in Part (a) to find the time at which $\dfrac{dN}{dt}$ was the largest.

  (c) What was the population at the time when $\dfrac{dN}{dt}$ was the largest? How does your answer compare with that in Part (c) of Exercise 6?

8. **Pacific halibut:** For a Pacific halibut population the value of r is 0.71 per year, and the environmental carrying capacity is 89 thousand tons of fish.[13]

  (a) Find the time it takes the population to grow from the optimum yield level to 90% of carrying capacity if the number b in the logistic growth equation, Equation (7.8), is 1.5.

  (b) Find the time it takes the population to grow from the optimum yield level to 90% of carrying capacity if the number b in the logistic growth equation, Equation (7.8), is 4.8.

  (c) Compare the times in Parts (b) and (c). Why could this result have been expected?

---

[12]From a study by M. B. Schaefer, as described by Colin W. Clark in *Mathematical Bioeconomics*, 2nd edition, 1990, John Wiley & Sons, New York.
[13]From a study by H. S. Mohring, as described by Colin W. Clark, *ibid.*

9. **Yeast growth rate:** The logistic model which best fits the yeast growth data in Figure 7.10 at the beginning of this section has $r = 0.54$ per hour and $K = 665$. But in Example 7.2 of the preceding section, when we treated the first 7 hours of growth as exponential we obtained $r = 0.46$ per hour. Explain the discrepancy and, in particular, why you would expect the value of r from the logistic model to be higher than that from the exponential model of the first segment of growth.

10. **Maximum growth rate:** We have seen that under the logistic model

$$\frac{dN}{dt} = rN \left(1 - \frac{N}{K} \right)$$

the growth rate $\dfrac{dN}{dt}$ is at its maximum when $N = \frac{1}{2}K$.

(a) Use these formulas to express this maximum growth rate in terms of r and K.

(b) Use your answer to Part (a) to find the maximum growth rate for the Pacific sardine population of Example 7.5.

11. **Logarithm of the logistic curve:** Consider the logistic curve sketched in Figure 7.15. Sketch, including an explanation, a graph of $\ln N$ against t. (Hint: The curve is nearly exponential at first. What is the logarithm of an exponential function?)

12. **Rate of change for logistic growth:** Consider the logistic curve sketched in Figure 7.15. Sketch, including an explanation, a graph of $\dfrac{dN}{dt}$ against t.

13. **Gompertz model:** One possible substitute for the logistic model of population growth is the Gompertz model, according to which

$$\frac{dN}{dt} = rN \ln \left(\frac{K}{N} \right).$$

For simplicity in this problem we take $r = 1$, so the equation of change for N is

$$\frac{dN}{dt} = N \ln \left(\frac{K}{N} \right).$$

(a) Let $K = 10$, and make a graph of $\dfrac{dN}{dt}$ versus N for the Gompertz model.

(b) Use the graph you obtained in Part (a) to determine for what value of N the function $\dfrac{dN}{dt}$ is at its maximum. Since this tells the population level for maximum growth rate, this is the optimum yield level under the Gompertz model with $K = 10$.

(c) Under the logistic model the optimum yield level is $\frac{1}{2}K$. What do you think is the optimum yield level in terms of K under the Gompertz model? (Hint: Repeat the procedure in Parts (a) and (b) using different values of K, such as $K = 1$ and $K = 100$. Try to find a pattern.)

14. **Another model:** Recall that the basic assumption of the logistic growth model is that each individual added to the population decreases the per capita growth rate by an equal amount. As Figure 7.17 shows, this means the graph of $\dfrac{1}{N}\dfrac{dN}{dt}$ against N is a straight line. In Figure 7.24 another relationship between $\dfrac{1}{N}\dfrac{dN}{dt}$ and N is suggested. Interpret this in physical terms, and discuss the impact on the growth curve, i.e., the graph of N against t.

Figure 7.24: *Alternative graph of* $\dfrac{1}{N}\dfrac{dN}{dt}$ *versus* N

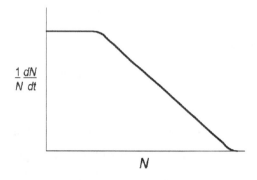

7.3 POPULATION STRUCTURE: SURVIVORSHIP CURVES

To study the structure of a population we need to know the number of individuals in each age category as well as the number of births and deaths within these categories. For human populations such statistics have long been recognized as an important subject of study. For hundreds of years, life insurance companies have set premium rates based on this information. However, the gathering and study of such statistics by ecologists for other populations is relatively recent. In this section we will see how the tools we have developed for analyzing functions and their graphs bear upon this study of the structure of populations.

Life tables and survivorship curves

Statistics on the age structure of a population are arranged into what is called a *life table*. Such tables usually contain several columns which give a variety of ways to present the basic data about births and deaths for individuals grouped by *cohort*, that is, according to their age.

For mortality statistics the following symbols are often used:

$$
\begin{aligned}
x &= \text{age at beginning of interval} \\
l_x &= \text{number of survivors at start of age interval } x \\
d_x &= \text{number dying during age interval } x \\
q_x &= \text{proportion of individuals dying during age interval } x .
\end{aligned}
$$

Note that l_x, d_x, and q_x are really functions of x. In ecology it is standard to put x as a subscript, writing l_x instead of $l(x)$, for example. Also, l_0 is the total number of individuals in the study, all born at about the same time. It is standard to rescale the numbers so that $l_0 = 1000$. With this convention, l_x represents survivors per thousand. The number q_x is usually called the *age-specific mortality rate*. It tells the percentage of a given age group that can be expected to die over the given time period.

A life table for the Dall mountain sheep is given in Table 7.2, where x is in years.[14] From the table, we see that of the initial $l_0 = 1000$ sheep in the study, $d_0 = 199$ died during their first year of life. This gives $1000 - 199 = 801$ for l_1, the number surviving past the first year of life. The fourth column is obtained by taking ratios between the third and second columns: $q_x = d_x/l_x$. Thus for the sheep we have $q_0 = 199/1000 = 0.199$. That is, 19.9% of newborn Dall mountain sheep do not survive their first year.

The various columns of a life table are redundant, in a sense: Once one column is known, the others can be derived from it using the definitions given above. Each column presents the basic information in a different way.

---

[14]From a study by A. Murie, as described by E. S. Deevey, Jr., in "Life tables for natural populations of animals," *Quart. Rev. Biol.* **22** (1947), 283–314.

Table 7.2: Life table for the Dall mountain sheep

x	l_x	d_x	q_x
0	1000	199	0.199
1	801	12	0.015
2	789	13	0.016
3	776	12	0.015
4	764	30	0.039
5	734	46	0.063
6	688	48	0.070
7	640	69	0.108
8	571	132	0.231
9	439	187	0.426
10	252	156	0.619
11	96	90	0.938
12	6	3	0.500
13	3	3	1.000

Survivorship curves are derived from life tables to aid in visualizing the data. The shapes of these curves convey characteristic features of the population, such as the incidence of juvenile mortality. A survivorship curve is nothing but a graph of l_x as a function of x. We will follow the common practice of looking at these graphs on a semi-logarithmic scale.[15] That is, we will graph $\log l_x$ against x. Here is the reason: Suppose for a moment that the mortality rate q_x is independent of age x, so a fixed fraction of the individuals in each age interval die. Then a fixed fraction of the individuals in each age interval survive, so l_x will be a decreasing exponential function. By the basic property of logarithms, $\log l_x$ will be a decreasing *linear* function of x, i.e., equal increases in x lead to equal decreases in $\log l_x$. In summary, a constant mortality rate yields a linear relationship between x and $\log l_x$, and this illustrates the value of examining survivorship data on a semi-logarithmic scale. It is noteworthy that on this scale the slope of the graph is roughly proportional to the mortality rate.

Classifying survivorship curves

The three general types of survivorship curves are shown in Figure 7.25. The type I curve, which is concave down, is relatively flat for a time, then drops off steeply. This indicates that mortality among juveniles is fairly low but that it increases rapidly for older individuals. This shape is strictly applicable only under rather special conditions, such as captive populations in which individuals benefit from close care in the juvenile stage and in the end die of old age. This general shape, however, models to some extent survivorship among humans and

---

[15]It is standard to use the common (base 10) logarithm for this instead of the natural logarithm.

indeed many mammals. The curve of type II is a straight line and represents a constant mortality rate (as discussed above). This roughly describes survivorship among most birds and some other animals (e.g., the hydra and some lizards). The type III curve, which is concave up, represents high juvenile mortality, followed by lower, relatively constant mortality thereafter. Such a curve is thought to be the most common in nature, valid for many fish, marine invertebrates, insects, and plants.

Figure 7.25: *Survivorship curves*

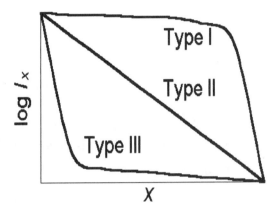

In reality, many species have survivorship curves representing a combination of the above types. High juvenile mortality (as in type III curves) is common. But often this is followed by a relatively constant mortality rate, so the middle portion of the curve is linear (at least over long intervals) as for type II curves. Towards the end the mortality rate increases, dramatically perhaps, as in type I. As an example of this combination of types, consider the survivorship curve for the Dall mountain sheep shown in Figure 7.26 (based on the life table in Table 7.2). This would be classified as a type I curve, but it has some characteristics of the other types.

Note: The calculator can be used to generate survivorship curves based on the data in a life table. You should refer to the *Keystroke Guide* to see the keystrokes needed.

Figure 7.26: Survivorship curve for the Dall mountain sheep

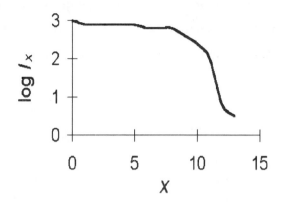

EXAMPLE 7.6   Survivorship Curve for Red Deer Stags

Consider the life table for red deer stags in Table 7.3, where x is in years.[16]   Graph the survivorship curve, classify as to type, and discuss the shape.

Table 7.3: Life table for red deer stags

x	l_x	d_x	q_x
0	1000	282	0.282
1	718	7	0.010
2	711	7	0.010
3	704	7	0.010
4	697	7	0.010
5	690	7	0.010
6	684	182	0.266
7	502	253	0.504
8	249	157	0.631
9	92	14	0.152
10	78	14	0.179
11	64	14	0.219
12	50	14	0.280
13	36	14	0.389
14	22	14	0.636
15	8	8	1.000

[16]From V. P. W. Lowe, "Population dynamics of the red deer (*Cervus elaphus* L.)," *J. Anim. Ecol.* **38** (1969), 425–457.

Solution: The calculator produces the graph shown in Figure 7.27. This is similar to a type I curve, in that it is relatively flat for a time and then falls off steeply. But note the drop at the very beginning, corresponding to the relatively large value of q_0. Also note that the middle part resembles a stair-step curve, with sections of linearity followed by somewhat steep drops: From year 2 until year 7 the curve appears to be linear, and in fact the value of q_x is quite small here. This section is followed by a steep drop to year 10, then another linear section in years 10 to 15, and then a final drop. Thus in addition to relatively high mortality at the beginning and end of the curve, there is a section of high mortality in the middle.

Figure 7.27: Survivorship curve for red deer stags

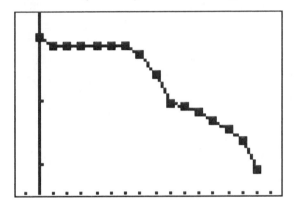

<u>Note</u>: It is not always clear why there are such drops in the middle of the curve. They may simply reflect complications in gathering data which accurately reflects the mortality rates in a population. They also may have to do with different stages in the life history of an individual. Often insects have different life stages, during which there are different susceptibilities to dying. Connections with life stages are explored in the following example and in Exercises 2 and 4.

EXAMPLE 7.7 *Survivorship Curve for the Spruce Budworm*

The spruce budworm infests forest conifers of eastern Canada and the northeastern United States.[17] During periodic outbreaks it kills many trees by feeding on flowers, buds, and needles. Here is a brief summary of its life history: A generation lasts about one year, during which the insect passes from egg to larva to pupa to moth. Eggs hatch in August, and shortly after emergence the larvae spin *hibernacula* (protective cases) in which they pass the winter. Upon emergence in May they begin feeding on the tree. In late June or early July the insect enters the pupal stage. In late July the moths emerge.

A survivorship curve for the spruce budworm is given in Figure 7.28. Here x represents half-months since August 15. Based on this curve, describe the variation in mortality of the spruce budworm as it corresponds to the life stages.

Figure 7.28: Survivorship curve for the spruce budworm

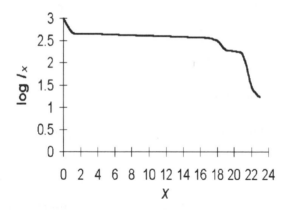

Solution: At the beginning of the curve there is a steep drop. This means that there is high mortality in the time between emergence from the egg and formation of the hibernaculum. Next comes a long section where the curve is nearly horizontal, so for several months the mortality rate is very low. This corresponds to the period when the larvae are in the protective hibernacula. After this is a drop in the curve around $x = 18$, which indicates high mortality in mid-May. This is the time of emergence from hibernacula. There is a final drop in the curve near $x = 21$, and so the mortality rate is high around the time of entrance into the pupal stage.

---

[17] This example is based on the work of R. F. Morris and C. A. Miller, "The development of life tables for the spruce budworm," *Can. J. Zool.* **32** (1954), 283–301.

For the first two drops in the curve, each after an emergence, it turns out that the main cause of death is dispersion, as some insects fall from the trees. The final drop is associated with mortality due to parasites and predators (such as birds).

Usefulness of survivorship curves

There are difficulties in gathering data for a survivorship curve. The greatest difficulty is associated with the young age classes. But often these difficulties do not greatly affect the general shape of the curve. It is this shape, we have stressed, which has much to say about the general characteristics of a population. Ecologists find it useful to compare curves between similar species or curves of the same species in varying environments. R. Dajoz[18] gives another reason why survivorship curves are studied: "They indicate at what age a species is most vulnerable." In the management of a population, whether in a game reserve or in pest control, it is important to know when the population is most vulnerable, for that is when intervention is most effective. For the spruce budworm described above, the stages of vulnerability occur after each emergence and around the time of pupation.

---

[18]*Introduction to Ecology* [transl. A. South], Crane, Russak & Company, 1976, p. 186.

Exercise Set 7.3

Age x is measured in years, unless otherwise specified. References are supplied at the end of the exercise set.

1. **Classifying survivorship curves:** For each of the following populations a survivorship curve is given. Classify it as to type, and discuss its shape.

    (a) African buffalo, in Figure 7.29.

    (b) Buttercup seedling, in Figure 7.30. (Here x represents the number of weeks since emergence.)

    (c) Warthog, in Figure 7.31.

Figure 7.29: *Survivorship curve for the African buffalo*

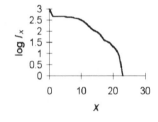

Figure 7.30: *Survivorship curve for buttercup seedlings*

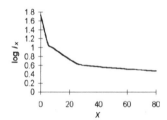

2. **Palm survivorship:** Consider the survivorship curve for a palm tree given in Figure 7.32.

Figure 7.31: *Survivorship curve for the warthog*

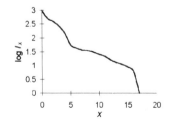

Figure 7.32: *Survivorship curve for a palm tree*

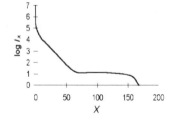

(a) Classify it as to type, and discuss its shape.

(b) Interpret your answer to Part (a) in light of the following table, which gives stages of the palm life cycle:

Stage	Seed	Seedling	Young tree	Mature, subcanopy	Canopy
Age in years	0	1	9	51	88

3. **Graphing and classifying survivorship curves:** For each of the following animals a life table is given. Graph the survivorship curve, classify as to type, and discuss the shape.

(a) The desert night lizard, in Table 7.4.

Table 7.4: Life table for the desert night lizard

x	l_x	d_x	q_x
0	1000	117	0.117
1	883	144	0.163
2	739	144	0.195
3	595	93	0.156
4	502	136	0.271
5	366	116	0.317
6	250	127	0.508
7	123	35	0.285
8	88	88	1.000

(b) The American robin, in Table 7.5.

Table 7.5: Life table for the American robin

x	l_x	d_x	q_x
0	1000	503	0.503
1	497	268	0.539
2	229	130	0.568
3	99	63	0.636
4	36	26	0.722
5	10	4	0.400
6	6	5	0.833

4. **Honeybee survivorship:** In Table 7.6 a life table for the European honeybee worker is given. (Here x is measured in days.)

Table 7.6: Life table for the honeybee worker

x	l_x	d_x	q_x
0	1000	42	0.042
3	958	137	0.143
8	821	10	0.012
20	811	15	0.018
25	796	16	0.020
30	780	14	0.018
35	766	27	0.035
40	739	149	0.202
45	590	221	0.375
50	369	256	0.694
55	113	76	0.673
60	37	32	0.865
65	5	5	1.000

(a) Graph the survivorship curve, classify as to type, and discuss the shape.

(b) Interpret the life table and survivorship curve in light of the following table, which gives stages of the honeybee worker life cycle:

Stage	Egg	Unsealed brood	Sealed brood	Adult
Age in days	0	3	8	20

(Here "unsealed brood" refers to feeding larvae and "sealed brood" refers to post-feeding larvae and pupae.)

5. **Sagebrush lizard mortality:** In Table 7.7 age-specific mortality rates for the sagebrush lizard are given, with the assumption that initially there were 1000 individuals.

Table 7.7: Partial life table for the sagebrush lizard

x	l_x	d_x	q_x
0	1000		0.770
1			0.378
2			0.469
3			0.553
4			0.618
5			0.462

(a) Fill in the blanks to give the life table through age 5. (Suggestion: To do this, use the formula $d_x = q_x \times l_x$. For example, to fill in the blank in the first row, we find

$$d_0 = q_0 \times l_0 = 0.770 \times 1000 = 770.$$

This means that 770 individuals died during their first year, so

$$l_1 = l_0 - d_0 = 1000 - 770 = 230$$

survived to the next year. This fills in the first blank in the second row.)

(b) Graph the survivorship curve and discuss its shape.

6. **Half die each year:** In a study of a certain animal cohort, each year half of those surviving to the start of the year died during that year.

(a) Make a life table with columns x, l_x, d_x, and q_x for this study. Assume there were initially 32 individuals in the cohort.

(b) What type of survivorship curve does this population have? Explain your answer.

7. **Experience of no use:** One ecologist interpreted the survivorship curve of a certain animal as saying that "experience of life is of no use . . . in avoiding death." Which one of the three types of curves do you think this animal has? Explain your answer.

8. **Grain beetles:** This exercise illustrates how the survivorship curves of two strains of grain beetles depend on environmental conditions. (Here x is measured in weeks.)

(a) In Figure 7.33 two survivorship curves for the small strain grown in wheat are given. For one the ambient temperature was 29.1 degrees Celsius (about 84 degrees Fahrenheit), and for the other the ambient temperature was 32.3 degrees Celsius (about 90 degrees Fahrenheit). Use the curves to decide at which one of these two temperatures you would store your wheat. Explain your answer.

(b) In Figure 7.34 two survivorship curves for the large strain grown at 29.1 degrees Celsius are given. For one the food was wheat, and for the other the food was maize. Use the curves to decide which one of these two grains the large strain would be more likely to infest. Explain your answer.

Figure 7.33: *Survivorship curves at varying temperatures*

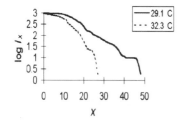

Figure 7.34: *Survivorship curves in varying grains*

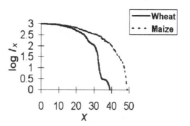

9. **Life insurance rates:** Consider the survivorship curves for males and females in the human population of the United States given in Figure 7.35.

Figure 7.35: *Survivorship curves for United States population*

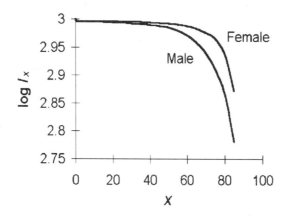

    (a) Based on these curves, explain why life insurance rates are higher for men than for women.

    (b) Based on these curves, explain why life insurance rates increase so rapidly at high ages.

    (c) What would you predict about life insurance rates if the survivorship curves were of type II? What about type III?

10. **Span of life:** Many scientists believe that, for humans at least, there is a natural span of life. That is, there is a limit beyond which human life cannot extend, regardless of advances in medical science. Currently the maximum life span in developed countries is around 120 years, and evidence suggests that it has been that way since Roman times. However, with improvements in medical care, the *average* life span has increased significantly during the past century. Draw a sequence of survivorship curves which illustrates increasing average life span with fixed maximum life span.

References

A classic reference here is E. S. Deevey, Jr., "Life tables for natural populations of animals," *Quart. Rev. Biol.* **22** (1947), 283–314. This is referred to simply as *Deevey* below.

African buffalo, warthog C. A. Spinage, "African ungulate life tables," *Ecology* **53** (1972), 645–652.

Buttercup J. Sarukhán and J. L. Harper, "Studies on plant demography: *Ranunculus repens* L., *R. bulbosus* L., and *R. acris* L.," *J. Ecol.* **61** (1973), 675–716.

Palm L. Van Valen, "Life, death, and energy of a tree," *Biotropica* **7** (1975), 259–269.

Desert night lizard R. G. Zweifel and C. H. Lowe, "The ecology of a population of *Xantusia vigilis*, the desert night lizard," *Amer. Mus. Novitates* **2247** (1966), 1–57.

American robin From a study by Donald S. Farner, as described in *Deevey*.

Honeybee worker S. F. Sakagami and H. Fukuda, "Life tables for worker honeybees," *Res. Popul. Ecol.* **10** (1968), 127–139.

Sagebrush lizard D. W. Tinkle, "A population analysis of the sagebrush lizard, *Sceloporus graciosus*, in southern Utah," *Copeia* (1973), 284–296.

Exercises 7 and 10 *Deevey*, p. 310.

Grain beetles L. C. Birch, "Experimental background to the study of the distribution and abundance of insects. I. The influence of temperature, moisture, and food on the innate capacity for increase of three grain beetles," *Ecology* **34** (1953), 698–711.

U.S. population *Demographic Yearbook of the United Nations*, 1992. Statistical Office of the United Nations (published annually), New York.

7.4 CHAPTER SUMMARY

Exponential and logistic functions as models for population growth have been used throughout the book. In this chapter we look more closely at their development and application to population dynamics. Logarithms and rates of change play key roles and are essential to an understanding of how populations change with time.

Exponential Growth

For any population, the rate of change in population is births per unit time minus deaths per unit time. The exponential model for population growth is a consequence of the assumption that both births and deaths per unit time are proportional to population size. This gives rise to the equation

$$\frac{dN}{dt} = rN,$$

the classic equation of change for an exponential function. As we saw in Chapter 6, the solution can be written in the alternative form

$$N = N(0)e^{rt}.$$

The letter r used here is standard notation. It is the measure of capacity for growth and is characteristic of a species in an environment. Ecologists know it as the species *r value*.

The *doubling time t_d* is the time required for a population to double in size, and for populations exhibiting exponential growth, the value of t_d does not depend on current population size. That is, the same period of time is required for such a population to grow from 100 to 200 individuals as from 1000 to 2000. The doubling time depends only on the r value and is given by

$$t_d = \frac{\ln 2}{r}.$$

Logistic Growth

Since exponential functions grow so rapidly, no population can maintain exponential growth indefinitely; ecological factors limit population size. The logistic model comes from the assumption that the population grows nearly exponentially at first but has an upper limit known as the *carrying capacity* of the environment. As carrying capacity is approached, the growth rate slows, yielding the classic S-shaped graph of logistic growth.

The basic shape alone of the logistic curve yields important information. In particular, the *maximum sustainable yield* theory of wildlife management says that a harvested population should be maintained at the point where the growth rate is the fastest; that is the steepest point on the logistic graph, which occurs at half of the carrying capacity.

As is often the case, logistic population growth is most easily understood by its equation of change

$$\frac{dN}{dt} = rN\left(1 - \frac{N}{K}\right).$$

When the population is small, the factor $1 - \frac{N}{K}$ is almost 1, and so the logistic equation of change is almost the same as the equation of change for the exponential function. This means that initially the population will grow exponentially. When the population N is near the carrying capacity K, the factor $1 - \frac{N}{K}$ is near 0, and so the rate of change is almost zero. This causes the population to level out near the carrying capacity.

In order to get the correct logistic equation from observed data, we rewrite the logistic equation of change as

$$\frac{1}{N}\frac{dN}{dt} = r - \frac{r}{K}N.$$

This tells us that $\frac{1}{N}\frac{dN}{dt}$ is a linear function of N, and we know how to fit a regression line to such data. This regression line can be used to recover the equation of change.

Survivorship Curves

A deeper analysis of population dynamics requires knowledge not just of the number of individuals in a population but also of the age distribution within the population. Individuals are grouped by *cohort*, that is, according to their age. The following symbols are standard in mortality studies:

$$
\begin{aligned}
x &= \text{age at beginning of interval} \\
l_x &= \text{number of survivors at start of age interval } x \\
d_x &= \text{number of deaths during age interval } x \\
q_x &= \text{proportion of individuals dying during age interval } x.
\end{aligned}
$$

A table providing values for these statistics is known as a *life table*, and a *survivorship curve* is the graph of l_x against x done on a semi-logarithmic scale. It shows how many individuals survive to a given age.

There are three basic types of survivorship curves. The type I curve is concave down and is characteristic of a population which has a low mortality rate among juveniles but a high mortality rate for older individuals. Rarely does a naturally occurring population exhibit true type I behavior, though humans and many other mammals show survivorship curves of roughly this shape. A type II curve is a straight line, which describes a species with a constant mortality rate. Type II curves apply to certain bird populations. The type III curve is concave up and represents high juvenile mortality with relatively constant mortality after that. This type of curve is thought to be the most common in nature.

BRIEF ANSWERS *to odd-numbered exercises*

Note: The answers presented here are intended to provide help when students encounter difficulties. Complete answers will include appropriate arguments and written explanations which do not appear here. Also, many of the exercises are subject to interpretation, and properly supported answers which are different than those presented here may be considered correct.

Prologue

1. (a) 63.21

   (b) 1.43

   (c) 1.69

   (d) 50.39

   (e) 3.65

3. $669.60

5. 3.45%

7. (a) 2; 3

   (b) 1.83

   (c) $9150

9. (a) 22.8%

   (b) $7368.00

   (c) $7520.41

11. (a) 6.28 feet

    (b) 6.28 feet

13. 3436 after 8 hours.
    51,458 after two days.

Chapter 1 Section 1

1. (a) 9.33

   (b) −2.21

   (c) 3.32

3. (a) 12 deer

   (b) $N(10) = 380$ deer (rounded to the nearest whole number).

   (c) $N(15) = 410$ deer.

   (d) 30 deer

5. (a) $C(800) = 4.54$ grams

   (b) 5730 years

7. (a) Higher

   (b) Continuous compounding

   (c) 0.0837 (or 8.37%)

   (d) $191.22 per month. It is 65 cents higher than if interest is compounded monthly as in Example 1.2.

9. (a) $P(350, 0.0075, 48)$; $14,064.67

   (b) $15,812.54

   (c) $19,478.33

Chapter 1 Section 2

1. (a) $P(1980) = 29\%$

   (b) Using averages $P(1975) = 26.5\%$

   (c) 0.5 percentage point per year

   (d) 30%

   (e) If growth rate from 1988 to 1990 continues at the same rate as from 1980 to 1988 then $P(1990)$ will be about 34%.

3. (a) Using averages $T(1.5) = 48.5$ degrees.

   (b) 0.35 degree per minute

   (c) 54.8 degrees

   (d) Answers will vary. Sometime near 6:30.

5. (a) Velocity will increase until air resistance causes it to stabilize. This is the terminal velocity of the parachutist.

   (b) Answers will vary. About 20 feet per second.

7. (a) Using averages $A(75) = 90.5$ degrees.

   (b) 5 degrees per minute

   (c) 0.63 degree per minute

   (d) The rate of cooling is more rapid when the difference between temperature of the aluminum and the air is greater.

(e) $A(73) = 91.81$ degrees

(f) $A(0) = 302$ degrees

(g) The temperature of the aluminum will cool until it reaches room temperature.

(h) Answers will vary. About 72 degrees

9. (a) $H(13) = 62.2$ inches.

   (b) (i) Units for growth rate: inches per year.

Period	Growth rate
0 to 5	4.2
5 to 10	2.5
10 to 15	2.4
15 to 20	1.3
20 to 25	0.1

   (ii) From age 0 to age 5.

   (iii) Answers will vary.

   (c) Answers will vary.

11. Answers will vary.

Chapter 1 Section 3

1. Answers will vary.

3. (a) $v(1960) = $10,000, v(1970) = 5000, $v(1980) = $35,000, v(1990) = $35,000$

   (b) Various acceptable graphs.

   (c) Answers will vary.

5. (a) Answers will vary.

   (b) Around 1970

   (c) $N(d) = 80$ thousand individuals

   (d) From 1955 to about 1970. Population is increasing at an increasing rate.

(e) From about 1970 to 1995. Population is still increasing, but at a decreasing rate.

(f) About 1970. This is the time when population is growing at its fastest rate.

(g) About 140,000

(h) Answers will vary.

7. (a) In 1991 there were about 70 tornados reported.

(b) In 1988 there were about 15 tornados reported.

(c) About 25 tornados per year

(d) About 40 tornados per year

(e) 0 tornados per year

9. Answers will vary.

11. Answers will vary.

Chapter 1 Section 4

1. (a) $N(3) = 186.56$ million

(b)

Year	Population
1960	180
1961	182.16
1962	184.35
1963	186.56
1964	188.80
1965	191.06

(c)

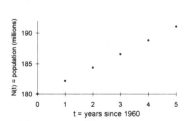

(d) Same numbers as in Part (b).

(e) 273.27 million.

3. (a) $64.00

(b) $C(d, m) = 29d + 0.06m$, where C is the rental cost in dollars, d is the number of rental days, and m is the miles driven.

(c) $C(7, 500) = 233.00

5. (a) $3.96

(b) $C = \dfrac{dg}{m}$, where C is the cost in dollars, d is the distance driven in miles, g is the cost in dollars per gallon of gas, and m is the mileage in miles per gallon.

(c) $C(28, 1.17, 138) = 5.77

7. (a) Two sides are 30 feet long and the other two are 50 feet long. The area is 1500 square feet.

(b) Each side is 40 feet long. The area is 1600 square feet.

(c) Yes, as is shown by Parts (a) and (b).

(d) $A(l) = l \times \dfrac{160 - 2l}{2}$

9. (a) Using k for the constant of proportionality, $t = kn$.

(b) k is the number of items each employee can produce.

11. (a) (i) $13 (ii) $C = \dfrac{150}{n} + 10$, where C is the amount charged per ticket (in dollars) and n is the number of people attending. (iii) $C(65)$, about $12.31.

(b) $P = \dfrac{250}{n} + 10$, where P is the amount charged per ticket (in dollars) and n is the number of people attending.

Chapter 2 Section 1

1. (a) Larger

   (b) Once each year, EAR is 10%. (This is the only circumstance under which the EAR and APR will be the same.) Monthly, EAR is 10.471%. Daily, EAR is 10.516% .

   (c) Monthly, $5523.55. Daily, $5525.80.

   (d) Continuously, EAR is 10.517%. The EAR for monthly and continuous compounding differ by less than 0.05 percentage point.

3. (a) $160.00 (If you rounded when you calculated r, your answer may be different by a few cents.)

   (b) We show the table only for months 18 through 24.

X	Y1	
18	937.74	
19	784.5	
20	630.05	
21	474.39	
22	317.5	
23	159.37	
24	0	

X=24

5. (a) $\left(\dfrac{Q}{2}\right) 850 + \left(\dfrac{36}{Q}\right) 230$

   (b) $4035

   (c) 4 cars at a time

   (d) 9 orders this year

(e) $80 per year per additional car

7. (a) 19.2 feet per second

   (b) The average rate of change in velocity during the first second is 16 feet per second per second. The average rate of change from the fifth to the sixth second is 0.005 foot per second per second.

   (c) 20 feet per second

   (d) With a parachute, about 3 seconds into the fall. Without a parachute, 25 seconds into the fall. A feather will reach 99% of terminal velocity before a cannonball.

9. (a) $P = 2h + 2w$ inches

   (b) $P = 2h + 2\left(\dfrac{64}{h}\right)$ inches

   (c) A square 8 blocks by 8 blocks with a perimeter of 32 inches.

   (d) $P = 2h + 2\left(\dfrac{60}{h}\right)$ inches. A rectangle 6 blocks by 10 blocks giving a perimeter of 32 inches.

11. Answers will vary.

Chapter 2 Section 2

1. (a) Horizontal 0 to 120, vertical 0 to 425.

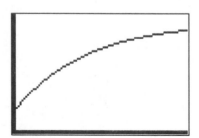

   (b) 75 degrees

(c) Temperature rose more during the first 30 minutes. Average rate of change over first 30 minutes is 4.89 degrees per minute. Over the second 30 minutes it is 2.68 degrees per minute.

(d) Concave down.

(e) After about 46 minutes.

(f) 400 degrees

3. (a) (i) $Q = \sqrt{\dfrac{800c}{24}}$

Horizontal 0 to 25, vertical 0 to 30.

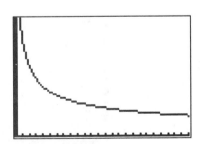

(ii) 14 items

(iii) The number will increase.

(b) (i) $Q = \sqrt{\dfrac{11200}{h}}$

Horizontal 0 to 25, vertical 0 to 125.

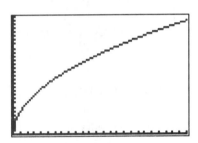

(ii) 27 items per order

(iii) It should decrease.

(iv) About -0.79 item per dollar.

(v) Concave up.

5. (a) $P = 200 \times \dfrac{1}{0.01} \times \left(1 - \dfrac{1}{1.01^t}\right)$

Horizontal 0 to 480, vertical 0 to 25,000.

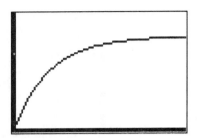

(b) $7594.79

(c) $13,940.10

(d) $20,000

7. (a) (i) $N = \dfrac{60}{2\pi} \times \sqrt{\dfrac{9.8}{r}}$

(ii) Horizontal 10 to 200, vertical 0 to 10.

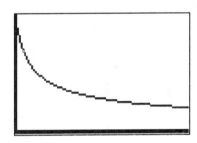

(iii) Decreases.

(iv) 2.44 per minute

(b) (i) $N = \dfrac{60}{2\pi} \times \sqrt{\dfrac{a}{150}}$

(ii) Horizontal 2.45 to 9.8, vertical 1 to 3.

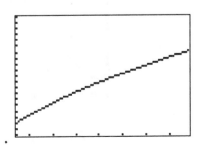

(iii) Increases.

9. (a) (i) $Y = 339.48 - 0.01535N - 0.00056N^2$

    (ii) Horizontal 0 to 800, vertical −50 to 400.

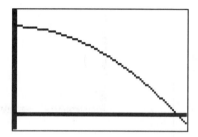

    (iii) Decreases.

    (b) (i) $Y = 260.56 - 0.01535N - 0.00056N^2$

    (ii) Horizontal 0 to 800, vertical −50 to 400.

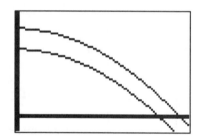

    (iii) Decreases.

11. (a) Horizontal 0 to 6; vertical 0 to 55

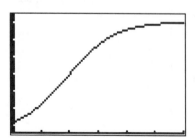

    (b) 19.67 thousand

    (c) Answers will vary.

    (d) Answers will vary.

    (e) 52 thousand.

13. (a) Horizontal 0 to 20; vertical 0 to 2000.

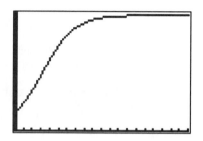

    (b) 299

    (c) In 1997.

    (d) 1961

    (e) Concave up from $t = 0$ to about $t = 3$; concave down afterwards.

Chapter 2 Section 3

1. (a) 288 miles

    (b) $m = \dfrac{d}{g}$
        (ii) 25.77 miles per gallon

    (c) (i) $g = \dfrac{425}{m}$
        (ii) Horizontal 0 to 30, vertical 0 to 75.

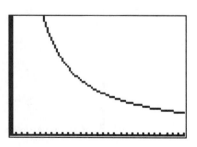

3. (a) $p = (d - c)n - R$ or equivalently $p = dn - cn - R.$

    (b) $2726.10

(c) $d = \dfrac{p + R}{n} + c$ or equivalently $\dfrac{p + nc + R}{n}$.

(d) \$9.43

5. (a) 303.15 kelvins

(b) $F = 1.8(K - 273.15) + 32$, or $F = 1.8K - 459.67$.

(c) 98.33 degrees Fahrenheit.

7. (a) $S(30)$; 3.3 centimeters per second.

(b) $T = \dfrac{S + 2.7}{0.2}$

(c) 28.5 degrees Celsius.

9. (a) $R = (Y + 55.12 + 0.01535N + 0.00056N^2)/3.946$

(b) $R = (55.12 + 0.01535N + 0.00056N^2)/3.946$

(c) Horizontal 0 to 800, vertical 0 to 120.

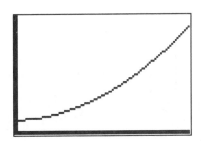

(d) Increases.

(e) 38.23 millimeters

(f) Die back.

Chapter 2 Section 4

1. (a) 30 foxes

(b) Horizontal 0 to 25. Vertical 0 to 160.

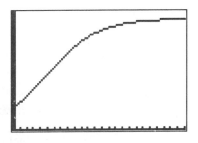

(c) 7.58 years

3. 12.68 seconds

5. (a) Horizontal 0 to 200. Vertical 0 to 220

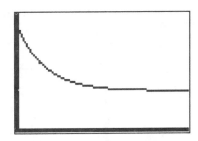

(b) 17.03 minutes

(c) 200 degrees

(d) 75 degrees

7. (a) (i) 6.89 grams (ii) 13.16 minutes

(b) (i) 4.12 grams (ii) 525 million years

9. 5.20%

11. (a) For sheep 2.8 pounds. For rabbits 0.2 pound.

(b) 287.82 pounds per acre.

Chapter 2 Section 5

1. (a) Horizontal 0 to 4, vertical 0 to 2.

    (b) 3.22 miles downrange

    (c) 1.39 miles high at 1.61 miles downrange

3. (a) Height= 4.77 inches. Area= 36.28 square inches.

    (b) Height= 0.19 inch. Area= 163.08 inches.

    (c) (i) Horizontal 0 to 4, vertical 0 to 50.

    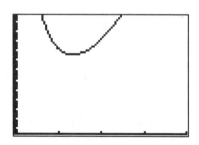

    (ii) 1.34 inches (iii) 2.66 inches

5. (a) $\dfrac{e^{10-32A^{-1}}}{A}$

    (b) Horizontal 0 to 60, vertical 0 to 500.

    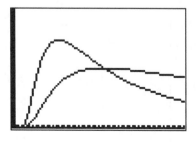

    (c) 32 years

    (d) The rotation age also occurs where the graphs of growth and mean growth cross.

7. (a) Horizontal 0 to 3, vertical 0 to 3.

    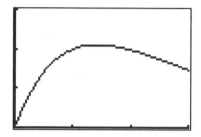

    (b) 2102 recruits

    (c) 802 spawners

9. (a) Horizontal 0 to 7, vertical 0 to 15.

    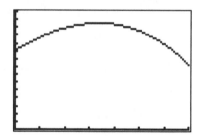

    (b) 3.36 years

    (c) 8.31 million dollars

11. (a) Horizontal 0 to 60, vertical 0 to 0.6.

    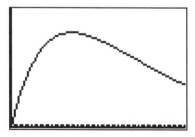

    (b) 0.48 gram

    (c) 0.39 gram

(d) 0 grams

Chapter 3 Section 1

1.

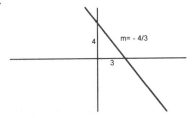

The slope will be negative. $m = -\dfrac{4}{3}$

3.

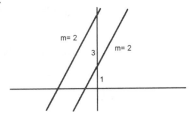

They do not cross. Lines with the same slope are parallel.

5. (a) 0.4 foot high

   (b) 6 feet high

7. (a) 0.83 foot per foot

   (b) 22.11 feet

   (c) 4.82 feet

9. (a) 276.67 feet per mile

   (b) 5513.35 feet

   (c) 22.30 miles

Chapter 3 Section 2

1. (a) $C = 0.56F - 17.78$ (rounding to two decimal places.)

   (b) slope $= 0.56$.

   (c) They are the same.

3. (b) Let S be the total amount of storage space (in megabytes) used on the disk drive and n the number of pictures stored. $S = 2.3n + 781$.

   (c) 270 pictures on the disk drive. There is room for 260 additional pictures.

5. (a) 0.37 pound per dollar.

   (b) 48.10 pounds

   (c) $33.30

7. (a) Slope: 3.5 pounds per inch.

   (b) Let h be height (in inches) and w weight (in pounds): $w = 3.5h - 85$.

   (c) 67.71 inches

   (d) Light.

9. (a) $0.5a$ dollars

   (b) g dollars

   (c) $0.5a + g = 5$

   (d) $g = -0.5a + 5$

Chapter 3 Section 3

1. (a) Let d be the number of years since 1988 and V the value. A table of differences will show a common difference of 2.2. The rate of change is constant, so the data is linear.

(b) Horizontal -0.4 to 4.4, vertical 3.4 to 15.2.

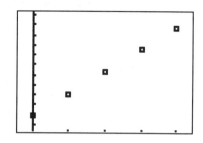

(c) $V = 2.2d + 4.9$

(d) Same scale as Part (b).

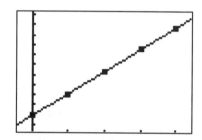

(e) 22.5 billion dollars

3. (a) Let d be the number of years since 1990 and T the tuition in dollars. A table of differences shows a common difference of 216 dollars.
$T = 216d + 2159$

(b) 216 dollars per year

(c) 821 dollars per year

(d) Tuition at public universities increased at a rate of $216 per year while tuition at private universities increased at a rate of $821 per year.

(e) Tuition at public universities increased by about 7.7% while tuition at private universities increased by about 5.9%.

5. (a) A difference table shows a common difference of 36.

(b) 1.80

(c) $F = 1.80K - 459.67$

(d) 310.15

(e) Increase of 1 kelvin causes an increase of 1.80 degrees Fahrenheit. Increase of 1 degree Fahrenheit causes an increase of 0.56 kelvin.

(f) -459.67 degrees Fahrenheit

7. (a) A difference table shows a constant difference of 1.05 ; $P = 2.1S - 0.75$.

(b) Horizontal 0 to 3, vertical 0 to 5.

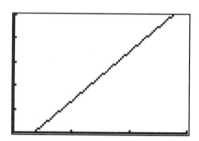

(d) 2.21 billion bushels

9. (a) Let V be the velocity (in miles per hour) and t the time (in seconds) since the car was at rest. A table of differences shows a common difference of 5.9 miles per hour.

(b) 11.8 miles per hour per second

(c) $V = 11.8t + 4.3$.

(d) 4.3 mph

(e) 4.72 seconds

Chapter 3 Section 4

1. (a) Let t be the time in hours since the experiment began and N the number of bacteria in thousands.

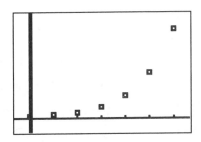

The data does not appear to fall on a straight line, and so a linear model is not appropriate.

(b) Let t be the number of years since 1986 and E the enrollment in millions.

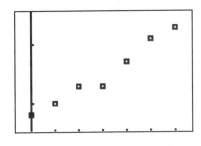

The data is not exactly linear, but it does nearly fall on a straight line. It looks reasonable to approximate the data with a linear model.

3. (a) Let t be the number of years since 1987 and T the number of tourists (in millions).

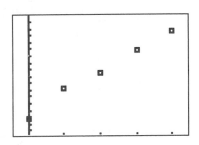

(b) $T = 3.16t + 30.28$

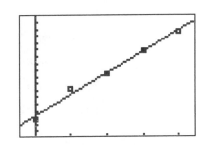

(d) The regression model gives an estimate of 23.96 million tourists.

5. (a)

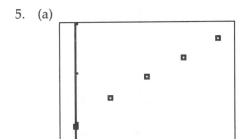

(b) $D = 0.88t + 51.98$.

(c)

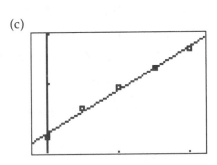

(d) The regression line is the thin line, and the true depth function is the thicker line.

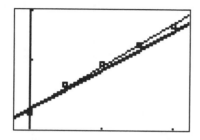

The two lines are very close, but the regression model shows a water level that is a bit too high.

(e) 54.4 feet.

(f) 54.62 feet.

7. (a) Let t be the number of years since 1992 and J the number of sales (in millions).

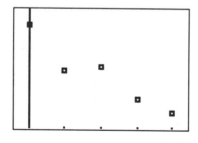

It may be reasonable to model this with a linear function, but there is room for argument here. There is enough deviation from a straight line to cast serious doubt on the appropriate use of a linear model. The collection of additional data would be wise.

(b) $J = -0.044t + 2.46$.

(c)

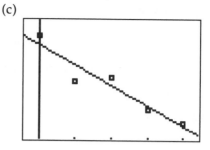

Answers will vary. To the eyes of the authors, this picture makes the use of the regression line appear more appropriate, but gathering more data still seems wise.

(d) The regression model gives a prediction of 2.24 million cars sold in 1997 and 2.20 million in 1998. Both of these estimates are higher than Commerce Department predictions.

9. The data points show that a larger number of churches yields a larger percentage of antimasonic voting. Furthermore, since the points are nearly in a straight line, it is reasonable to model the data with a linear function.

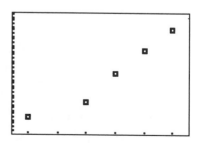

The formula for the regression model is

$M = 5.89C + 45.55$. Its graph is added to the data below.

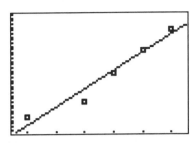

This picture reinforces the idea that the data can be approximated by a linear function.

11.   (a) $E = 0.34v + 0.37$

       (b) 0.34

       (c) Higher cost; less efficient.

       (d) 0.37; higher.

Chapter 3 Section 5

1. If $C =$ chips and $S =$ sodas, then $2C + 0.5S = 36$ and $S = 5C$. You buy 40 sodas and 8 bags of chips.

3. If $c =$ crocus bulbs and $d =$ daffodil bulbs, $0.35c + 0.75d = 25.65$ and $c + d = 55$. You buy 39 crocus bulbs and 16 daffodil bulbs.

5. In 7.55 years there will be about 504 foxes and 504 rabbits.

7. At 160 degrees Celsius the Fahrenheit temperature is 320 degrees.

9. $D =$ number of dimes and $Q =$ number of quarters. $D + Q = 30$ and $0.1D + 0.25Q = 3.45$. 27 dimes and 3 quarters.

11. The graphs lie on top of one another. Every point on the common line is a solution of the system of equations.

13. $N =$ number of nickels, $D =$ number of dimes, and $Q =$ number of quarters.
    $0.05N + 0.1D + 0.25Q = 3.35$
    $N + D + Q = 21$
    $D = N + 1$.
    5 nickels, 6 dimes, and 10 quarters.

Chapter 4 Section 1

1. 23×1.4^t. Horizontal $t = 0$ to 5, vertical 0 to 130.

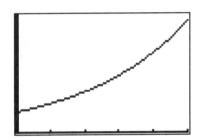

3. 3×1.023^t (in millions)

5. (a) 32%.

    (b) Monthly decay factor 0.968. Monthly percentage decay rate 3.2%.

    (c) Percentage decay rate per second 0.00000122%.

7. 4 half-lives, or 11.32 years

9. 23.45 years.

11. (a) $N = 67.38 \times 1.026^t$, where N is population in millions and t is years since 1980.

    (b) $t = 11.28$. Sometime in 1991.

13. 9.22×10^{18} grains.

About 35.6 trillion dollars.

Chapter 4 Section 2

1. The data shows a common ratio of 1.04 rounded to two decimal places, and thus it is exponential. $f = 3.8 \times 1.04^t$.

3. The successive ratios are not the same.

5. (a) $1750

(b) Let t be the time in months and B the account balance in dollars. The data shows a common ratio of 1.012 , rounded to 3 decimal places. $B = 1750 \times 1.012^t$

(c) 1.2%

(d) 15.4%

(e) $23,015.94

(f) 58.1 months to double the first time. 58.1 months to double again.

7. (a) The data shows a common ratio of 0.42, and so it is exponential. $D = 176 \times 0.84^t$

(b) 16%.

(c) $V = 176 - 176 \times 0.84^t$

(d) About 26.41 seconds into the fall.

9. (a) Let t be the time in minutes and U the number of grams remaining. In the display, the data appears to be linear.

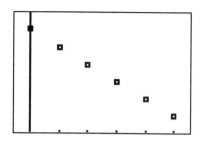

(b) Rounding the calculated parameters to 3 decimal places,
$U = -0.027t + 0.999$.

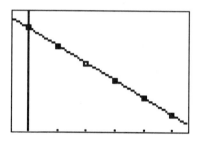

(c) 18.5 minutes.

(d) That there would be -0.621 grams remaining.

Chapter 4 Section 3

1. (a) 4

(b) 6

3. (a) Let t be years since 1976 and C the percent with cable.

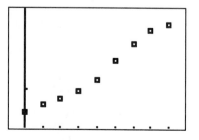

(b) $\ln C = 0.144t + 2.631$

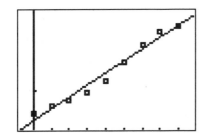

(c) $C = 13.888 \times 1.155^t$

(d)

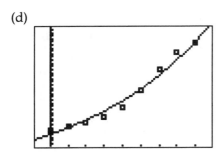

(e) 15.5%

(f) Yes, the model predicts 67.77%.

5. (a) Let t be years since 1950 and H the costs in billions of dollars.

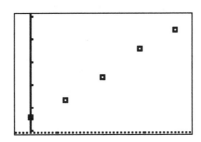

(b) $\ln H = 0.099t + 2.426$

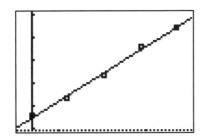

(c) Yearly growth factor 1.104. 10.4% per year.

(d) $H = 11.314 \times 1.104^t$

(e) 1.592 trillion dollars

7. (a)

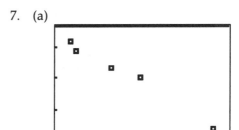

(b) $\ln D = -0.012V - 1.554$

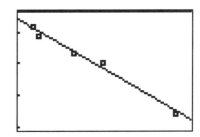

(c) $D = 0.211 \times 0.988^V$

(d) $A = 0.18 - 0.211 \times 0.988^V$

(e) 203.89 pounds per acre (depends on rounding in previous parts)

9. Table A is approximately linear, with the model $f = 19.84t - 16.26$. Table B is approximately exponential, with the model $g = 2.33 \times 1.56^t$.

Chapter 4 Section 4

1. 1.046

3. $263.48

5. (a) APR: 3.5%; EAR: about 3.56%.

   (b) EAR: 3.5%; APR: 3.44%.

7. (a) \$7.75

    (b) 0.044 per year

    (c) 1.045

    (d) $W = 7.75 \times 1.045^t$

    (e) 4.5%

9. (a) Let t be time (in years) and C the circulation (in hundreds).

    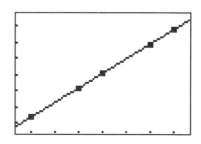

    $$\ln C = 0.901t + 2.474$$

    (b) $C = 11.87e^{0.901t}$

    (c) 0.901 per year

    (d) 0.075 per month

    (e) 1.078

Chapter 5 Section 1

1. Larger positive values of c make the power function with a positive power increase faster.

3. About 6.5 times as fast.

5. (a) 118.63 feet

    (b) Yes.

    (c) Between 42 and 43 miles per hour.

7. (a) When the distanced is halved, the force is 4 times larger. When the distance is one quarter of its original value, the force is 16 times greater.

    (b) $c = 1.8 \times 10^{11}$. When $d = 800$ kilometers, $F = 281,250$ newtons.

    (c) Horizontal 0 to 1000, vertical 0 to 20,000,000.

    When the asteroids are very close, the gravitational force is extremely large. (In practice, physicists expect that when massive planetary objects get too close, the gravitational forces become so great as to tear the objects apart.) When the asteroids are far apart, the gravitational force is so small that it has little effect.

9. (a) By a factor of 125,000,000.

    (b) By a factor of 250,000.

    (c) By a factor of 500.

Chapter 5 Section 2

1. (a) $\ln V = 0.50 \ln p + 2.34$

    (b) $V = 10.4p^{0.5}$

    (c) Answers will vary.

3. (a) Yes.

    (b) $F = 20.70L^{0.34}$

(c) Horizontal 0 to 300, vertical 0 to 150.

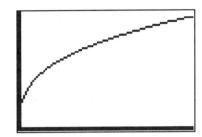

(d) Concave down.

(e) 2.19 times faster.

5. (a) $B = 40.04W^{0.75}$

    (b) (ii) 2.14; 37.79

    (iii) 0.08; 0.07

7. (a)

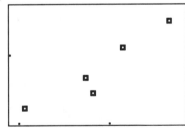

    (b) $\ln h = 0.66 \ln d + 3.58$

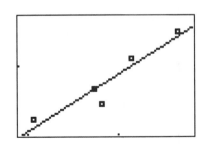

    (c) Plains cottonwood.

    (d) $h = 35.87d^{0.66}$

    (e) (ii) No.

9. (a) Answers will vary.

    (b) $w = 4770p^{-1.48}$

(c) Weight increases by a factor of 2.79.

(d) Yield increases.

11. (a) Decreases, generally.

    (b)

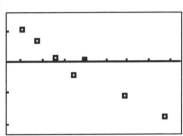

    (c) $\ln C = -0.40 \ln W + 2.15$

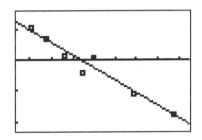

    (d) Answers will vary.

    (e) $C = 8.58W^{-0.40}$

13. (a) $W = 0.0076L^3$

    (b) About 8 times as heavy.

Chapter 5 Section 3

1. Horizontal 0 to 144, vertical -5 to 5.

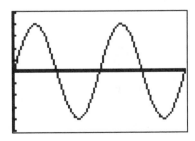

Period 72; amplitude 4.

3. Horizontal 0 to 40, vertical −6 to 6.

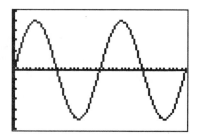

4. .05 feet above sea level

5. Horizontal 0 to 60, vertical −6 to 6.

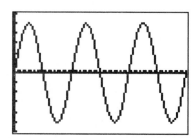

0 centimeters

7. 412.12 feet

9. 10.41 degrees

Chapter 5 Section 4

1. The second-order differences are all 4.

3. $Q = 3x^2 - 2x + 5$.

5. (a) The second-order differences are all 0.5.

   (b) $H = 0.25D^2 + 1$

   (c) 3.46 inches

   (d) 7.25 inches

7. 7.44 minutes

9. (a) Horizontal 0 to 15, vertical 0 to 100.

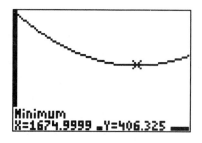

   (b) Answers will vary.

   (c) 6.3 feet perpendicular, 12.7 feet parallel

Chapter 6 Section 1

1. Answers will vary.

3. (a) Horizontal 0 to 2. Vertical 0 to 20.

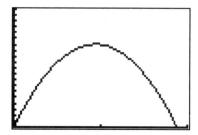

   (b) 14.06 feet.

   (c) 1.88 seconds after it is tossed.

   (d) Answers will vary.

5. Answers will vary.

7. Answers will vary.

9. Answers will vary.

11. Answers will vary.

Chapter 6 Section 2

1. Horizontal -3 to 3. Vertical -10 to 10.

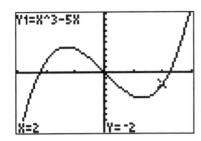

(a) Positive.

(b) $x = 0$, or any number between -1.29 and 1.29.

3. (b) Negative.

5. Answers will vary.

7. (a) Horizontal 0 to 5, vertical -5 to 3.

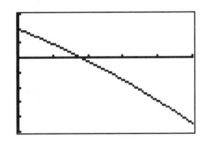

(b) From 0 to 1.82 years.

(c) 1.82 years

9. Answers will vary.

11. (b) It will be positive.

(c) $\dfrac{dN}{dt}$ would be smaller than before.

(d) $\dfrac{dN}{dt}$ will be near zero.

13. Answers will vary.

15. Answers will vary.

Chapter 6 Section 3

1. (a) $V = -32t + 40$

(b) 1.25 seconds

(c) 2.5 seconds

3. (b) $B = 10,000e^{0.07t}$

(c) $B = 10,000 \times 1.073^t$

(d) 9.90 years

5. (a) $\dfrac{dN}{dt} = 0.04N$

(b) $N = 30e^{0.04t}$ or $N = 30 \times 1.04^t$

(c) About 13 years.

7. (a) Linear.

(b) $\dfrac{dW}{dt} = 5.5$

(c) $W = 5.5t + 13.5$

9. (a) -0.05 per day

(b) $A = 3e^{-0.05t}$

(c) 13.86 days

Chapter 6 Section 4

1. (a) Horizontal 0 to 225, vertical -20 to 40.

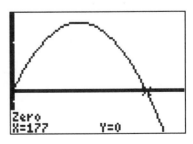

Equilibrium solutions $N = 0$ and $N = 177$.

(b)

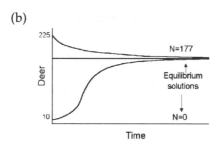

(c) $N = 6$

3. (b) 8

5. (a) $F = 0.15$ million tons per year.

(b) (i) Horizontal 0 to 3, vertical -0.15 to 0.15.

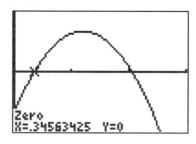

Two equilibrium solutions, $N = 0.35$ million tons and $N = 2.05$ million tons. (ii) The biomass increases when N is between 0.35 and 2.05 million tons. It decreases if N is less than 0.35 million tons or more than 2.05 million tons. (iii)

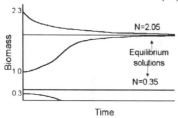

7. (a)
$$\frac{dN}{dt} = -0.338N \left(1 - \frac{N}{0.8}\right) \left(1 - \frac{N}{2.4}\right)$$

(b) Horizontal 0 to 3, vertical -0.25 to 0.25.

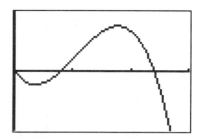

The equilibrium solutions are the threshold $N = S = 0.8$ and the carrying capacity $N = K = 2.4$.

(c) Increasing when N is between 0.8 and 2.4. Decreasing otherwise.

(d) The population will eventually die out.

(e) $N = 1.77$ million tons.

9. (a) Yes.

(b) All bacteria are of type B. There will never be any type A bacteria.

(c) Eventually all type B bacteria will disappear and all the bacteria will be of type A.

11. (b) Horizontal 0 to 3, vertical -0.01 to 0.05.

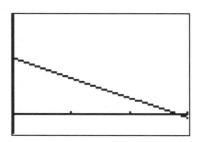

(c) 2.8 pounds

Chapter 6 Section 5

1. (a) 423 reindeer per year

   (b) 1524

   (c) Too large.

3. (a) -18.98 deaths per 100,000 per year

   (c) 689.86 deaths per 100,000

   (d) -4.36 deaths per 100,000 per year

5. (a) Horizontal 0 to 10, vertical 0 to 200.

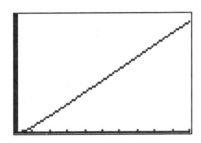

   (b) 28 feet

   (c) 19.20 feet per second

7. (a) Horizontal 0 to 45, vertical 0 to 300.

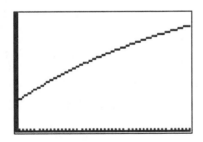

   (b) 5.32 degrees per minute

   (c) 3.57 degrees per minute

9. (b) Negative.

   (c) −8000 gallons per minute

   (d) 1,960,000 gallons

Chapter 7 Section 1

1. (a) 42; 14,939

   (b) 0.18 year; 0.35 year

   (c) 60

3. $r = \dfrac{\ln 2}{t}$; 0.099 per year

5. 0.35 per year

7. (a)

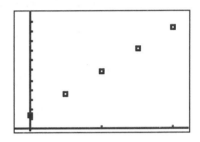

   (b) 4.50 per year; 0.15 year

   (c) $N = 4e^{4.5t}$

   (d) 117

9. Collard

11. Horizontal 0 to 5, vertical 0 to 500

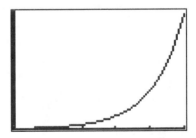

   486

13. Population declines; population is unchanging.

Chapter 7 Section 2

1. Answers will vary.

3. $r = 0.37, K = 14,800$

5.   (a) 11.39 years

    (b) 17.91 years

7.   (a) Horizontal 0 to 5, vertical 0 to 160.

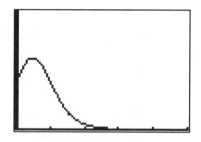

    (b) 0.49 years

    (c) 74 thousand tons

9.   Answers will vary.

11.   Answers will vary.

13.   (a) Horizontal 0 to 12, vertical -1 to 6.

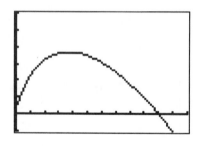

    (b) 3.68

    (c) About $0.368K$.

Chapter 7 Section 3

1.   (a) Type I.

    (b) Type III.

    (c) Type II.

3.   (a) Horizontal -0.8 to 8.8, vertical 1.77 to 3.18.

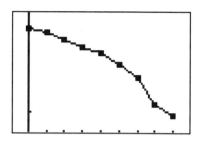

Type I.

    (b) Horizontal -0.6 to 6.6, vertical 0.4 to 3.38.

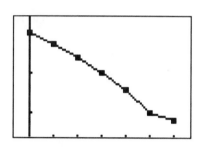

Type II.

5.   (a)

x	l_x	d_x	q_x
0	1000	770	0.770
1	230	87	0.378
2	143	67	0.469
3	76	42	0.553
4	34	21	0.618
5	13	6	0.462

    (b) Horizontal -0.5 to 5.5, vertical 0.79 to 3.32.

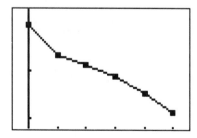

7.   Type II.

9.   Answers will vary.

Index

Index of applications (continued from first page of book)